Drums A'beating
Trumpets Sounding

Drums A'beating,

Trumpets Sounding

ARTISTICALLY CARVED POWDER HORNS
IN THE PROVINCIAL MANNER
1746–1781

by WILLIAM H. GUTHMAN

HARTFORD · CONNECTICUT
THE CONNECTICUT HISTORICAL SOCIETY
MCMXCIII

THIS CATALOGUE WAS PREPARED FOR THE EXHIBITION
Drums A'beating, Trumpets Sounding
ORGANIZED BY THE CONNECTICUT HISTORICAL SOCIETY

Exhibition Schedule

HERITAGE PLANTATION OF SANDWICH, MASSACHUSETTS
MAY 9, 1993 – OCTOBER 24, 1993

THE CONNECTICUT HISTORICAL SOCIETY, HARTFORD, CONNECTICUT
NOVEMBER 16, 1993 – MARCH 27, 1994

THE CONCORD MUSEUM, CONCORD, MASSACHUSETTS
APRIL 19, 1994 – NOVEMBER 15, 1994

Illustrations

TITLE PAGE: DETAIL, NATHANIEL SELKRIG HORN, NO. 24
COPYRIGHT PAGE: DETAIL, ZEBULON WATERMAN HORN, NO. 27
DEDICATION PAGE: DETAIL, DAVID BALDWIN HORN, NO. 14

ISBN 1-881264-05-X
LIBRARY OF CONGRESS CATALOGUE NUMBER: 92-075868

BOOK DESIGN BY CHRISTOPHER KUNTZE
PRINTED BY THE STINEHOUR PRESS IN LUNENBURG, VERMONT
COPYRIGHT © 1993 BY WILLIAM H. GUTHMAN

THIS BOOK IS DEDICATED TO

The Carvers

PARTICULARLY JOHN BUSH AND JACOB GAY

The Recorders

PARTICULARLY RUFUS A. GRIDER

AND STEPHEN V. GRANCSAY

The Collectors

PARTICULARLY NATHAN L. SWAYZE

AND THOMAS J. SEGAL

The Believers

PARTICULARLY ROBERT F. TRENT,

ELIZABETH STILLINGER GUTHMAN, AND

ELIZABETH PRATT FOX

Contents

Foreword *by Robert F. Trent*	9
Preface *by Elizabeth Pratt Fox*	13
Acknowledgments	15
Author's Reflections	17
Powder Horns Carved in the Provincial Manner, 1746–1781	19
Eighteenth-Century American Forts	61
Catalogue	71
Color Plates *(following page 72)*	
Bibliography	213
Index	221

Abbreviations and Short Titles

Antiques, 1978: William H. Guthman, "Powder Horns of the French and Indian War 1755–1763," *The Magazine Antiques*, Vol. CXIV, No. 2 (August, 1978), 312–331.

ASAC: American Society of Arms Collectors.

Clements Library: "Powder and Ball: The Colonial Soldier in the 18th Century," exhibition and catalogue, William L. Clements Library, University of Michigan, May 17–July 27, 1984.

duMont: John S. duMont, 1978, *American Engraved Powder Horns*, Canaan, N.H., Phoenix Publishing.

Grancsay: Stephen V. Grancsay, 1946, *American Engraved Powder Horns*, N.Y., Metropolitan Museum of Art.

Grider, NYHS: Collection of powder-horn drawings made by Rufus S. Grider in the 1880s and 1890s and now in the possession of the New-York Historical Society.

KRA: Kentucky Rifle Association.

MHS: Massachusetts Historical Society.

NYHS: New-York Historical Society.

Swayze: Nathan L. Swayze, 1978, *Engraved Powder Horns of the French and Indian War and the Revolutionary War Era*, Yazoo City, MS, Gunhill Publishing Co.

Drums A'beating, Trumpets Sounding

Here we have two monumental, polarized versions of the folk artist that, in my opinion, reflect the needs and wants of the commentators, rather than those of any folk artist, living or dead. The first view—the "naive" stance—is today reinforced by the enthusiasms and investments of dealers and collectors, while the second—the "traditional" stance—depends upon the momentum of scholarly studies dating back 150 years. The relative merits and demerits of both are obvious. Both assume a great deal about the mentality of the artist without much proof, other than psychological insights based upon compositional analysis. Anyone with more than a passing acquaintance with art historical literature knows how half-baked and mechanical that kind of approach can be. The naive view stresses individual creativity to a ridiculous degree, while the traditional approach coerces the artist into a straightjacket of repetition. Both are alarmingly condescending and provide little insight into the act of creation, an intersection of mind and matter.

The powder horns in this catalogue do not reflect either model. They do not reflect even the most plausible compromise positions between these two antagonistic views. As you read through the essays and object entries, this will become evident. The horns seem to spring from an aristocratic European tradition of carved staghorn powder receptacles or other engraved military and hunting accoutrements, but no proof of the presence of such transmitters in the colonies exists. The probable design sources seem quite diffuse, and just about any graphic source encountered by functionally illiterate New Englanders may have been regarded by them as being of equal interest. This is a convenient theory, but people who rely more on concepts and memory than on written sources absorb, analyze, and recombine graphic images in ways that go far beyond immediate religious and political concerns. I'm particularly startled by the free-form interpretations of space *and* iconography seen on the horns. Where did they learn to depict reality in this unprecedented manner? Did these ready juxtapositions of animals, humans, and mythological or ideological beings reflect a genuine, almost phenomenological perception of reality? I ask these questions with all due allowance for the humor found on many horns.

The context in which the horns were made is troubling, too. The unstable psychological climate among a group of men deprived of female companionship fosters art forms whose meaning rarely is clear; a classic example is scrimshaw and other whaling memorabilia. The military campaigns of the successive French and Indian Wars are not quite in the same category. The boredom of life in the camps cannot adequately explain the surprising stylistic uniformity of the engraving on the horns, nor is it evident how men with strong calligraphic skills mastered the academic conventions of engraving with the burin. I am willing to suppose that the masters of powder horn carving set up what amounted to shops in the camps where they concentrated on engraving horns that were prepared for them by others. Only consistent marketing of the horns could have fostered consistent demand, and competition among the best carvers may have been more overt than we think. Too many good horns survive for carving to have been an intermittent, ephemeral activity.

If this was the case, what happened when the carvers returned to their homes? My suspicion is that some dropped carving altogether, but that the best carvers studied professional lettering and gathered print sources that they drew upon in formulating new formats and decoration. In this sort of highly regulated activity, marketing and demand reinforce

Foreword

AMERICAN POWDER HORNS AND THE PROBLEM OF FOLK ART

I'M EXTREMELY HONORED to have been asked by Bill Guthman to provide a brief commentary on his beloved powder horns. After working on them with Bill for ten years or so, I feel I have only the barest intimation of where they fit in the general scheme of eighteenth-century North American material culture.

First and foremost, the task of the observer is to rid his or her mind of presentist foibles and prejudice before addressing these strange objects. In the 1990s, most Americans still adhere to a definition of folk art that harks back to the 1930s. When you ask the average person with no particular interest in collecting or museum-going what the definition of folk art is, his or her response is usually "something artistic and unique made by a gifted amateur." When it comes to artistic production, Americans have some extremely odd ideas, many of which are thought to reflect democracy. Where art is involved, academically-trained artists are distrusted, because they are thought to be bohemian intellectual snobs and too heavily influenced by "foreign" ideas. For an object to be "art," it must be unique, otherwise it might be merely "craft," or worse, mass-produced. It follows that the highest American art must spring from the mind and hand of the untutored, preferably the impoverished untutored. No sophisticated concepts or highly developed skills must intervene between the authentic revelation and its realization as artwork. An ideal instance of this might be, say, a former slave in Alabama carving a walking stick with a jack-knife. The assumptions are that the slave had no skills, no first-hand knowledge of African art, nothing beyond some undifferentiated "spiritual" yearning and the most rudimentary tools and materials with which to express it.

The classic understanding of folk art among academics in Europe and America is the polar opposite of this widespread American one. Professional folklorists are obsessed with placing the individual in a tradition of design and workmanship that is identified with a specific group and place. Academic study of folk art can be a branch of art history or even psychology, but is most often classed as part of ethnography. In these disciplines the influence of the individual artist seldom is emphasized; most of the time the artist or artisan is submerged in a professional master-and-apprentice line of descent that can be traced back for decades, if not centuries. In this case, an ideal example might be a Lithuanian maker of painted wagons whose shop has existed since the 1700s. His pride is based on a high degree of skill, exacting repetition of a precise design found in no other village, and decoration whose original content and meaning may have been forgotten. Creativity, if it ever enters into the work of such an individual, is restricted to minor ornamental variations that are part of the overall tradition.

Foreword

consistency of performance, but boredom on the part of the artist and the introduction of novelties to stimulate new purchases and fend off competitors reinforce creativity.

To return to this idea of self-education and self-improvement, if we see the major carvers in this light, it's striking how rarely the carvers used ideologically-charged images from official publications or art prints. The vast majority of the motifs have no obvious high style or popular antecedents. Nevertheless, I shrink from the tempting conclusion that the carvers formulated most of their motifs on the basis of empirical observation. Engraving is a copyist's art par excellence. In those cases where specific motifs were drawn from immediate experience, they swiftly must have become stylized, for ease of execution and also for memorization.

Given these factors (most of which are speculative), how does one approach the mentality of the carvers? I, following Henry Glassie, would suggest that their outlook represented a balance of elite, popular, and folk culture. All these qualities are relative, especially in colonial societies, but New England was literate and literary at this time. We might question if any New Englander was all that folklike, save for his or her narrow-minded religious convictions. Most of the horns were carved by men from heavily-populated areas of Connecticut and Massachusetts, areas that were not marginal, not poor, not unaware of world events. I would tend to think that only elite and popular cultures had an appreciable impact on the horns, both thematically and artistically. That the heyday of horn production extended from 1746 to 1780 or so further suggests that horns passed from fashion because they were a reminder of an immediate colonial past that was both inspiring and embarrassing. I believe that the horns only began to be regarded as folklike in retrospect, perhaps during the lifetimes of their original owners, when professional engraving in the Neoclassical taste became readily available.

With this well-researched, strong-minded catalogue, the status of powder horns as major eighteenth-century American artworks is secured. What began as a conviction on Bill Guthman's part has turned out to be a historical truth. While some new horns of the first importance may come to light in the future, I doubt that any new examples will seriously alter the canon established here. My hope is that future iconographic work may give us a more detailed notion of the sources the carvers drew upon. With that knowledge, we will be better able to assess the relative merits of their concepts.

Robert F. Trent
CURATOR AND IN CHARGE OF FURNITURE
THE HENRY FRANCIS duPONT WINTERTHUR MUSEUM

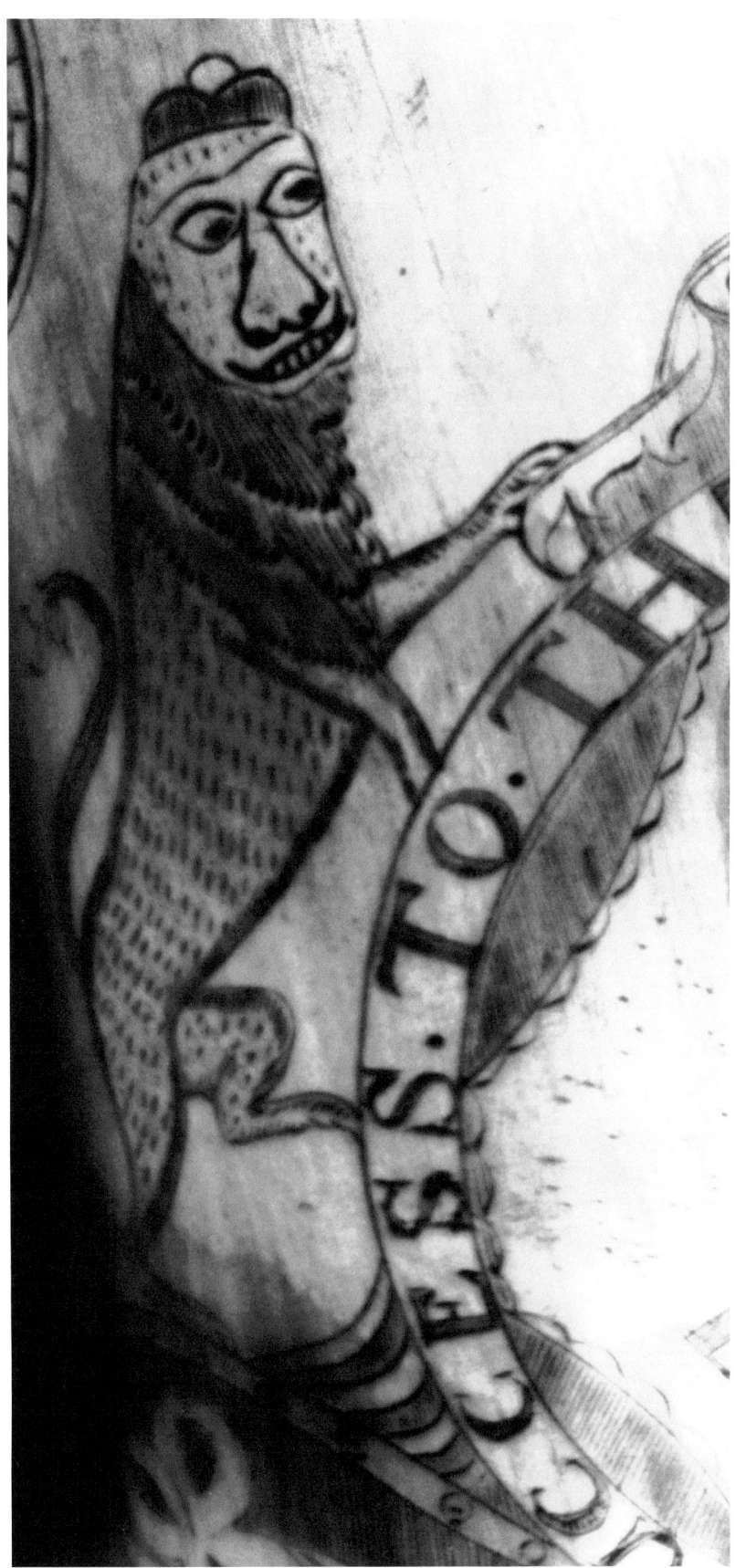

Preface

THIS CATALOGUE AND EXHIBITION, *Drums A'beating, Trumpets Sounding: Artistically Carved Powder Horns in the Provincial Manner*, was in the preliminary planning stages when I became Curator of The Connecticut Historical Society in 1988 and it has been a delight to work with William H. Guthman on both the exhibition and the catalogue. Drawing upon his fifteen-year study of the North American powder horn, Mr. Guthman undertook this definitive work as a labor of love, traveling to look at every major horn in the country, obtaining many of the photographs which you see in this catalogue, and researching each and every clue as to the possible carver of a horn. Mr. Guthman's approach was to consider the horns in the context in which they were made and used, and by doing so he has drawn an important connection between the maker and the owner. As he points out in the text, powder horns have too often been considered relics of the important people who owned them or the wars in which they were used. This study considers the important role they played in the transformation of an English culture to an American culture.

Unlike scrimshaw objects, which were made by whalemen during long periods of inactivity between catches, powder horns were sometimes made by professional carvers who, following the troops to various forts and battlefields, carved the horns in those locales. Their work depicts the fortifications, battle scenes, and other frontier environments familiar to the customer. Unlike scrimshaw, the horns were often carved quickly under pressure and most of them have no print sources. Many of the carvers used images found only on other powder horns, reinforcing Mr. Guthman's thesis that there was a community of carvers observing one another's production. A few extraordinary powder horns were also carved with scenes of historic events or political cartoons not recorded in any other medium. Examples are an American chasing a Hessian soldier after the Battle of Trenton and a hog and a naked soldier representing the oppressive taxation of the British.

We are delighted to have as hosts of the exhibition two other museums. The first site, Heritage Plantation in Sandwich, Massachusetts, is known for its strong military and folk art collections. It is highly appropriate that this exhibition, which focuses on both of these areas, should open there. The other site is the Concord Museum in Concord, Massachusetts, which has a long history of collecting and interpreting the material culture of the battles which marked the beginning of the American Revolution.

We are especially grateful to the Kentucky Rifle Association and the American Society of Arms Collectors, who through their publication funds supported this project from the beginning. Without these organizations' support the catalogue would not have been possible.

Elizabeth Pratt Fox
CURATOR, THE CONNECTICUT HISTORICAL SOCIETY
PROJECT DIRECTOR

Preface

THIS CATALOGUE AND EXHIBITION, *Drums A'beating, Trumpets Sounding: Artistically Carved Powder Horns in the Provincial Manner*, was in the preliminary planning stages when I became Curator of The Connecticut Historical Society in 1988 and it has been a delight to work with William H. Guthman on both the exhibition and the catalogue. Drawing upon his fifteen-year study of the North American powder horn, Mr. Guthman undertook this definitive work as a labor of love, traveling to look at every major horn in the country, obtaining many of the photographs which you see in this catalogue, and researching each and every clue as to the possible carver of a horn. Mr. Guthman's approach was to consider the horns in the context in which they were made and used, and by doing so he has drawn an important connection between the maker and the owner. As he points out in the text, powder horns have too often been considered relics of the important people who owned them or the wars in which they were used. This study considers the important role they played in the transformation of an English culture to an American culture.

Unlike scrimshaw objects, which were made by whalemen during long periods of inactivity between catches, powder horns were sometimes made by professional carvers who, following the troops to various forts and battlefields, carved the horns in those locales. Their work depicts the fortifications, battle scenes, and other frontier environments familiar to the customer. Unlike scrimshaw, the horns were often carved quickly under pressure and most of them have no print sources. Many of the carvers used images found only on other powder horns, reinforcing Mr. Guthman's thesis that there was a community of carvers observing one another's production. A few extraordinary powder horns were also carved with scenes of historic events or political cartoons not recorded in any other medium. Examples are an American chasing a Hessian soldier after the Battle of Trenton and a hog and a naked soldier representing the oppressive taxation of the British.

We are delighted to have as hosts of the exhibition two other museums. The first site, Heritage Plantation in Sandwich, Massachusetts, is known for its strong military and folk art collections. It is highly appropriate that this exhibition, which focuses on both of these areas, should open there. The other site is the Concord Museum in Concord, Massachusetts, which has a long history of collecting and interpreting the material culture of the battles which marked the beginning of the American Revolution.

We are especially grateful to the Kentucky Rifle Association and the American Society of Arms Collectors, who through their publication funds supported this project from the beginning. Without these organizations' support the catalogue would not have been possible.

Elizabeth Pratt Fox
CURATOR, THE CONNECTICUT HISTORICAL SOCIETY
PROJECT DIRECTOR

Acknowledgments

IN 1981 ROBERT F. TRENT, then Curator of the Connecticut Historical Society and now Curator and In Charge of Furniture at the Winterthur Museum, encouraged me to expand upon the 1978 article on powder horns as an art form that I had written for *The Magazine Antiques*, in order to present an exhibition at the Historical Society. With Trent's guidance and enthusiastic support and with the additional encouragement of Director Christopher P. Bickford and Associate Curator Elizabeth Pratt Fox, I began this catalogue. In 1988, when Trent left the Society for his present post at Winterthur, Ms. Fox became Curator and continued to support the project. The entire staff at the Society, in fact, has been encouraging and helpful throughout. Of particular assistance have been Registrar Richard C. Malley, Head of Library and Crofut Curator of Rare Books and Manuscripts Everett C. Wilkie, Jr., Reference Librarian and Genealogist Judith E. Johnson, and Curatorial Assistant Elizabeth I. Blakelock.

Many others helped with this project: Arthur Vitols of Helga Studios produced most of the excellent photographs for the catalogue; Robert Bitondi produced the photographs used in the introductory essay; Elizabeth Stillinger Guthman not only encouraged the project but has capably edited the manuscript; Christopher Kuntze of The Stinehour Press oversaw the design and production of a beautiful book; David S. Hansen, Rudolph W. Gleichman, Frank Nocilla, and the membership of the Kentucky Rifle Association provided consistent support and did me the honor of including this publication in their Kentucky Rifle publication series; and the board of directors, publications committee, and membership of the American Society of Arms Collectors also offered support in the form of a Joint Venture Publication Participation Conditional Grant. James Dresslar, William Myers, James Routh, and Carol Segal have offered much appreciated moral support, and have graciously lent horns for the exhibition.

Others without whose help this project could not have come to completion are: Anne E. Bentley of the Massachusetts Historical Society; Lisa Broberg Quintana, now Curator at the New Haven Colony Historical Society; Dr. Edmund Carpenter; Wendy Cooper, Curator of Decorative Arts, Baltimore Museum of Art; Brian Cullity, Curator of the Art Museum, Heritage Foundation of Sandwich, Massachusetts; the late Anthony D. Darling; Sharon Darling, former Curator, Chicago Historical Society; Dennis Fiori, Director, Concord Museum; Alice Cooney Freylinghuysen, Associate Curator, American Decorative Arts, Metropolitan Museum of Art; Jay Gaynor, Curator of Mechanical Arts, Colonial Williamsburg; James L. Gavin, former Curator, New Hampshire Historical Society; Kathy Griffen of the Massachusetts Historical Society; Thomas Grinslade; Thomas A. Graves, Jr., retired Director, Winterthur Museum; George R. Hamell, Senior Museum Exhibits Planner in Anthropology, New York State Museum; Elizabeth Hammel of the Maine Historical Society; Anne Coffin Hanson of the Yale University Art Gallery; Diana M. Hawes; Jay E. Hopkins, MD; Charles F. Hummel, former Deputy Director of Collections, Winterthur Museum; James R. Johnston; George Juno, Sr.; Patricia E. Kane, Curator of the Mabel Brady Garvan and Related Collections, Yale University Art Gallery; Richard J. Koke, former Curator, New-York Historical Society; Don Ladd; Jane M. Lape, former Curator, Fort Ticonderoga Museum; Barbara R. Luck, Curator, Abby Aldrich Rockefeller Folk Art Center, Colonial Williamsburg; Warren T. Lewis; Wilhelmina V. Lunt, former Curator, Old Newbury Historical Society; Olivia Mahoney of the Chicago Historical Society; Richard C. Meyer; Elizabeth Miller of the Maine Historical Society; Crosby Milliman; Robert E. Mulligan, Jr., Associate Curator of History, New York State Museum; Walter O'Connor; the late John H. G. Pell, former President, Fort Ticonderoga Museum; Mrs. John H. G. Pell; Stuart W. Pyhrr, Curator of the Department of Arms and Armor, Metropolitan Museum; Richard F. Rosenberger; Beatrix Rumford, Vice President of Museums, Colonial Williamsburg; John O. Sands, Manager and Administrator of the Collections Department, Colonial Williamsburg; John A. H. Sweeney, former Assistant to the Director, Winterthur Museum; Ted Trotta; Deborah D. Waters, Curator of Decorative Arts, Museum of the City of New York; Carolyn J. Weekley, Director of the DeWitt Wallace Gallery, Colonial Williamsburg; James R. Weitzel; Robert K. Weiss, former Curator, Essex Institute; David Wood, Curator, Concord Museum; Philip Zea, Curator, Historic Deerfield; and Catherine Zusy of the New Hampshire Historical Society.

Very special thanks, too, to Robert C. Anderson for his genealogical search for Jacob Gay; to Elaine Bush Prince for her genealogical search for John Bush; and to Wendy Shadwell, Curator of Prints at the New-York Historical Society, for her search for printed source material powder-horn carvers may have used.

Author's Reflections

CARVED POWDER HORNS represent to me the last vestige of an unspoiled era when American antiques were available in the marketplace in "as found, unspoiled" condition. Much like the disappearing rain forests in the Amazon, American antiques that have not been "improved" or desecrated are a rare commodity.

When I began to collect in the late 1940s, before turnpikes and superhighways had been built and before auction houses had become merchants to the masses, a collector could follow a variety of routes in Pennsylvania, New York, Connecticut, Rhode Island, Vermont, Massachusetts, New Hampshire, and Maine, and find dozens of untouched treasures in shops along the way. Furniture, ceramics, pewter, paintings, silver, weapons, accoutrements, and other such objects were easily found in the many shops that lined the sides of the roads. There were auctions held on the front lawns of old family homes, too, the stuff just pulled out, room by room, and sold "as is." The collector's prizes were usually not sophisticated Queen Anne and Chippendale pieces from prominent estates, but honest country antiques of those same periods.

The "look" and the "smell" of things, as well as the "feel," had great appeal for those of us who hunted antiques in those days. It was the object that was important, and although condition was a factor, it never got in the way of a decision to buy something because of what it was and where it had been. When, as often happened, parts and paint were missing, we concentrated not so much on what wasn't there as on what *was*, and that determined our decision.

Not so today, when condition is almost everything. And because it is, most objects appear on the market in a remarkably fine "original" state—so much more so today than in those good old days when objects were much more plentiful, but not always pristine. Carved powder horns, however, still appear "as found." The best part is that the carved powder horn represents a true art form, unmistakably identifiable and indigenous to North America. I am thankful for the handful of curators, collectors, and dealers who have been able to find in the carved powder horn those qualities that make them so desirable.

Powder Horns Carved in the Provincial Manner 1746–1781

HERE, for the first time to my knowledge, powder horns are being treated as art objects and not as military accoutrements. This is an art exhibition, and thus it departs from horn-collecting criteria that Rufus Grider first suggested in the late nineteenth century and that most collectors and curators have adhered to ever since. The undocumented "historic" legends and traditions that accompany many horns have been discarded, and only those that aid in placing a horn chronologically or that offer genealogical information have been retained in this study. Not one of the horns in the present exhibition would be included if it did not possess the required artistic qualities or embody features that were to become distinctive characteristics of the schools of powder-horn decoration that are described here.

Understanding historical periods, historical facts, and historical environments is certainly imperative in comprehending why the various schools of carvers chose the particular formats and motifs that are the hallmarks of their styles. Indeed, certain scenes and vignettes *must* be viewed in the light of historic events to be understood. But it is the art that resulted from a given historical background that is the true focus of this exhibit.

The interesting and visually appealing facet of New England material culture that carved powder horns embody was, until recently, totally ignored by both academics and most collectors. The style of horn decoration that we are dealing with is indigenous to North America, and fewer than twenty percent of the powder horns carved during the specified period qualify for inclusion. The examples selected were created by several different groups of extremely talented craftsmen who employed distinct calligraphic and design styles that today enable us to identify their various schools of carving. These artists engraved the horns on a particular stretch of the frontier that included northern New England, upper New York State, the eastern Great Lakes, and Canada. By selecting and analyzing only the best examples, we can establish critical standards of quality and appreciation for American carved powder horns. Two other important goals are identifying trends within the development of the art form and setting forth characteristics of the carvers who spearheaded the trends.

Because of the traditional bias toward historical and associative rather than artistic criteria on the part of both the public and horn collectors, fakers have flourished. They may be responsible for more powder horns with "historic" maps, scenes, and personages than the carvers who were working during the periods in which the commemorated events took place!

It is also my belief that some horns with Revolutionary War dates were carved from the 1820s through the 1840s in order to convince officials that the owner was a veteran and was entitled to pension land. Other horns may have been carved during the late nineteenth-century Colonial Revival to establish the fact that a person or family had participated in the American Revolution and was therefore eligible for membership in certain patriotic organizations.

Fortunately, the growing study of individual powder-horn engravers' products and those of carvers in the schools they founded, influenced, or simply belonged to, will make it much more difficult for fakers to market their dishonest wares. At the same time, sensitive collectors of both military artifacts and early American art of other kinds will, I hope, increase their understanding and enjoyment of the beauty of this long neglected art. It is my hope that this exhibition will set a standard for both collectors and institutions to use in judging future powder-horn acquisitions. Unlike other facets of American folk art, in which European and English objects have often been represented as North American, powder horns carved in the provincial manner have well documented places of origin. It is difficult to confuse a North American carved powder horn of the 1746–1781 period with a horn carved elsewhere.

We can only speculate as to how much influence the frontier environment, with its constant threat of confrontation with the French and their Indian allies, had on the new style of horn carving. It seems obvious, however, that since a major change in approach occurred during this period, those trying circumstances *did* affect horn development. The French and English had been enemies in North America since the early part of the seventeenth century, but it wasn't until the Louisburg Expedition of 1745 that major expeditions of Provincial (and British Regular) troops were planned and carried out. And it wasn't until these expeditions took place that the carved powder horn appeared.

Among the experiences that affected carvers' choices of designs and rhymes were the rigors of frontier living, the brutal effect of frontier warfare, the impact of Indians the soldiers faced in combat, and the homesickness that each soldier felt while serving at a remote frontier fort. Other influences may have included the insecurity of being away from a wife or girl friend for a long period of time and (since most Provincial soldiers were farmers) the worry of being away from the farm during prime farming season, when most campaigns were fought. A good example of frontier influence is embodied in the horn of Connecticut soldier Jonathan Hobart. It bears the rhyme, carved on August 3, 1757, "I now at Number Four Remain / Tho Tis Agin My Will / I hope I shall no Enemy Meet / But What I Wound or Kill." Number Four was Fort No. 4 on the Connecticut River at what is today Charlestown, New Hampshire. The rhyme tells us that Hobart was not in the Connecticut regiment of his own free will but was conscripted, and that he was living in fear of combat (horn not illustrated).

The horns that are the object of this study are decorated with intaglio designs and verse. We can identify roughly eighty to ninety percent of the New England soldiers who carried carved powder horns and at least eighteen New England carvers (not all by name) who decorated them and whose distinctive styles others copied. The decorative motifs, rhymes, formats, and themes of the horns we've chosen are indisputably characteristic of the region that encompasses Lake George; Lake Champlain; Shrewsbury, Massachusetts; the Connect-

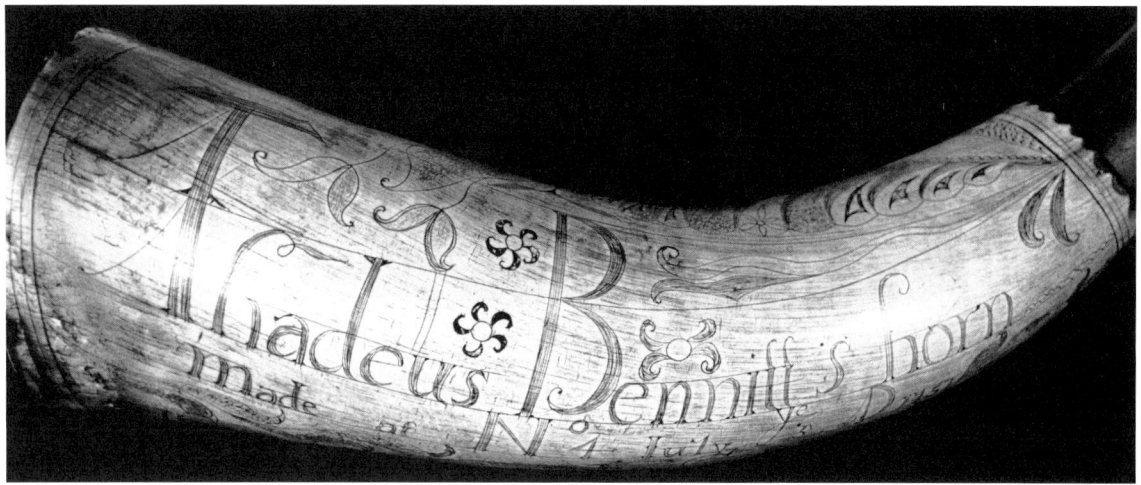

Intaglio lettering and designs (detail, Thaddeus Bennett horn, No. 19).

Revolutionary War motif (detail, Howland or Rowland horn, No. 105).

icut River Valley; and the Lake Ontario/Niagara area. They were made during the King George's War (1744–1748); the French and Indian War (1755–1763); the pre-Siege of Boston period, when anti-British sentiment began to emerge (1765–1774); the Siege of Boston period (1775–1776); through the end of the American Revolution (1776–1783) and on to 1787.

During the years just prior to the American Revolution, horn decorations and formats continued to incorporate earlier ingredients but also introduced new rhymes and decorations expressive of New England's anti-British sentiments. In the early stages of the Revolution, themes reflected events and circumstances in and around the military camps outside Boston. As a result, a new and final set of horn-carving criteria evolved.

I began a preliminary study of carved powder horns more than thirty years ago when I acquired several French and Indian War examples that were almost identical in style and design. Further searching turned up other similar horns, all dated within a one- to five-year period and located in the Lake George region. Eventually I discovered enough horns to document a group of carvers working between 1755 and 1761 at the French and Indian War forts in the areas of the Mohawk, Hudson, and Connecticut Rivers; Lake George; and the Lake Ontario/Niagara region. I concluded that there was a definable school of horn carvers working for the New England troops during the French and Indian War.

Subsequent study and searching led to the discovery of another definite school of carvers working between the end of the French and Indian War and the beginning of the American Revolution (1764–1774) and still another group working during the Revolution at the time of the Siege of Boston (1775–1776). The influence of these later carvers carried through on a few horns until after the Revolution.

Occasionally a signed horn has turned up, enabling the identification of other examples by the same hand. Since most horns are not signed, however, carving characteristics must serve as the basis for identifying anonymous carvers of two or more horns. Eighteen distinct hands are identifiable and another dozen can be closely associated with one of the eighteen because of similarities of style and format. As a result of these findings, I wrote the first article on the subject, which appeared in *The Magazine Antiques* for August, 1978.

My study continued, and in 1982 former Connecticut Historical Society Curator Robert F. Trent and I began to formulate plans for an exhibition of beautifully engraved powder horns of the period 1746–1781, to be hosted by the Society at Hartford, Connecticut. The selection process began with a search for the best available horns carved during those years, because in my opinion the handsomest North American horns date from this brief period. Naturally, some horns worthy of investigation were made later, but none bears a date prior to 1746. Furthermore, of the fine horns carved during the late eighteenth and first half of the nineteenth centuries, none is comparable to those of the thirty-five-year period of this exhibit. Upon Trent's departure from the Connecticut Historical Society in 1989, Elizabeth Pratt Fox became Curator and, with the support of Director Christopher P. Bickford, has assumed a major role in the completion of this exhibition.

Having identified regions and carvers, I proceeded to organize the powder horns selected for exhibition on two graphs: the first accommodated examples that date from 1746 through 1766, the second those made from 1767 through 1781. The graphs showed years and months, and each entry had the date, place, carving school, and, when there was a signature, the carver. Some horns from King George's War (1744–1748) and a few made during the years between that conflict and the French and Indian War (1749–1754) do not have the consistent artistic quality of later examples, and it was therefore necessary to compromise on standards of quality when we included them. I felt it was absolutely necessary to include those examples, however, because they were the first to manifest indigenous North American designs and layouts. They are extremely important in enabling us to perceive the critical transition away from traditional Continental and English decorative motifs and formats and toward a unique American style.

The earliest horns, carved during King George's War, retain traditional European designs and calligraphy. At the same time, examples by a few carvers provide evidence of a transition in which styles, designs, and subjects are selected and applied with much more freedom than was characteristic of European examples. The break from tradition became increasingly apparent during the French and Indian War and eventually a unique new style emerged. (However, a few isolated examples retained European characteristics throughout the thirty-five-year period covered by this exhibition).

Again for the first time, this exhibition shows horns in groups rather than as individual objects, for in general their similarities are more important than their differences. When regarded as part of a group, powder horns take on both historic and artistic significance from the points of view of chronological and regional identification and of quality. Only when they are seen as products of the same period and milieu can their symbolism be most fully understood and the level of their artistry assessed.

Schools of carvers are therefore listed within each time period. I first presented this idea of schools of carvers in the 1978 *Antiques Magazine* article, having by then noticed that lesser horn carvers copied the most fashionable or popular decorations and designs of the leading carvers of their respective periods. This became even more obvious as soon as the horns were arranged chronologically on graphs. Almost identical types, designs, verses, and calligraphy are executed with varying levels of skill within each time period and school.

The graphs proved that the majority of skillfully carved horns were produced during the French and Indian War and that the most productive year was 1758 (survival rates may obviously have a bearing on these findings). There were, of course, many horns carved during this period that had no artistic merit, but these were not included in the tabulation. The next step was to attempt to isolate the specific characteristics of each carver, for even though most horns are unsigned, the work of individual carvers who made more than one surviving horn can usually be recognized.

*Typical J. W. geometric device
(detail, Robert Baird horn, No. 32).*

Drums A'beating, Trumpets Sounding

Constant danger from the French and Indians had been a fact of life for English colonists living on the New York and New England frontiers during the seventeenth and first half of the eighteenth centuries. Both sides raided the trading posts, blockhouses, and settlements of the enemy, encouraging their Indian allies to do a large percentage of the damage, but neither the British nor the French royal government supervised these military actions. In 1665, as part of his ambitious expansion plans, Louis XIV became involved in North American colonial affairs when he sent regular troops to Nova Scotia and Canada to construct fortifications at strategic points along the rivers. In 1669, France established the Ministry of the Marine to oversee its colonies. Troops that were enlisted in France were encouraged to settle in Canada when their tour of duty expired. For the most part, officers were native Canadians who were well acquainted with the terrain and with the type of warfare waged in North America.

In 1686, Holland, Spain, Sweden, several German states, and England formed the League of Augsburg, whose purpose was to fend off Louis's encroachments. The American phase of the League began during King William's War (1689–1697). Small detachments of British engineers, infantry, and artillery were sent to strengthen and man North American military posts. It wasn't until Queen Anne's War (1702–1713) that major shipments of troops recruited in England, Ireland, and Scotland reached North America. Provincial regiments, raised to augment the regular British regiments, were often forced to supply their own accoutrements and were thus not so well equipped as the regulars.

One essential item usually not available to Provincial troops was the cartridge box, a container for the pre-fixed cartridges used in flintlock muskets. Provincial soldiers relied on hollowed-out cow's-horn containers to carry the powder with which they made their own cartridges. These were the most popular powder carriers on the frontier.

The widespread use of powder horns was therefore itself a distinctly American practice, for by the eighteenth century most European and English soldiers used cartridge boxes. These consisted of a leather-covered wooden block in which ten to thirty holes had been drilled to contain paper-wrapped measures of powder and lead balls. It was speedier and more efficient to have pre-fixed paper cartridges, so before each battle artificers, the work force of European and British artilleries, prepared paper cartridges in a part of the powder magazine called the laboratory. The cartridge box also provided an effective means of control over the distribution of powder and ammunition: before each battle a set number of cartridges could be issued to each soldier and after the battle was over and the soldier returned to the fort, unused cartridges could be collected.

Writing this essay has allowed me to view the entire history of horn carving in a brand-new and exciting light. It led me to place carvers in chronological order and, for the first time, to realize how they acquired their styles, developed them, and passed their ideas on to other carvers. It proved that the various carvers were not working in isolation, but were frontier craftsmen capitalizing on the fashions and trends of the world around them. The horns they produced have an almost rigid conformity that was reinforced by other artists' work and by their customers' expectations—the artistic boundaries of the craft can therefore be said to have been established by the tastes and perceptions of both carvers and customers.

Horn carvers' products may not be as supremely perfect as the work of craftsmen who spent years in apprenticeships. However, those more perfect works were not accomplished on round, curved surfaces under what must have been, most of the time, conditions of extreme hardship. Unpleasant weather and working environments could make the horn carver's task almost impossible.

In selecting horns for exhibition, I felt it was important to include as many as possible that were carved by known hands. This occasionally presented difficulties because the work of a single carver can vary dramatically. For example, a carver like James Greenfield engraved remarkable designs and ships, but they are offset by his amateurish calligraphy, and this raised serious doubts about including his work. In the end, his masterful engraved designs and Siege of Boston themes outweighed his calligraphic deficiencies. Actually, the uneven quality of Greenfield's work led to speculation that he may have been functionally illiterate and struggled with spelling and letter formation. This is an interesting possibility and may help explain similar disparities in the work of other carvers.

The horns of Jacob Gay, a carver who worked over a longer period of time than any other, also fluctuate widely in quality. This posed a question: did Gay execute all his own work, or did he partially carve some horns for others to complete later on? (I am led to believe that horn carvers sometimes left blank spaces to be filled in later by the fact that I have seen one horn by the Memento Mori Carver on which the cartouche for the owner's name and date has never been filled in.) After considering this question, I decided to include all available Gay horns, even inferior ones. Several modern fake Gay horns that I have encountered convinced me that representing the entire range of his work was important from a connoisseurship standpoint.

Even master carver John Bush's work varied during the two years he engraved horns. But the J. W. Carver and the Selkrig-Page Carver appear to have turned out consistently good work, as did the Bedford Carver. The latter's work is included in order to represent an important Revolutionary hand, even though he was working well after the Siege of Boston. It would be wrong to infer that because the quality of several carvers' work varies, all of their output is not of a superior quality. It is, and that explains why, after all candidates for inclusion had been considered, most of the master carvers' work was retained in the exhibition. I would estimate that eighty percent of the horns that survive from the French and Indian War and the Revolution are undecorated or crudely carved, amateurish, and stereotyped.

Some of the horns I chose were cut down in length during their period of active use, probably because one end was damaged. After improvising a new plug or spout, their owners continued to use them. I selected, as well, a few others whose surfaces are damaged by worm infestation or moisture or are so worn by hard use that the decoration is only faintly visible. These impaired horns are still significant, for horns of high quality are so rare that even damaged examples are sometimes of great study value.

Rufus Grider (1817–1900), the first great student of powder horns, was attracted to the subject by a fragmentary horn he found in 1886 (No. 104). His interest in powder horns aroused, Grider went on to draw in fine detail over five hundred American powder horns. Grider's first horn, a fine Siege of Boston example, is included here despite its missing half. As it happened, this is the only cut-down horn he depicted, suggesting that he soon lost sight of the artistic and imaginative qualities of horns and concentrated on recording as many historically oriented examples as he could find. The result was that Grider's sketches include good, mediocre, and poor examples, as well as some fakes.

Charles Winthrop Sawyer, a respected early writer on American colonial arms, included a passage in a 1929 article that epitomizes the attitude of later scholars who were attracted to powder horns:

> Military horns, according to the nature of their engraving, are classed as "map horns" and "culch horns." Culch, in this application, does not mean rubbish, but merely second in quality or interest among American military horns—and this frequently for the sole reason that the fad of the moment has raised the map horn to first place. Indeed, a culch horn may be a work of art, a map horn may be a crudity; and yet, because of fashion's dictates, the culch horn is less esteemed. In general, however, a culch horn is really a second grade horn from the standpoint of art. And there is a reason for this; map horns were often ornamented with designs pertinent to military affairs, wrought by men skilled in drawing and engineering.*

This prejudice in favor of map horns persisted from Sawyer's time until 1978, when my article appeared. Since that time, a small group of collectors have carefully examined the pictorial "culch" horns produced by skilled artisans. For them, military significance has become secondary to the quality of the engraving and of imaginative vignettes and verses. Most collectors, however, continue today to adhere to Sawyer's point of view. Among this conservative group, a glamorous military association and a detailed map that extends from New York to Albany and thence up the St. Lawrence River to Montreal or up the Mohawk River to Oswego, are still the hallmarks of excellence. This persistence of what amounts to a Victorian historical prejudice is made all the more ironic by the facts that most map horns were not carved on the frontier and that many of them were made in England long after the campaigns they commemorate. A great many map horns are, in fact, outright fakes. This is also true of the entire genre of horns carved to commemorate the 1762 Havana Expedition, which were professionally engraved and have little in common with the kind of frontier art under examination in this exhibition.

An obsession with historical events and topography is out of the question for today's informed collector or curator, who must exercise judgment with respect to talent, skill, and creativity. Studying powder horns for evidence of these latter qualities will lead, eventually, to connoisseurship and an informed appreciation of these objects as works of art, albeit of an extremely specific type.

Students who want to understand the criteria used here for selecting horns should handle and analyze a horn of the best quality, for connoisseurship of these intimate and complex works is not quickly acquired. A first step is to carefully examine the horn in daylight and under strong magnification. Its physical characteristics should be immediately apparent: horns are light in weight, but heavier than they look; they tend to be at room temperature, but swiftly absorb heat from the holder's hands or sunlight; their surfaces are smooth, but

---

* *The Magazine Antiques*, October 1929, p. 283.

are often abraded in strategic areas. Horn color is a complex phenomenon, for it is made up of the inherent color of the horn plus surface dirt, varnish, or stain; original pigment in the engraving; and the effects of reflected light.

It is difficult to decide on the most desirable horn color. No horn with masterful carving should be rejected on the basis of color alone, although to modern eyes a mellow honey color is probably the most appealing. Some dark horns of a greenish hue that were shaved to make them translucent are superbly carved, however. In some cases horns retain an original varnish that has turned dark brown or greenish-yellow and obscures the decoration. This presents difficulties, for if the varnish is removed to reveal the engraving, part of the original treatment is lost. Except for wiping away surface dirt with warm water and Kleenex, I tend to leave horns as they are. New enzyme treatments for removing oils and varnishes may open up conservation possibilities in the future.

In the past, curators and collectors who thought that powder horns were supposed to be bone-white brutally scrubbed a great many examples. The use of caustics and the practice of scrubbing with abrasives and bleaches were not uncommon. The result was irreversible damage, for not only were the inherent color and original artificial pigments removed from horns so treated, but the structural integrity of the proteins of which horns are composed was attacked. After stripping, these zealous "restorers" often emphasized the carving with black pigments that caricature subtler period effects and are highly offensive. The worst effects of stripping can be reduced in some cases by applications of lanolin. While I have never worked with a conservator at recreating appropriate coloration on a stripped horn, I believe it is possible that techniques may be found to rehabilitate some examples.

Because so many factors are involved in the "patina" or apparent color and surface condition of a horn, this is not easy to fake. Therefore, most fakers use old horns, because plain genuine horns are abundant.

A horn's plug and the pins that secure it, the engraving of the body, and all incidental cracks and crevices should be consistent and free of accumulations of recent pigment. Condition is important, but I believe that minor surface problems, the loss of a plug, or a chipped or split spout should never prevent a collector or curator from acquiring a choice example of a major carver's work or a horn that clearly relates to one of the recognized schools. I prize highly the five or six cut-down horns in my collection because they are magnificent and informative fragments that are worthy of being preserved for present and future study. This also holds true for worm-damaged horns. A powder horn's size is important because each design layout is directly related to the amount of space available. Large horns are often "easier" to look at, but small horns often elicited extremely fine, small-scale engraving from the carvers.

If the horn's color, surface, and fittings seem convincing, the next step is to see if the style and calligraphy of its engraving are appropriate to the date of its inscription. The schools of carving presented in this study differ markedly. However, as noted above, even the work of known carvers varies extraordinarily, and establishing the authenticity of important signed horns is a painstaking procedure.

A knowledge of period calligraphic styles is an important aspect of connoisseurship that can be acquired through the study of modern monographs and of period writing books and

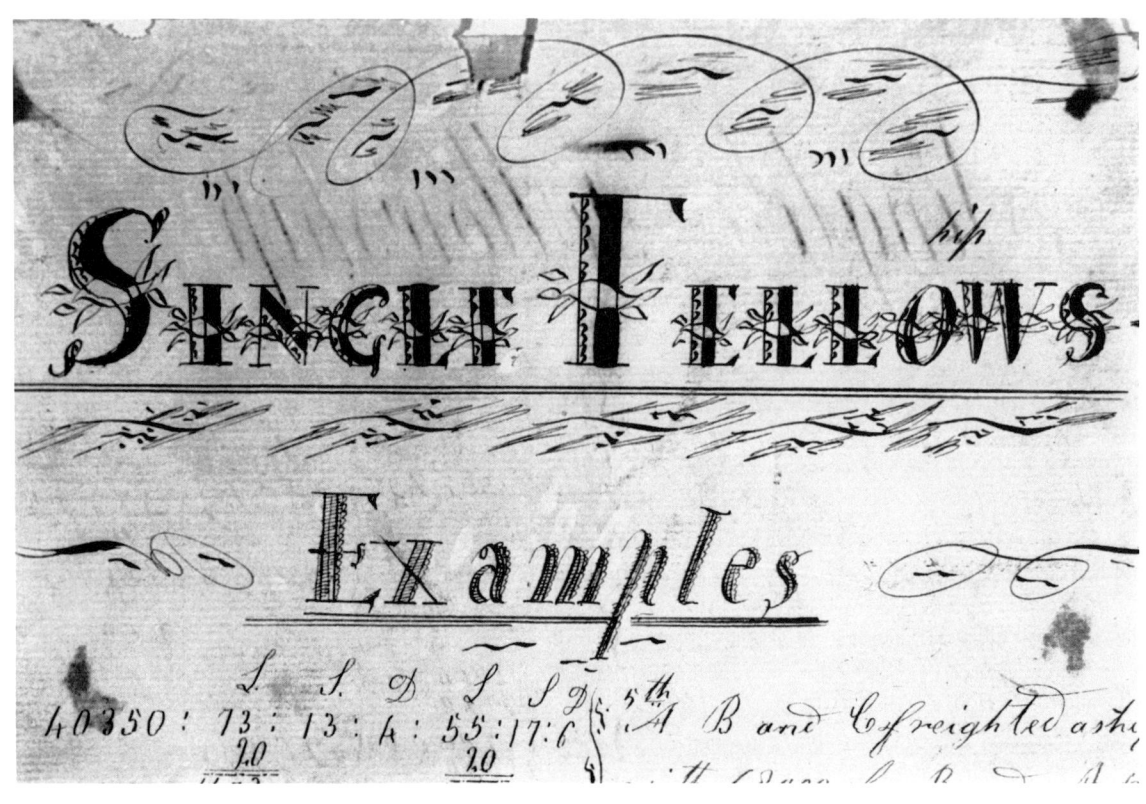

Schoolboy's exercise book, top, c. *1770*, bottom, c. *1760*.

Powder Horns Carved in the Provincial Manner

manuscripts. Horn carvers were consulting material whose calligraphic sources dated from the 1500s up to their own era, so that eighteenth-century scripts in the copperplate style are not the only ones we encounter. Today's student can acquire a more immediate idea of period orthography, ampersands, abbreviations, and phonetic spellings, as well as the ability to weed out spurious or later inscriptions by reading diaries and letters and perusing the jottings and scribblings inside eighteenth-century book covers and exercise books. Such research can also familiarize us with the thoughts, feelings, and concerns of the soldiers, whose letters and diaries indicate that they were often cold, hungry, homesick, and terrified. The motifs, scenes, and verses they chose to engrave on their horns have meaning and poignancy.

Calligraphy manuals like John de Beauchesne's and John Baildon's *A Booke Containing Divers Sortes of Handes* (London, 1571), John Seddon's *Penman's Paradise* (London, 1695), John Ayre's *A Tutor to Penmanship* (London, 1698), or George Bickham's *Penmanship in its Utmost Beauty* (London, 1731) are obvious possibilities as inspirational sources for horn decoration, although their use by New England writing masters is difficult to substantiate. More logical sources would include the Bible; engraved trade cards; labels on trunks; engravings on official documents like commissions, bonds, maps, currency, and deeds; prints; magazines; and title pages of books. It is important to remember, however, that while investigation of such materials does provide insight into sources for certain motifs, the mentality of barely literate soldiers often transformed these in surprising ways. After the initial period of experimentation exemplified by the simple calligraphy of King George's War horns, a fairly consistent style developed in the 1750s, which undoubtedly became standard through imitation in the camps and forts where horns were made.

Our admiration for the powder horn as an art form increases when we understand the psychological climate in which it was made. While some might find fault with the concept and execution of these "crude" objects, modern American connoisseurs can appreciate more and more the uniqueness of this outstanding and ephemeral art form.

Detail, Massachusetts Bay military commission engraved by Nathaniel Hurd, Boston, 1771.

Drums A'beating, Trumpets Sounding

HOW POWDER HORNS WERE CREATED

THE SOURCE of the horns that were carved during the period of this exhibition has never been documented. They were a by-product of the cow, oxen, and steer industry that slaughtered animals for their meat and hides and packed these for shipment.

In order to hold enough powder—probably a pound for most examples in this exhibition—a horn had to be at least sixteen to twenty-four inches long in its raw state. Early writers on powder horns such as Charles Winthrop Sawyer and Stephen V. Grancsay claimed that horns of this size were found only on cattle descended from Spanish stock and raised in South America. These writers also asserted that the majority were imported by the Bird family of tanners and wholesale leather merchants who settled in Dorchester, Massachusetts, in the seventeenth century. The horns, which were shipped with the hides to North America, are said to have cost the Bird family about one cent each in the rough. Early articles and books give no documentation for this story and it is my suspicion, again undocumented, that many horns were acquired by soldiers when cattle were slaughtered for food during the campaigns.

No period documents describe the method used to prepare horns to serve as powder containers. Since Grider's time, however, many writers have added their own conjectures to his theories. Because I respect Grider's early research on carved horns, his ability to identify the work of various carvers, and his attempts to copy their characteristics in his drawings, I rely on his theory about the preparation of horns to receive powder. It may be that Grider spoke with farmers or mechanics who had prepared cow horns to be used as powder horns, drinking vessels, scoops, or haircombs. He wrote:

> A cured horn possessed all the requisites for carrying gunpowder by the soldier in wartime. It was easily obtained, light in weight, quickly prepared and very strong. It neither molded nor decayed, even if buried in the earth. When carefully fitted, the powder kept dry even if the horn were carried in the rain for days.... A horn could also be floated on water without harm to the powder within, which was of great importance when an army had to ford streams. When a horn was fresh, the pith inside was removed by soaking or boiling it in water; then it was scraped, cleaned and fitted with a wooden end. [Grider did not mention that several inches were trimmed off the butt end to provide an even opening into which the plug could be fitted.] The point was shortened by being sawed off, and was then bored to effect an opening. Next came polishing and engraving. Crude engraving could be done with a pocket knife, but that method fractured the surface and the lines.
>
> Some horns were neatly marked by pricking the surface with a needle, but the best work was done by means of a graver [or burin], the decoration being first outlined with a lead pencil.
>
> All those [horns] made by professionals were given a bath in some yellow or orange dye which imparted an amber color. To show the engravings, the surface was rubbed with wet or dry brown color which filled the lines made by the graver. When that was dusted and cleaned, the process ended by a polish with emery and oil.
>
> Sometimes vermillion was used in giving prominent features, [such] as forts, roofs

of buildings, etc. The yellow dye used was enduring; water and soap had no effect upon it and it could only be removed by scraping. What was used to produce such a lasting dye could not be ascertained; some thought it was due to boiling in water with copperas or butternut rind, but experiments did not reveal the process, hence it will have to be classed among the lost arts.*

Many of the better horns I have seen retain the old varnish Grider describes, and even when it has turned dark with age, the original engraving underneath it is preserved in almost perfect condition.

The wooden plugs of most horns are fastened with wooden pegs, although brass or iron pins, nails, or even friction fitting were also sometimes employed. It was not unusual to paint plugs to waterproof them, and often some kind of sealer was placed between the plug and the horn before the plug was inserted. Wooden stoppers were used at the spout end. Some are beautifully carved, but most have long since been replaced with crude substitutes.

HORN NOMENCLATURE

Neither period documents nor early researchers supplied a comprehensive set of terms for referring to the various parts of a powder horn. I have regularly used the following terms in this work:

Plug: The rounded wooden insert at the butt end of the horn.
Plug or butt end: The wide end of the horn, where it was attached to the cow's head.
Body: The main portion of the horn where most of the engraving is placed.
Border: Decoration encircling the horn at the butt end or near the recessed portion.
Recessed portion: An inset area near the spout where more layers of horn have been removed to provide a collar for attaching one end of the carrying strap.
Throat: An alternative name for the recessed portion, or narrow end of the horn.
Raised ring: Rounded rings carved at the tip or within the recessed portion to retain the carrying strap.
Spout: The end of the recessed portion, or narrow end, plus the tip.
Tip: The end of the horn where the pouring aperture for the powder is located.
Stopper: A separate plug of wood or bone that closes the tip.
Extension lobe: A piece of the butt end of the horn that extends beyond the main body of the horn, pierced with holes for securing one end of the carrying strap.
Staple, rivet, or nail: Horns that later lost the extension lobe or did not have one originally are often provided with alternative suspension devices at the butt end.
Carrying strap: A leather or textile strap attached to the horn at both ends, worn over the shoulder.

---

* Rufus A. Grider, "Powder Horns, Their History and Use," edited and annotated by A. J. Wall, *The New-York Historical Society Quarterly Bulletin*, 15, No. 1 (April 1931), 4.

Funnel: A small, separate funnel, often made of a tip of another horn, used for filling the horn with powder.

Flintlock: The mechanism used to fire the muskets, rifles, or pistols for which gunpowder was provided; it consists of a cock (a steel clamp or jaws that held a piece of worked flint), a battery (a steel piece that the flint struck to create sparks that ignited the powder), and a pan (a cuplike container between the cock and the battery, which held priming powder set alight by the sparks, thus entering the barrel through a small hole, igniting the main charge, and propelling the bullet).

RUFUS A. GRIDER AND HIS DRAWINGS

RUFUS ALEXANDER GRIDER was born in Lititz, Pennsylvania, on April 13, 1817, to a family of Swiss extraction. His father was an ardent member of the Federalist Party and named his son after Rufus King, a United States senator from New York. Grider attended Beck's School for Boys in Lititz, where he became interested in colonial history. He married Elizabeth Skirving of Germantown in 1864 and they had two daughters before Mrs. Grider died in 1875. Grider belonged to the Moravian Church of Bethlehem.

In 1883, Grider moved to Canajoharie, New York, where he taught drawing in the schools of that town and of nearby St. Johnsville. As R.W.G. Vail stated in a 1943 article,

> From 1886 to about 1900 Grider travelled up and down his beloved Mohawk Valley, with occasional excursions to the Cherry and Schoharie valleys or to Lakes George and Champlain, in search of historic buildings, battlefields, the sites of ancient forts, the relics of the Indian and pioneer, all of which he transferred to the pages of his pictorial albums. . . .*

Vail went on to assert that Grider's "hobby was to leave a pictorial record of the historical spots of the State for the use of later historians who would not have his opportunities or his talent with pen and brush."

Grider's surviving papers, now at the New York State Library, occupy nine volumes that contain over 1,000 pieces, including 623 watercolor sketches, 42 watercolor copies of miniatures, 7 original manuscripts, 169 tracings of original manuscripts, 71 tracings of maps, 81 engravings, 23 photographs, and 25 colored drawings of engraved powder horns. The New-York Historical Society owns the major collection of 500 Grider drawings of powder horns. One hundred forty-three of the drawings were exhibited at the United States Geographical Exhibition at the Chicago World's Fair of 1893 at the request of the Secretary of the Interior. In 1894, part of the drawings collection was exhibited at the Smithsonian Institution in Washington, D.C., and in 1896, some were displayed at the University of Pennsylvania upon the occasion of a visit by President McKinley.

Isaac J. Greenwood, a trustee of the New-York Historical Society, purchased the collection of powder-horn drawings in 1895. Before selling his collection to Greenwood, however,

---

* Vail, New York State Librarian, wrote this for the *Journal* of the Schoharie County Historical Society, p. 1.

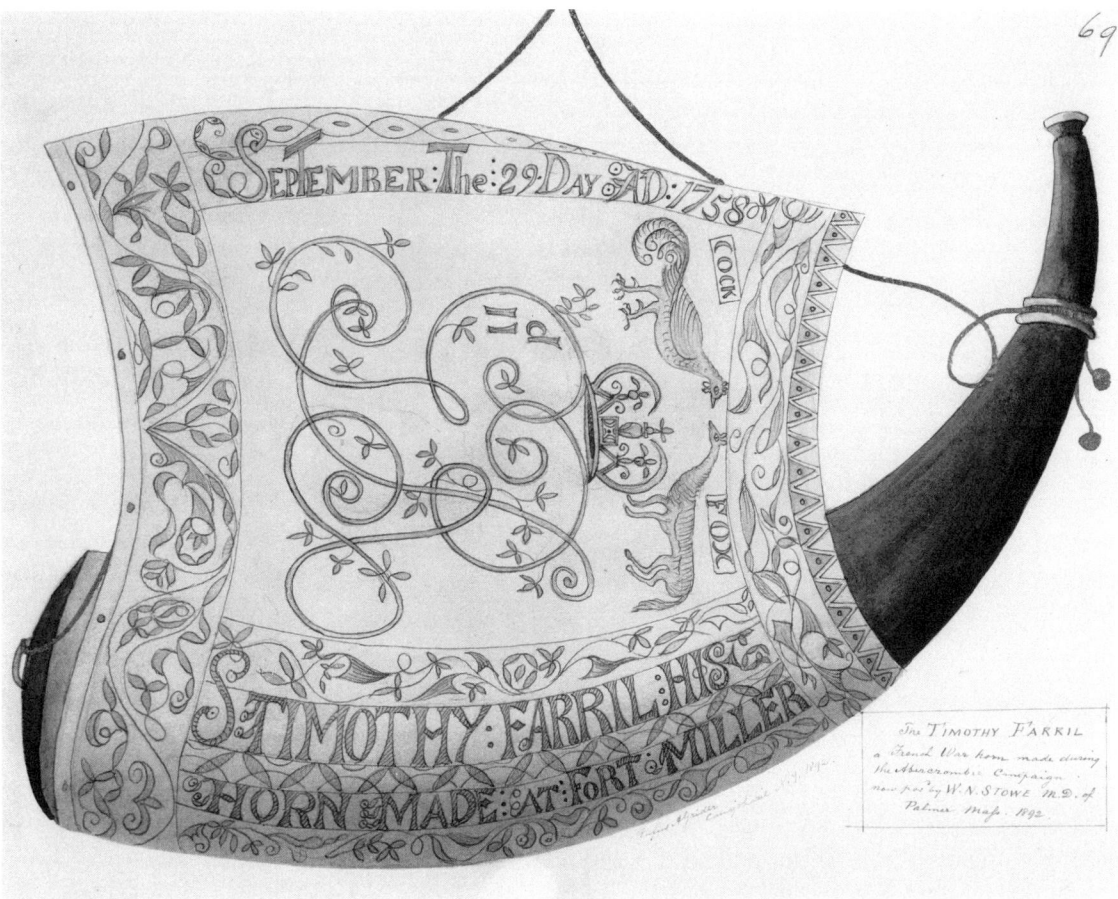

Typical Rufus Grider roll-out powder-horn drawing, 1892. New-York Historical Society.

Grider compiled an annotated catalogue, and this forms the basis for most of what is known about the artist's research activities. Grider died on February 7, 1900, and in 1907 Greenwood presented the drawings to the Society.

Grider's pioneering drawings of and observations about North American carved powder horns underlie all subsequent work on the subject. Much of the literature listed in the bibliography that was produced after Grider's initial efforts merely echoes his thoughts and theories. An exception is the work of the late Stephen V. Granscay, Curator of Arms and Armor at the Metropolitan Museum of Art. In his *American Powder Horns*, Grancsay drew upon Grider's work and upon the collection of J. H. Grenville Gilbert, which was donated to the Metropolitan by Gilbert's widow. The horns in this collection, like those recorded by Grider, were of every type and were valued for their historical interest. It is apparent that both men lacked connoisseurship skills with respect to powder horns, for both recorded or collected examples that are obvious fakes. Grancsay copied Grider's method of drawing horns, utilizing the services of Works Projects Administration artists. He expanded Grider's checklist of horns, adding examples collected after the latter's death.

Scholars today would be severely handicapped without the work of these two early compilers. Although many of the horns in Grider's drawings and Grancsay's book are of slight artistic value or are in fact spurious, many others are of the first importance and are unfortunately unlocated today. Grider, in particular, was extremely careful to note the work of individual carvers that he recognized, and his vast compilations have proven invaluable to me in the years of research that have led up to this study.

CHRONOLOGY OF THE CARVING

CARVED POWDER HORNS decorated in the provincial manner that is the subject of this exhibition began to be made during King George's War. Almost all the horns selected appear to have been carved by New Englanders for New England soldiers. No genuine American powder horn that dates before 1746 is known, and consistent development of both decorative schemes and calligraphy suggests that most horns were made at the time they are dated. However, some examples inscribed with dates earlier than 1746 may have been carved later and back-dated.

Where did this carving style originate? The evidence suggests that there was no single source, for a great number of motifs and compositional strategies seem to have converged in the North American product. Europeans had been decorating metal powder flasks and cow- or stag-horn powder containers for almost two hundred years before the American genre came into being. The designs that accomplished craftsmen at European courts carved or chased on horn and metal probably originated in the workrooms of French master gun designers, particularly those working for Louis XIII and Louis XIV between 1630 and 1710. Because elaborate firearms were popular royal gifts, European noblemen who had seen or received them probably recommended these elevated prototypes to their local gunsmiths.

It is highly unlikely that any of these royal treasures ever reached North American forts. More likely, craftsmen working in other mediums copied gunsmiths' designs and *their* products made their way to North America. Mountings on lesser-quality officers' swords and pistols, designs on officers' insignia, printed documents embellished with illuminated flourishes or enriched typefaces, calligraphically inscribed official manuscripts—all were available for inspection by those involved in the militia system. Other possible sources include gunnery and drafting instruments, decorated silver and ceramic objects, picture books and magazine illustrations, architectural design books, decorated maps and drums, and, of course, the Bible.

By 1700, Continental and English craftsmen were producing simpler incised powder horns, but their designs are still far closer to high-style precedents than were those of North American horns. Europeans eventually produced horns with simplified floral decoration and hunting scenes, some of which may have been brought to North America, but even these reflected earlier traditions.

The decorative style that emerged in New England and New York during King George's War represented a distinct break with this earlier heritage and embodied a freedom of expression that reflected the frontier environment. By the time of the French and Indian War the distinctive frontier style of powder horn reflected the fantasies and feelings of both officers and soldiers. The tradition persisted into the Revolution, and isolated schools of carvers carried the style far into the nineteenth century, but the high point of artistic accomplishment in horn carving passed halfway through the Revolution, when cartridge boxes superseded powder horns. All horns carved after that time were more or less souvenirs rather than a necessary piece of equipment, except those used by riflemen and for priming artillery.

Horns of the King George's War Period (1744–1748)

Suddenly, in 1746, a definite style of carved powder horn appeared in several examples made at North American frontier forts. Talented carvers working during King George's War developed a recognizable manner that combined traditional European designs with innovative new motifs and compositional strategies. The initial style, which formed the basis for the great developments of the 1750s, has the following index features:

> (a) Calligraphy that is a composite of Gothic and German Renaissance lettering. Stylized, double-line letters are rigidly incised and accented with diagonal dashes or dots between the lines. A sharp "V" is formed at the junction of branches of letters, as well as at the ends of letters and of most date numerals.
> (b) Elaborate scrolled decorations of foliate or geometric types that are applied to the branches of many letters.
> (c) Geometric designs that are often incised between words or between the letters of one word.
> (d) Elaborate geometric or foliate borders that are placed above and below inscriptions.

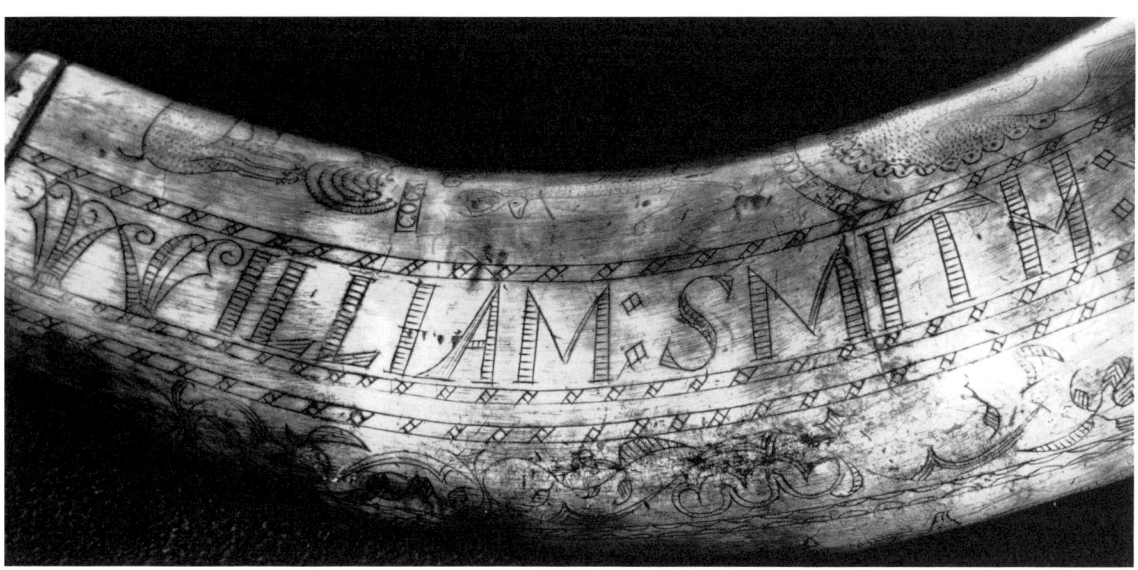

Calligraphy in the King George's War style (detail, William Smith horn, No. 1).

Relatively few horns survive from this formative period. In contrast to horns from the two later wars, horns from King George's War lack campaign locations and the owner's military rank. Some original owners can be traced in enlistment records, but existing rosters are incomplete and many early horn owners' names are unknown. It is obvious, however, that the unidentified horns are not random efforts but are closely related to examples that can be documented to specific military encampments.

If we accept the King George's War School of carvers as the source of later schools, we can detect in several examples characteristics remarkably similar to those of horns of the Lake George School of the 1750s. For example, the David Willson horn (No. 4) and the Nathaniel Willson horn (No. 5) conform strictly to King George's War styling. A second group contains four other contemporary horns that combine King George's War traits with calligraphy of the kind associated with the carver John Bush (working 1755–1756) and decorative features seen in the work of the carver Jacob Gay (working 1758–1787). These are the 1746 William Smith horn (No. 1), the 1747 David Fletcher horn (No. 2), and the 1747 Meshach Taylor horn (No. 3). Their elaborate calligraphy, geometric devices, and animated depictions of people, animals, birds, and fish clearly influenced the decoration of horns made later.

Fantastic animals characteristic of King George's War horns (detail, Jonathan Cook horn, No. 6).

The third group that developed during the later part of King George's War extended into the early 1750s. The earliest horn of this group (No. 6) is inscribed with three different names and dates: "Jonathan Cook 1747," "James Melvin 1746," and "J M 47." The first inscription, in mediocre Gothic and block lettering, is incised within a single-line border and is the only one of the three groupings that obviously was planned for the space it occupies. The second inscription is indifferently scratch-carved, but "J M 47" is cleanly carved in the King George's War manner. The inference is that the Cook inscription is that of the first owner and that Melvin acquired the horn soon after it was made, possibly back-dating it. Profusely carved with realistic and fantastic animals that are appealing but not well executed, the horn also displays geometric and vine borders and meanders of the same pedestrian level of accomplishment. A doe with a nursing fawn is the most important motif, for it appears on other King George's War horns and persists into the next period. The importance of this otherwise undistinguished horn and its background is that in exhibiting the free compositional spirit associated with the Lake George School of the 1750s and 1760s it represents a complete break with European formal traditions. Records indicate that a Jonathan Cook served at Fort Dummer in 1749, so perhaps he was already serving in 1747. Fort Dummer was built near Brattleboro in 1724 on the site of present-day Vernon, Vermont, as part of a line of forts the Massachusetts provincial government constructed to protect the frontier.

Fort Dummer is a two- or three-day march from Shrewsbury, Massachusetts, which was probably the birthplace of the later Lake George School of carvers. Shrewsbury was definitely the birthplace of the carver John Bush (1725–1757), some of whose Lake George and Fort William Henry horns dated 1755 and 1756 are among the finest ever made. It is also significant that a horn carved for Samuel Crosby and inscribed "Shrewsbury, 1748" (No. 7) exhibits the most radical departure from the restrained and simple King George's War style. Its features are a composite of those typical of many Lake George horns and the Crosby horn is thus key in linking the new approach to horn decoration, Shrewsbury, and John Bush. An elaborate scrolled border above and below the inscription has great freedom of execution, as does a whimsical character in period garb. Fully-rigged sailing ships sport cartoonlike figureheads, and one has a rooster weathervane atop the mainmast. A variety of geometric devices separate the words of the inscription and stylized lunettes form borders at both the spout and plug ends.

A second Shrewsbury horn is inscribed "Shrewsbury * November * The * 9 * 1749 / ASA * HAPGOOD * HIS HORN" (No. 8). The asterisks stand for a diamond-shaped device Lake George School carvers of the 1750s often used. Other forward-looking features are the neat but simple foliate and geometric borders and the shaded, double-line block lettering.

A third Shrewsbury horn is engraved "Shrewsbury * January * 14 * A DOMINI / LEVI * WHITNEY * HIS * HORN * 1750" (No. 9). Family tradition relates that the horn descended from Lieutenant Samuel Whitney of Stratford, Connecticut, to an Elisha Foote of Simsbury, Connecticut, whose two

Whimsical characters on a Shrewsbury horn (detail, Samuel Crosby horn, No. 7).

militia commissions accompanied the horn when it was purchased from a descendant near Albany, New York. The first commission, issued by the State of Connecticut, is dated 1793, while the second, issued by the State of New York, to which Foote moved, is dated 1798. The connection appears to be that Samuel Whitney's daughter Sarah married Daniel Foote of Newtown, father of the Elisha Foote who later owned the horn.

However, records do not confirm this traditional history, and it has been suggested that Whitney was born in Westminster, Massachusetts, in 1719, served in the Massachusetts militia, and held a commission of lieutenant in that province. This has obvious implications for a probable Shrewsbury origin for the horn.

The calligraphy of the Whitney horn is almost identical to that of the Crosby horn, with simpler diamond devices between words. A snake labeled "BRAZEN / SERPENT" runs the length of the horn. Schools of fish and flocks of birds surround a grotesque creature labeled "Divel." Other motifs include weedlike borders underneath the snake, a geometric design composed of shaded scrolls, and three sunbursts. The reference to "Brazen Serpent" is from the Bible, John 3:14–15, and is a symbol of the redemptive death of Christ.

These three horns, which may be directly related to John Bush's early work, demonstrate that many important features of the Lake George School almost certainly had their inception in Shrewsbury, a relatively isolated frontier town through which companies of troops marched on their way to forts farther in the interior.

John Bush himself deserves further scrutiny. His only known horns are dated 1755 and 1756, when he was roughly thirty years old. He was born in the North Parish of Shrewsbury (now Boyleston) about 1725 or 1726 and, from 1747 up to and including 1757, is listed as serving as a clerk in a Shrewsbury company during some of the campaigns he enlisted in. This indicates that he could keep accounts and had a legible hand. If his horns are any indication, he probably wrote an excellent copperplate script.

John Bush's mother's name is not known, but his father, Georges Bush, was born in South America or perhaps in the Caribbean and emigrated to the Massachusetts Bay early in the century. The father was a free Black who died in 1767 at the age of eighty, leaving personal property and land to five surviving children. Three of his sons died while serving in the Massachusetts militia: Georges Bush, Jr., the oldest, died at Lake George on September 25, 1755; Joseph Bush died on April 8 during the 1756 campaign; and John Bush, the carver, was captured by Indians in 1757 and was never heard from again.

John Bush had prepared a will prior to enlisting in the 1757 campaign. Dated April 25, 1757, it was probated October 14, 1758. A clerical docket attached to the will describes Bush as the Negro son of Georges Bush, Sr., husbandman.* John Bush left his father and his son three pounds each and his four sisters the sum of twenty pounds to be divided among them. He makes no mention of a wife. He left the residue of his estate to his brother, Benjamin Bush, who was named executor.

Georges Bush, Sr., wrote a letter to Governor Pownall of Massachusetts on September

---

* Worcester probate case #9402, dated April 25, 1757. Elaine Bush Prince very kindly supplied this information.

14, 1758.* He asked for help in locating his son John, whom he described as a "Mulatto Fellow, about 30 years of age," which suggests that John Bush's mother or perhaps a maternal grandparent was white or Creole. Bush explained that his son had been taken prisoner by Indians allied with the French under General Montcalm at the surrender of Fort William Henry and had last been seen alive while being led off as a prisoner. Bush asked that the governor inquire whether his son was still alive in a prison camp outside Quebec, and solicited help in securing his release. So far as we know, nothing came of these requests.

Unanswered questions about Bush are many. Where and when did he start to carve horns? Did he originate the Shrewsbury carving style, or did he learn it from an older man? The crudity of Shrewsbury horns made in the late 1740s bears no comparison with the excellent work Bush did in the mid-1750s. At the same time, at least two of his 1756 horns are of lesser quality. Did severe weather conditions or physical illness inhibit him some of the time?

John Bush's graceful copperplate calligraphy (detail, William Williams horn, No. 11).

Bush's calligraphic styles, his formats, and his decoration became the basis for the Lake George School of the 1750s. All were copied by the Selkrig-Page Carver, the J. W. Carver, and the Memento Mori Carver during the campaigns of 1757 through 1761. Thus an otherwise obscure Black farmer can with some justification be regarded as one of the founders of an entire tradition of American folk art.

---

* Mass. Archives docket, Vol. 84, p. 310.

Horns of the Lake George School (1755–1763)

Horns belonging to the Lake George School range in date from 1755 to the mid-1760s. Most were made at the string of forts from Albany to Lake Champlain that witnessed the most important battles between the English and the French for the control of North America. It is important to remember that of the hundreds of horns that survive from this period only about one-fifth received the spectacular engraving and carving now identified with a small group of artists that are the subject of this catalogue.

The index features of the Lake George School are far more elaborate than those of the King George's War School:

(a) Gothic, German Renaissance, and copperplate calligraphic styles are freely combined, sometimes augmented by block lettering, illuminated initials, and decorative devices like finials on letters. The engraving is looser and more accomplished than it was in the 1740s.

(b) Inscriptions follow fairly consistent formats that can include military rank, name, place, and date; rhymes; captions on forts or animals; brief accounts of battles; and mottoes on coats-of-arms.

(c) Lettering sometimes incorporates pictorial devices like animals, birds, and human and grotesque faces.

Pictorial devices characteristic of John Bush's work (detail, David Baldwin horn, No. 14).

(d) Detailed borders appear above and below inscriptions and at throats and plugs. Designs include chevrons and dogtooth, sawtooth, zigzag, diamond, or triangular and shield-shaped patterns, shells, and scrolls.

(e) Representations of people, animals, caricatures, or grotesques cover a wide range of subjects, including soldiers, Indians, women, ship figureheads, sea monsters, mermaids, Loreleis, gargoyles, and cherubs. Often figures are combined in vignettes.

(f) Soldiers are often depicted performing the evolutions of warfare or drill. When cavalry is portrayed, horses and riders often have the same facial expressions.

(g) Other common motifs include scenes of towns, diagrams of forts, ships under sail, birds in flight, schools of fish, implements of war, entrenching tools, and smoking and drinking paraphernalia.

(h) Ornament distinct from borders includes scrollwork, floral and vine designs, and baroque cartouches.

Powder Horns Carved in the Provincial Manner

The Lake George School Carvers

John Bush

Two different hands carved the four earliest horns in the full-blown Lake George style. The Thomas Williams horn is the only known horn that John Bush signed (No. A), while the William Williams horn (No. 11) is attributed to Bush. The Rufus Hill (No. 10) and Sol Tyler (not in the exhibition) horns are by an unknown hand (henceforth the Hill-Tyler Carver). Both carvers employed the combination of Gothic and German Renaissance lettering characteristic of the King George's War period, but dispensed with the rigid structure of that early style. They introduced whimsical birds, fishes, and "funny faces," as well as beautifully executed floral and geometric designs that filled the spacial voids inherent in the surfaces of horns.

The Hill and Thomas Williams horns bear explanations of the Battle of Lake George, while the Tyler horn displays the caption "Carrying Place," an alternative name for Fort Edward. The Hill and Thomas Williams horns are dated at the time of the battle, and the William Williams horn most likely dates from then as well, but the Tyler horn dates two weeks prior to the battle. The rhymes engraved on the Thomas Williams horn, and the narratives on that and the Hill horns, as well as the date and place engraved on the Tyler horn, illustrate standard patterns for most Lake George horns that follow.

Superbly executed letter "A" on Bush horn (detail, David Baldwin horn, No. 14).

The David Baldwin horn (No. 14), carved at Fort William Henry and dated October 18, 1756, is one of the finest attributed to Bush. It has all of his best features, including superb illuminated lettering of the word "WAR"; incised chevron, scrolled floral, and geometric borders; fine copperplate-script calligraphy with winglike serifs; a four-line rhyme; and a scalloped throat embellished with cherubs and crosses.

The Selkrig-Page Carver

Because Bush lived in a center of horn carving, he is most likely to have introduced key aspects of the style. But Bush disappeared from view after his capture on August 9, 1757. The Samuel Lounsbury horn (No. 23), dated June 20, 1757, follows the Bush tradition closely but is not by his hand. The Lounsbury horn does, however, relate directly to a group of five horns carved in 1758 that differ from it in depicting formations of soldiers.

These six related horns are named after the two most dramatic examples, those made for Nathaniel Selkrig (No. 24) and Aaron Page (No. 25). They display the word "WAR" in magnificent illuminated lettering, sawtooth carving, floral-scroll borders, and popular rhymes, but their most outstanding feature is the line of soldiers. The troops are arranged in formations taken from formal drill as taught in military manuals, and the accurate uniforms are those of various units like light infantry, rangers, and artillery. The figures have amusing animated facial expressions, and while they purportedly show the horrors of war, they have a distinctly comic (or at least ironic) aspect.

All the owners of Selkrig-Page horns were Connecticut soldiers in the same regiment; some were in the same company and several were rangers. The locations given on the horns are Fort No. 4 and Lake George. The carver was most likely a Connecticut soldier.

The J. W. Carver

Another carver whose engraving is of the highest quality sometimes signed his horns "JW—his pen." His work dates from 1758, when he executed a horn for Robert Baird (No. 32), to November of 1761, when he made the Samuel Whitaker example at Crown Point (No. 37). Unlike John Bush and Jacob Gay, J. W. consistently produced high-quality work. On several horns, most notably those made for Whitaker and for Enoch Cooper (No. 35), he approached the uppermost limits of the art's possibilities.

J. W.'s calligraphy, which closely resembles that of John Bush, is meticulous. It is carefully and evenly spaced and is executed in a superb script that always adheres to the rule of copperplate "thick-and-thin" alternation, never deviating from horn to horn. Many of the letters in his inscriptions are embellished with winglike serifs similar to Bush's, a trademark that helps in identifying his unsigned horns. These winglike devices also appear elsewhere on J. W.'s horns, sometimes attached to design motifs, sometimes floating in space. Equally light and graceful are his animals and birds, which appear floating in space or as part of scrolls. The long, gracefully turned necks of these creatures complement J. W.'s floral designs. His scrolls often serve as borders around inscriptions or as a setting for a central geometric device that also functions as an index feature of J. W.'s work. His carving is probably the most delicately executed of the later Lake George School. His design motifs and his clever rhymes, which are often abstract, prompt the viewer to consider the meaning of J. W.'s composition as a whole.

Jacob Gay

Only one carver was more prolific than J. W., because he carved over a much longer span of time. His name is Jacob Gay, sometimes spelled Gauy or Guay or simply indicated by the initials "JG." His identity is still a mystery. Genealogist Robert C. Anderson does not believe that Jacob Gay belonged to the large Massachusetts family that descended from John Gay of Watertown, Massachusetts, but has found deeds that identify a Jacob Gay or Gauy who gave his residence from 1760 to 1781 as Allenstown, New Hampshire.*

A possible clue to Gay's origins lies in a horn he signed that now belongs to a small

---

* Robert C. Anderson of Salt Lake City generously did the research on the Gay family for this exhibition.

Powder Horns Carved in the Provincial Manner

Formation-of-soldiers motif introduced by Lake George School carvers (detail, Nathaniel Selkrig horn, No. 24).

Carver J. W.'s signature, J W his Pene.

Typical J. W. calligraphy (middle and bottom details, Robert Baird horn, No. 32).

historical society near Boston. Local tradition relates that the carver of the horn presented it to an inhabitant of the town who had befriended him in 1755, when he came with 6,000 other French Acadian refugees who were expelled from Nova Scotia and settled in British colonies along the Eastern Seaboard. Gay or Guay is a name associated with these French settlers, many of whom dropped the "u" from their names later on. The Hamilton Davidson horn (No. 69) of 1772 is signed "Jacob Guay," as are some of the Allenstown deeds. At this point, the most likely theory is that Jacob Gay and the New Hampshire resident are one and the same. Jacob Gay of New Hampshire could easily have traveled to Lake George and Lake Champlain during the French and Indian War and to Boston and New York during the Revolution.

The variable quality of horns in the Gay style is a definite problem. Gay's best animals are beautifully carved. The facial expressions on his animals and soldiers, the curled snouts of his animals, and the jaunty posture of his figures are all readily recognizable. A few signed horns, however, have animals, soldiers, and calligraphy of indifferent quality. A number of examples demonstrate how difficult it is to explain this situation: the John Pemberton horn (No. 39) and the William Goding horn (No. 40), for example, were both made at Fort Edward in March of 1759. They are nearly identical, with inferior carved animals and soldiers. The Pemberton horn is signed "Jacob Gay / Hand Writ" and confirms a Gay attribution for the unsigned Goding horn. Yet the Jotham Bemus horn (No. 41), made at Stillwater and dated September, 1759, is signed "JACOB GAY han" and is a superlative example of the artist's finest carving of animals and soldiers. Why should such a tremendous variation in Gay's production have occurred in the course of seven months? Similarly, the Amos Bostwick horn (No. 95), carved seventeen years later, has non-Gay calligraphy and an atypical lion and unicorn, but exhibits other animals that are without doubt peculiar to Gay's work.

Several theories for the variation in Gay's production arise. One is that he carved a group of horns and left blank spaces so that others could later fill in the inscriptions. Another is that his work was so highly esteemed that others imitated his style and forged his signature. The wild variations in Gay's carving could also be explained by the existence of a drinking problem or arthritis.

During the early years of the Revolution Gay carved bust portraits within elaborate cartouches that included patriotic slogans. These resemble portraits on the William Smith horn (No. 1) and the David Fletcher horn (No. 2), illustrating the fact that the influence of the earlier King George's War School endured to some extent in later work.

The Memento Mori Carver

Another major carver is notable for his sardonic humor. He is called the Memento Mori Carver because most of his horns include that phrase as a decorative device. His known work dates from 1756 through 1760 and was executed at Lake George, Lake Champlain, and Lake Ontario (Oswego). All his horns have a border circling the plug end consisting of a series of either dotted or cross-hatched lunettes or triangles. The cartouches that enclose his inscriptions are similarly embellished with chevrons or zigzags.

The Memento Mori Carver's design motifs fill the empty areas of most of his horns, save

Powder Horns Carved in the Provincial Manner

Typical Jacob Gay calligraphy (detail, John Mills horn, No. 43).

A Gay signature, Jacob Gay hand writ *(detail, John Pemberton horn, No. 39).*

A decorated cartouche by the Memento Mori Carver (detail, John Rockwell horn, No. 31).

for several instances in which he used only a simple inscription and cartouche. He used a linear vinelike design on many horns, as well as drinking and smoking accessories like decanters, goblets, punch bowls, ladles, and pipes; sometimes he used drums and trumpets.

An especially intriguing motif that occurs on many Memento Mori examples is a profiled head with horns. Occasionally the face has a pipe in its mouth. The head is often made to utter the phrase "fools such as we," and the presence of the horns, signifying a cuckold, suggests that while the soldier is off fighting on the frontier, his wife or girlfriend is being unfaithful. A variant shows an Indian's head with the same label, implying perhaps that the soldier is a fool to be fighting in the wilderness at all. The Memento Mori Carver's diagrams of forts and the surrounding countryside are highly detailed and of almost cartographic excellence.

Motifs characteristic of the Memento Mori Carver, including a cuckold at upper left (detail, John Rockwell horn, No. 31).

The Cleaveland-Carril-Cotton Carver

Although his horns do not reflect the freedom and creativity of the carvers just discussed, this carver deserves inclusion here (Nos. 39, 50, and 51). His execution is above average, but his designs are far more rigid than those of his gifted contemporaries and conform to older European prototypes. His hand is precise and his deeply incised, sometimes almost raised, carving testifies to his skill. He displayed a wide knowledge of architectural motifs.

The Spencer-Hitchcock-Walker Carver

The Josiah Walker Horn (No. 46), the Hobart Spencer horn, carved at Crown Point on November 1, 1759 (No. 47), and the Ebenezer Hitchcock horn, carved at Crown Point in 1762 (No. 48), are of extraordinary quality. Two of the three owners served in Nathan Whiting's Connecticut regiment, but in different campaigns. The incomplete enlistment rolls do not indicate that these men were stationed at Crown Point together, although this may well have been the case.

Each of the three horns has floral designs that resemble those of J. W. and formations of soldiers reminiscent of Jacob Gay's less accomplished work. This melding of two different styles says less about the Spencer-Hitchcock-Walker Carver than it does about the two more significant artists. One begins to speculate that J.W. and Jacob Gay may have worked together and borrowed one another's motifs. The other, less plausible, explanation is that minor carvers saw the work of both masters and selected motifs from each.

The Lyme, or Miller-Tribble, Carver

Another anonymous carver produced two splendid horns in September of 1758 for soldiers at Fort Edward. The owners were from Lyme, Connecticut, and both horns are inscribed "Lyme," but only Thomas Miller (No. 44) is listed as serving in the 1758 campaign; he was a drummer in a company recruited at Lyme. Both owners served in the 1759 campaign.

The Lyme Carver's horns are of the highest quality and could be mistaken for the work of J.W. because of their use of winglike serifs and other motifs found on J. W. horns. The Lyme Carver also employed the deeply incised chevron design used by John Bush and the Selkrig-Page Carver, as well as animals, birds, and fish similar to those of Jacob Gay. His copperplate calligraphy is a combination of the styles of J. W. and Jacob Gay, while his floral and vine designs bear a generic similarity to those found on all Lake George School horns.

A mysterious face motif appears on both of the Lyme horns. One is a face with wings and a fan of feathers, while the other is composed of two faces of a slightly more baroque gravity. Both horns have unusual, tightly dotted, serpentine borders at the recessed portion near the spout, and each has an amusing rhyme. Once again, the confluence of styles associated with different masters tempts the observer to attribute these two horns to one of them, but the idiosyncrasies of the two horns are so consistent that they must represent the work of another individual.

The Lyme, or Miller-Tribble, Carver's mysterious face motif (detail, Thomas Miller horn, No. 44).

Richardson Minor

Another skilled carver who produced only two known horns was the silversmith and clockmaker Richardson Minor of Stratford, Connecticut. He made his own horn, inscribed LaGallette and dated August 29, 1760 (No. 52), at which time he was serving as an armorer in the Second Connecticut Regiment. He carved the Stephen Tambling horn (No. 53) at Crown Point in 1761.

These two horns are exceptional in every way. As might be expected of a skilled metalworker who engraved clockfaces, Minor developed formats and motifs that are far more academic in character than those of other horns. His scrollwork is deeply incised and informed by a working knowledge of baroque flourishes. His copperplate calligraphy is equally magnificent. The Tambling horn is distinguished (if that is the appropriate term) by a pornographic verse of a sort that is extremely rare. Perhaps Minor suggested it.

The Goldthwait-Smith-Diamond Carver is represented by Nos. 16 and 17.

Lesser Hands of the Lake George School
A number of talented carvers are known to us by only one or two beautifully carved horns. These carvers were as skillful and imaginative as the more prolific makers, and as additional examples come to light, they may one day be identified.

Singleton Horns
As we have already seen, a great many horns are unique but nevertheless bear a striking resemblance to the work of one of the identified masters, suggesting the possibility that their owners made them while working under the direction of a master. Horns of this sort include that of Thaddeus Bennett (No. 19), Lieutenant Christopher Palmer (No. 55), Lieutenant Joseph Smith (No. 56), and Captain Nathaniel Porter (No. 59).

However, a number of unsigned horns fall into another category, for they are mavericks. They can be classified as of the Lake George School, but their carvers were not so skilled nor so imaginative as the masters. The great number of surviving horns in this category strongly suggests that every soldier with some talent tried his hand at engraving what was unquestionably a fashionable accoutrement.

Between the Wars (1763–1774)

The momentum of the fad for carved powder horns carried over into the interval between the end of the French and Indian War and the outbreak of the Revolution. During this period there was practically no need to raise regiments to fight beyond the confines of each colony, although the Siege of Havana of 1762 and Pontiac's Rebellion of 1763–1764 produced a few horns. The existing militia system was probably responsible for a few more, but cartridge boxes were becoming far more readily available. A small number of horns may have been made for hunting purposes and some must also be the result of the build-up of martial spirit that accompanied New England's resistance to the Sugar and Currency Acts of 1764, the Stamp and Quartering Acts of 1765, the Boston Massacre of 1770, the Boston Tea Party of 1773, the Boston Port Act of 1774, and the British troops' seizing of provincial powder supplies at Charlestown, Massachusetts, in 1774. Establishment of the Committees of Correspondence, the minutemen, and the Continental Congress was also partially responsible for horns of this interim period.

These horns display a range of characteristics. That made for Giles Barnes (No. 64) remains strictly within the confines of the Lake George School, while those of Tilton Bennett

(No. 66), Titus Cave (No. 67), Thomas Holt (No. 68), Hamilton Davidson (No. 69), and Jonathan Clark Lewis (No. 70) prefigure characteristics of the Siege of Boston School.

The Barnes horn (No. 64), which has a number of innovative features that were ideological in origin, resembles examples by several Lake George School masters. A raised plaque at its spout depicts the king's arms with the three vines of Connecticut's arms substituted in the shield. A rare view of New Haven Green and the brig *Lively* afloat in the harbor also appear.

The Nicholas Edgecomb Pickett horn (No. 65) belongs to the spontaneous and imaginative "maverick" category that includes many Lake George School horns. It features charming and ingenious images drawn from illustrations about the West Indies trade in travel books and magazines. The serio-comic depictions of Pickett's three children approach the satirical manner of the Selkrig-Page Carver. Pickett was an officer in the Massachusetts militia before the Revolution.

Four horns are attributed to Hugh Tolford of Chester, New Hampshire. Tolford carved the Tilton Bennett horn (No. 66) in 1770 and the Titus Cave horn (No. 67) in 1772, both for men who lived in Chester, New Hampshire. Carver Jacob Gay, who lived in nearby Allenstown, New Hampshire, may very well have trained Hugh Tolford, because their styles are remarkably similar. Tolford also carved the Thomas Holt horn, which is inscribed "Chester, NH, 1772" (No. 68), and his own horn, which is not illustrated.

Just as Gay's work is characterized by distinctive calligraphy and highly unusual animals and decorative devices, so is "H. T.'s" work, for he employed consistent features that were often dependent on Gay's prototypes. The abstract British coat-of-arms on both the Bennett and Cave horns and the lion and unicorn used in conjunction with them are almost identical, although they were carved two years apart. Tolford's calligraphy in the Gay style is good and his animated people and animals, among them a turtle snapping at a snake, are delineated with fairly advanced graining and stippling.

Neither Titus Cave nor Tilton Bennett served in the Revolution and it is odd that no Hugh Tolford horns are known that date from the Revolutionary period. The horn of Ichabod Robie (not included in this exhibition), who did serve in the Revolution, which was made in Chester, New Hampshire, in 1766, remotely resembles Hugh Tolford's work and may represent an earlier effort on his part. An interesting fact to consider in connection with the question of whether or not Hugh Tolford studied with Jacob Gay is that Gay was active between the wars.

The 1772 Hamilton Davidson horn (No. 69) and the 1773 Jonathan Clark Lewis horn (No. 70) are high quality examples of Gay in a creative mood. The Davidson horn is a rare, authentic political horn with a Boston Massacre scene copied from Paul Revere's print of the same subject (most horns with political vignettes or cartoons derived from famous print sources are Victorian or modern fakes).

Outstanding on the Davidson horn is the persistence of the Lake George School idiosyncrasy of serio-comic engraving. Although Gay followed Revere's print quite closely, even to the extent of employing engraver's hatching and shading, his people are the same animated types he used to represent soldiers in the 1750s and early 1760s. It is doubtful that either Jacob Gay or the Selkrig-Page Carver was being openly sarcastic about what was obviously

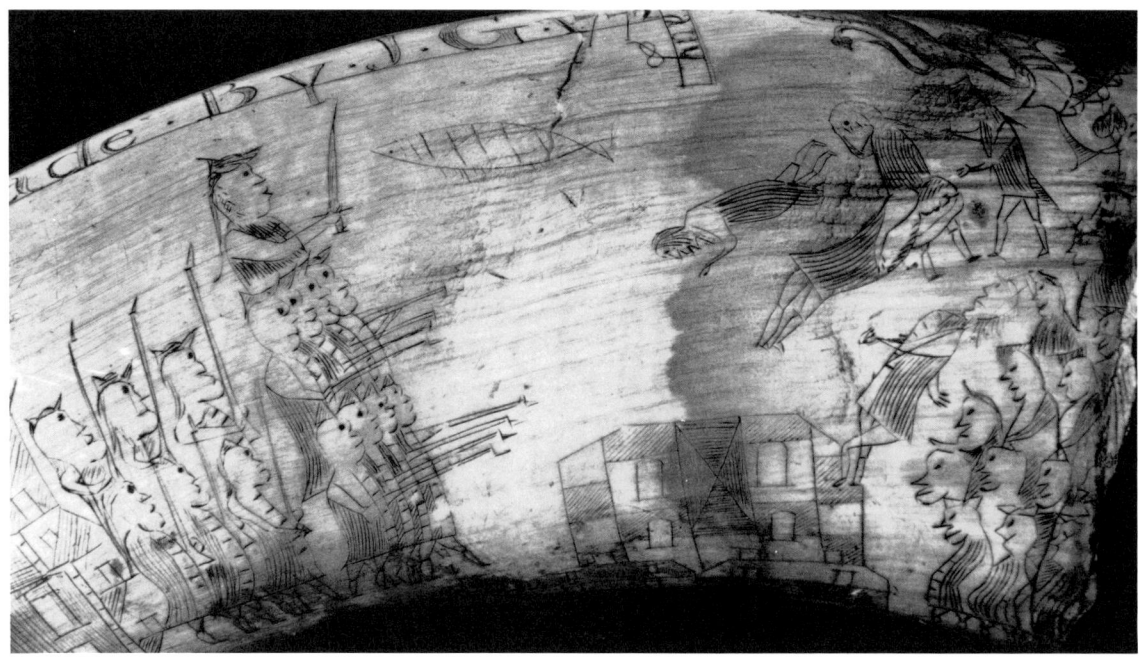

Jacob Gay's version of Paul Revere's Boston Massacre *print
(detail, Hamilton Davidson horn, No. 69).*

a tragic event. Rather, it seems that the injection of a comic aspect into depictions of war or political confrontations was intended to lighten the terror and hardship such events imposed. Many of Gay's Siege of Boston horns exhibit an explosive New England patriotism, so to view him as detached or alienated would be incorrect.

Gay's Davidson horn makes a number of important stylistic and ideological points about continuity between the Lake George and Siege of Boston Schools, and this establishes it as a major monument for reasons quite apart from its unique subject matter and high quality. The many diaries of soldiers who served in both conflicts convey little about the emotional traumas they experienced; they are instead dreary recitals of marches, duty, weather, punishments, and bare descriptions of battles. Such diaries, unlike many other seventeenth- and eighteenth-century examples, were not intended to record emotional or religious experiences but were probably merely records of service that could later be used in disputes about payment or pensions. Powder horns, on the other hand, depict scenes and facial expressions that serve as sensitive barometers of soldiers' everyday preoccupation with homesickness, fear of death, or betrayal on the home front. They also exhibit ambiguous satires on drill, vices, and the lack of creature comforts, and record battles in a way that often seems more resigned than celebratory. Horns reveal, too, that the carvers themselves interacted with their clients in subtle ways that do not suggest manipulative marketing. Horn carvers seem to have been under considerable pressure to please their customers.

Bob Trent has pointed out that Jacob Gay's Jonathan Clark Lewis horn (No. 70) is an even more spectacular instance of an ideologically-inspired scene and it has, to my knowledge, no published precedent. It was made for a militia officer who later, during the Revolution, became aide-de-camp to Brigadier General Oliver Prescott. Unusually profuse and detailed geometric and scroll ornament complements one of the oddest political cartoons to

Powder Horns Carved in the Provincial Manner

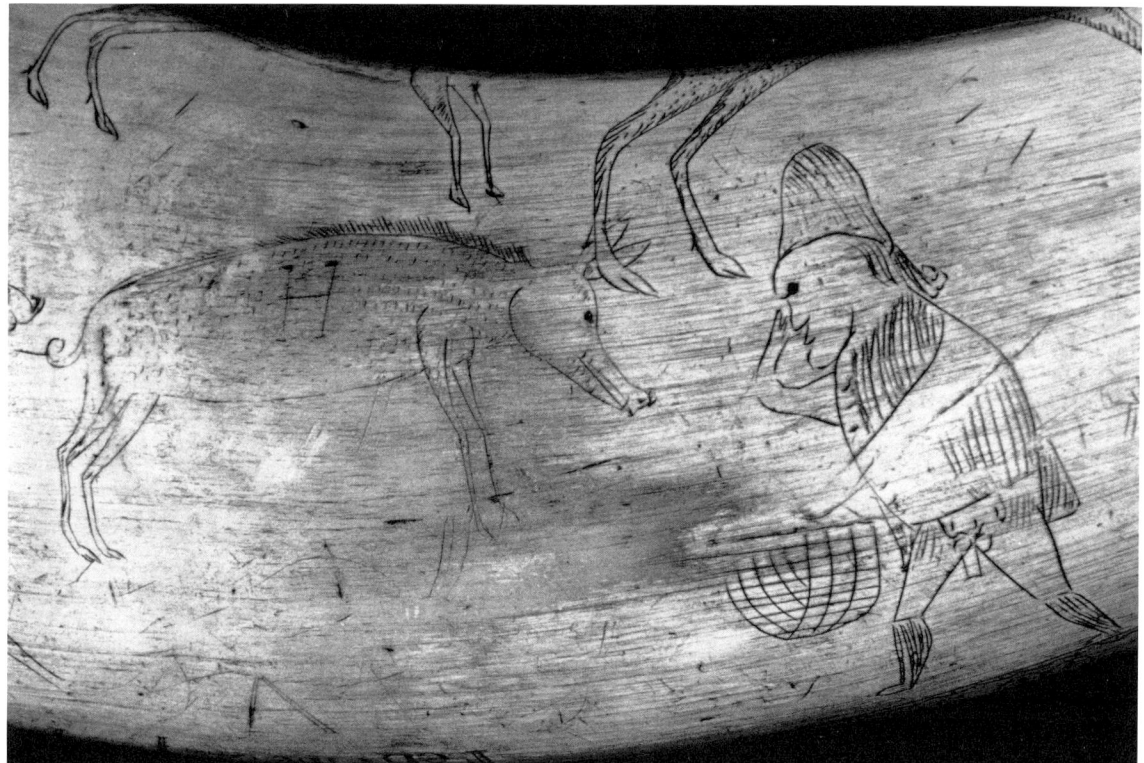

Jonathan Clark Lewis's pre-Revolutionary political cartoon (detail, Lewis horn, No. 70).

emerge from the Revolutionary era. This consists of the full figure of a man dressed in a peculiar hat, coat, and boots, but without pants and with his genitals depicted in detail. He is saluting a hog with one hand and offering it a vessel of food with the other; the hog appears to be snorting in the direction of the food. Other animals above and below this vignette do not appear to be taking part in it. The implication of the scene seems to be that the hog represents the oppressive taxing power of the British imperial government, which is seizing so much of the farmer's income that he has literally lost his pants and is reduced to servility. This was undoubtedly Lewis's own image, executed by Gay at his request. As such, it deserves to be ranked in importance with the better-known engraved political cartoons and broadsides of the period.

The Siege of Boston School (1775–1777)

A new school of carvers emerged in and around Boston after the battles of Lexington and Concord on April 19, 1775. We have seen that the carver Jacob Gay continued to work in a style that dated back to 1758, the height of the flowering of the Lake George School. When a new need for troops arose, Gay subtly changed his format to include expressions of American loyalty. Others who had carved horns during the French and Indian War probably also worked in the camps around Boston, but no surviving horns substantiate their presence.

Early in 1775, when Massachusetts created minutemen organizations out of a third of its existing militia units and other colonies followed suit, many of those enlisted had to provide their own equipment. The various New England legislatures passed temporary resolutions allowing recruits to carry ammunition in either cartridge boxes or powder horns with leather shot bags if cartridge boxes were not available. Most colonies required that each soldier in the militia also provide his own musket and bayonet (or sword or tomahawk) unless he could not afford them, in which case the town where the recruit enlisted was obligated to supply them. Some military gear may have been available as a result of local militias' increased levels of preparedness in the years immediately preceding the war, while some was undoubtedly personal hunting equipment.

It was far quicker and cheaper to supply powder horns than cartridge boxes, which were not readily available during the first two years of the war. It follows that the most productive years for the manufacture of Revolutionary powder horns were the first ones, before contracts for cartridge boxes were filled, and certainly the finest carved horns date from 1775 and 1776. Specialized troops, notably riflemen, the artillery (who were on hand to prime cannon), the newly formed light infantry, and the ranger units, continued to use horns, but the majority of men carried cartridge boxes during the Revolution. After 1777, horns were produced only in small numbers as mementoes, and consequently the quality of the work often declined. Another factor in the decline of the powder horn is that the main armies became far more mobile than formerly, and horn carving could not flourish without stable camp existence.

When the Alarm was called on April 19 after the Battles of Lexington and Concord, many men hastily inscribed their horns for identification purposes just before they marched for the siege camps around Boston. The brief inscriptions on most horns of this period reflect the immediate need for manpower and the short period of service, for the Alarm lasted only a few days. Although most examples have a name and date scratched into their surfaces, some have inscriptions that were merely scribbled on scraps of paper and pasted onto the ends of the plugs, which are flat, slightly recessed, and protected by a piece of glass or thinly shaved horn.

Immediately after the Lexington Alarm and before the Battle of Bunker Hill (June 17), the various New England militia units that answered the Alarm call joined the already encamped Massachusetts Army in fortifications around Boston, forming the New England Army. The skirmishes at Lexington and Concord prompted a wave of patriotism comparable to that created by Pearl Harbor on December 7, 1941. Patriotic feeling swept through New England and enabled the consolidation of the command of the army under General Artemas Ward of Massachusetts. Although each state raised and governed its own forces—and each had, since the formation of the colonial militia system in the 1630s, jealously guarded its power to regulate its military forces—the momentum of events and the hatred of British troops miraculously created a strong, if temporary, unity of purpose.

This patriotism was expressed in many mediums, including that of decoration on carved powder horns. After the Battle of Bunker Hill there was a lull, and the troops had a great deal of time on their hands before their enlistments expired. A great many Siege of Boston horns were produced between June, 1775, and March, 1776, when the men were not fully occupied and when new enlistments and reenlistments began to serve for one year. Connecticut enlistments expired December 10, and the majority of the men went home despite urgent requests that they reenlist. Most other New England colonies then had to use their own militia to fill the gaps left in their ranks by departing Connecticut troops, but by the end of 1775 all enlistments expired and many soldiers chose to return home.

The Siege of Boston School was thus operative from just after the Battles of Lexington and Concord through the New York campaign, April, 1775–October, 1776. True Siege of Boston horns began to appear when the men were settled into camps around Boston at Roxbury, Charlestown, Cambridge, and other fortifications. During the summer of 1775 carvers began producing horns that bore the name of the owner, the location of the encampment, and the complete date; they sometimes also added the name of the regiment and often included topographical diagrams of the fort, the town of Boston, and surrounding landscapes.

Some horns depict Washington, whom the Continental Congress had appointed Commander in Chief on June 15, 1775, but none is known that reflects the fact that the Continental Army was established on June 14 and that the New England Army had been placed under the jurisdiction of the Congress. Nor do Siege of Boston horns reflect the Northern Campaign of General Benedict Arnold's abortive expedition into Canada or the earlier captures of Fort Ticonderoga and Crown Point.

The Siege of Boston School produced horns that retained many Lake George School characteristics and between-the-wars innovations, but also added new features that swiftly altered the overall appearance of decorated horns. Regional allusions appear often in both the calligraphy and the decoration of these horns. There are views of local buildings and entire towns—sometimes these are the soldiers' hometowns, rather than the towns around Boston in which they were stationed—as well as captions identifying them. This appearance of regional features reflects the fact that Washington had been unable to fill the recruitment quotas allowed by Congress for 1776, and had had to supplement his undermanned army with drafts from the various state militias. This poor showing was principally the result of homesickness, a need to be back helping with farm chores, and a sense that the emergency had passed.

The most consistently encountered motifs on Siege of Boston horns, however, are topographical sketches of the encampments around Boston Harbor. These often dominate the compositions and many are engraved in detail, just as the town views are often precisely rendered. Recognizable ships are shown in the harbor and on the Charles River and the waterways. The combination of town views in profile, views in plan or skewed perspective, and shipping activity on the waterways prompted a new genre of formats and ingenious combinations. Examples include the depiction of Charlestown Neck on the neck of a horse, or men-of-war floating over the steeples of Boston. Only two carvers, Jacob Gay and James Greenfield, signed horns during this period.

The index features characteristic of Siege of Boston horns may be summarized as follows:

(a) Neatly executed and regularly spaced calligraphy that is not as important a compositional feature as it is on Lake George School horns except in a few instances, like those of the George Morley and Simeon Smith horns (Nos. 81 and 82) and of Jacob Gay's works (Nos. 88–99).
(b) Topographical sketches of fortifications, encampments, and waterways, and town views, buildings, and ships in harbors.

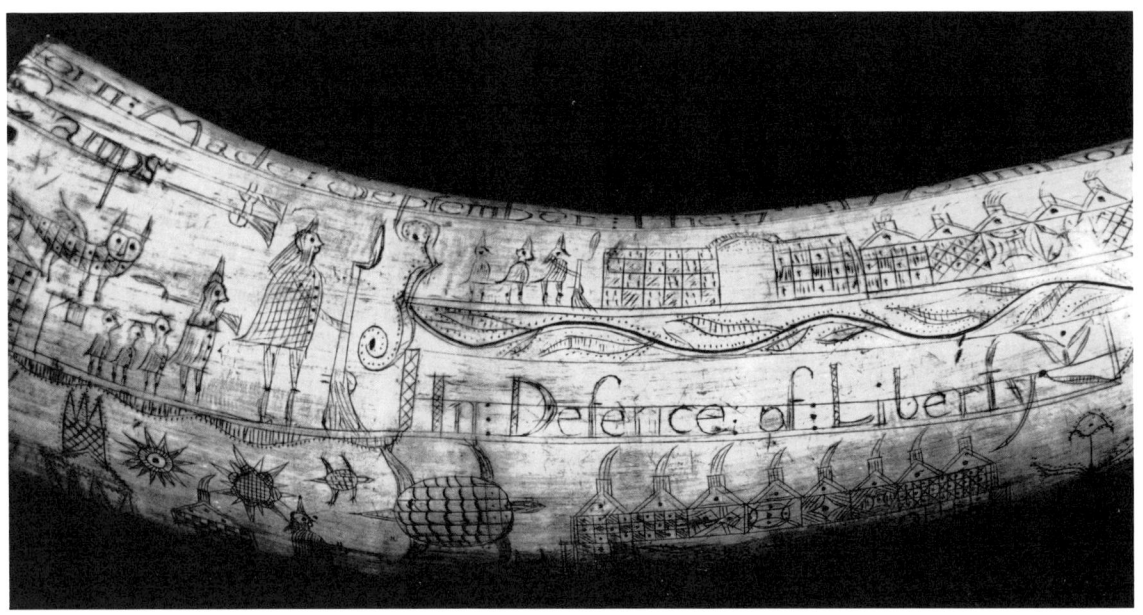

Patriotic slogans and vignettes (details, top: Frederick Robbins horn, No. 83, and bottom: Edward Sherburne horn, No. 91).

POWDER HORNS CARVED IN THE PROVINCIAL MANNER

Formations of soldiers characteristic of Jacob Gay's Siege of Boston horns (detail, Hull Curtis horn, No. 90).

(c) Imaginative animals, monsters, fish, birds, mermaids, and Loreleis.
(d) Formations of soldiers.
(e) Patriotic vignettes, portraits, and slogans, sometimes satirical in nature.
(f) Elaborate cartouches and borders.
(g) Ambitious architectural devices framing vignettes and slogans.
(h) Weapons and accoutrements used as fillers.

Decorations on period maps and prints seem to have provided many compositional strategies and individual motifs used in powder-horn decoration. These were not functional field maps but maps used in book and magazine illustrations.

THE SIEGE OF BOSTON CARVERS

John Parker

The John Parker horn is dated April, 1775 (No. 71). Although there is no documentation other than the date "1775" on the horn itself, it is here attributed to Captain John Parker, who was in charge of the detachment of the Lexington militia that saw action at the Battle of Lexington. There were, however, eight other Massachusetts militiamen named John Parker who answered the alarm on April 19, and New Hampshire records list a ranger captain named John Parker in 1775. Connecticut rolls list twenty-one John Parkers who served at various times during the Revolution, and undoubtedly other John Parkers served whose records have been lost.

For the purposes of this exhibition, the John Parker horn will be interpreted as if it did in fact belong to the captain of the Lexington militia. It is a large horn, seventeen inches long, and neatly if not artfully inscribed in large block letters, "JOHN PARKER HIS HORN

1775." This simple style of inscription is typical of most horns dating from immediately after the Lexington Alarm; nearly all of these are undecorated and of documentary interest only. The Parker horn, however, is engraved with an elaborate war canoe filled with eight feathered Indians. Two of the warriors are holding halberds, five are paddling, and one is standing in the bow holding a telescope. The canoe is decorated with Woodlands Indian designs along the hull and with several large eagle or turkey feathers that dangle from the beak of a bird effigy on the bow.

This unusual American Indian war canoe was copied from a woodcut in the September, 1773, issue of *The Gentleman's Magazine*, published in London. The caption to the woodcut reads "the Savages of the South Seas" and depicts the identical scene with New Zealand natives instead of American Indians and a canoe decorated with authentic Oceanic motifs, exotic feathers, and other ornaments specific to that culture. The engraving was prepared from drawings made during one of Captain James Cook's exploratory expeditions to the South Pacific in the early 1770s.

English woodcut from which the John Parker horn's Indian war canoe was copied (see No. 71 and Pl. 10).

The Parker horn, which could be classified as a between-the-wars horn were it not for its Lexington Alarm-type inscription, is unusual for the degree to which its maker altered his source to conform to a New England context.

James Greenfield
James Greenfield, one of the most significant identified carvers of the period, signed one of five extant horns thought to be by his hand. Apparently Greenfield carved horns for men who served in his company during the 1775–1776 campaign at Boston. Two horns were carved at Roxbury in July and September of 1775 and one is dated March, 1776. The other two show the encampments at Roxbury but were carved in February and May of 1777 after

the Siege of Boston, probably at Peekskill, New York, and one was carved in Lyme (No. 103) after the Siege (all, however, conform to Siege of Boston standards).

Greenfield and two of his customers lived in Lyme, Connecticut, and the other two horn owners lived in towns nearby. Greenfield's calligraphy is painstakingly but not expertly executed. His great ability was for engraving ships, animals and Loreleis with human faces, people, and wonderful decorative devices, often laid out with a compass or dividers. His distinctive lions and Loreleis have identical expressions and exotic crownlike hairdos. Chip-carved compasswork is another of his characteristic features. All his horns, with the exception of his own, have fort outlines with gun emplacement details. His own horn has a fine profile view of Boston in place of the forts.

Jacob Gay

The enigma of Jacob Gay, whose career spans the interval from 1758 to 1787, has already been discussed. Two of Gay's best Siege of Boston horns, both deeply engraved and embellished with colored pigments, are dated January, 1776. In these two horns, one made for Hull Curtis, a Connecticut soldier (No. 90), and the other for Edward Sherburne, a New Hampshire officer (No. 91), Gay appears to have reached a pinnacle of imaginative power and technical ability, perhaps because the Revolutionary cause inspired him. None of his earlier horns are so profusely carved and well executed, nor are his later horns as accomplished. His enthusiasm for the Siege of Boston work is evidenced in the carefully planned and imaginative designs, ranging from borders to pedestals to backgrounds, that decorate the Curtis and Sherburne horns.

Jacob Gay's unicorn and patriotic cartouche (detail, Edward Sherburne horn, No. 91).

Two quite similar horns are those made for Colonel Samuel Connor (No. 94) and Samuel Webster (No. 96), both of New Hampshire. Both examples have elaborate cartouches. The adaptation of Paul Revere's American soldier with upraised sword and the legends "Success to the American Army" and "Liberty" underneath the soldier on the Connor horn, as well as the rampant lion standing above the word "Success" on the Webster horn, are compelling images. The Webster horn's cartouche is composed of rococo scrolls and a crest enclosing a flintlock musket, while that of the Connor horn has elaborately shaded rococo scrolls embellished with flowers and supported by a lion and a unicorn. Both horns have floral borders and a few animals. These two examples are in excellent condition, and the Webster horn retains its original protective varnish. The extraordinary quality of both suggests that they were expensive souvenirs.

Even though Ephraim Moore (No. 89) served in Samuel Webster's company in 1777, his horn was carved at Boston in 1775. It is closely allied with the

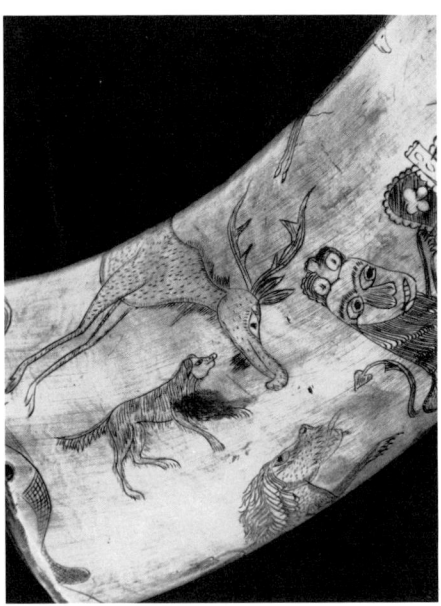

Jacob Gay animals, including a moose and a lion (detail, Hull Curtis horn, No. 90).

horns of Hull Curtis and Edward Sherburne. These three are the most ambitious horns Gay ever undertook, and the preservation of the original varnish on the Sherburne horn suggests that they, too, were souvenirs that never saw extensive use in the field.

The John Noyes horn (spelled "Noyce" in the New Hampshire rosters; No. 93) is a combination of the earlier Curtis, Sherburne, and Moore horns and the later Webster and Connor horns. Noyes served in Colonel Timothy Bedel's New Hampshire regiment during the Northern Campaign of 1776. His horn displays an extraordinary complement of motifs: grand calligraphy, a heart-shaped cartouche with a bust of Washington and the slogan "LIBERTY," typical Gay lion and unicorn supports with the slogan "Success To America," formations of soldiers, scroll borders, and a grenadier with a raised sword taken from Revere's currency illustration.

These histories and the locations cited strongly suggest that Gay was on the move. After working the Boston camps from late 1775 to early 1776, he might have gone to a staging area like Fort No. 4 to work for New Hampshire troops headed for the Northern Campaign. Then he went down to New York and back to New Hampshire. Some of the horns he made at that time, especially those dated 1776 and later, may have been carved well after the campaign and date that appear on them. Collaboration seems indicated because of inconsistencies in calligraphy and decoration. Notable examples are the Amos Bostwick horn (No. 95) ostensibly made in New York on May 21, 1776, and the Elijah Bradbury horn (No. 97), identified as having been made at West Point on December 21, 1778.

Unidentified Minor Hands of the Siege of Boston School

Just as thought-provoking as the signed and attributed horns are finely carved, carefully planned, and visually stimulating examples by a scattering of unidentified individuals. While these are clearly by many different carvers, they nevertheless have similar formats, design motifs, calligraphic styles, and quality. The amazing consistency of this group of horns can be explained only by the theory of a community of workmen who were observing each other's production closely and perhaps collaborating from time to time.

Topographical Horns

Horns with views of the camps surrounding Boston seem to have been extremely popular. Those made for Jonathan Goff (No. 72), Captain John Pennoyer (No. 73), Jabez Arnold (No. 74), Obadiah Johnson (No. 75), Elisha Crain (No. 77), John Arnold (No. 78), and Gershom Mott (No. 101) belong to this group. All were carved by different hands except for the Crain and Arnold horns, which are almost identical. Most have decorative devices and views that had symbolic significance for their owners but that we no longer understand.

Powder Horns Carved in the Provincial Manner

Buildings in the town profile views represent actual structures, but in many other instances carvers placed imaginary buildings in views or rendered views in an abstract manner. Fortification plans tend to be strictly literal and were undoubtedly intended as souvenirs and conversation pieces.

Horns Depicting Formations of Soldiers

The horns made for Elijah Case (No. 79), Frederick Robbins (No. 83), Jabez Gooddel (No. 102), and John Abbott (No. 104) all show soldiers engaged in various maneuvers of warfare; they also show decorative arrangements made up of weapons, accoutrements, and wildlife. The Caleb Johnson Hall horn (No. 106) with a sword- and tomahawk-wielding soldier also belongs in this category. These are similar to Lake George School horns but have patriotic inscriptions.

Horns Depicting Satirical Cartoons

Both the Christopher Andrus horn (No. 76) and the "—WLAND" horn (No. 105) show a harried General Thomas Gage leaving Boston in a hurry. General Gage was relieved of his command of British forces in North America in September of 1775 and left in October, whereupon he was succeeded by General William Howe. The fragmentary "—WLAND" horn is superior in quality, and was the example that attracted antiquarian Rufus Grider to the subject of powder horns.

Horns Depicting Architecture

Five horns included here—those made for George Morley (No. 81), Simeon Smith (No. 82), Reuben Hosmer (No. 99), Nathaniel Hosmer (No. 100), and Gershom Mott (No. 101)—display a new form of layout found in only a few other examples. This was accomplished by recessing both the spout end and a portion of the butt end of the horn. The magnificent engraving of the Morley horn is so professional that one wonders if its maker were not a clockmaker or a silversmith; its gravestone-shaped cartouche engraved in two lines with "LIBERTY" seems to be a rebuslike play on the slogan "Liberty or Death."

Late Siege of Boston, Revolutionary War, and Later Horns

The Aaron Foot horn (No. 108), carved on December 25, 1776, on the Pennsylvania side of the Delaware River after the Battle of Trenton, is testimony to the soldiers' lack of concern about specific events in the war. The carver's only reference to the battle, one of the most important of the Revolution, is the figure of an American chasing a Hessian soldier; the American is almost lost amidst the surrounding calligraphy and animals, birds, and flowers.

The Stephen Newell horn (No. 109), carved at Springfield, Massachusetts, on September 24, 1777, was undoubtedly made at the supply depot which later became the Springfield Armory. It continues the typical Siege of Boston vocabulary of formations of soldiers and patriotic mottoes.

Horns carved late in the war as mementoes could certainly have been used for hunting. Also, the decoration on some of them could have been executed after the war and back-dated to a time of service. The Jacob Forman horn (No. 111), carved at Bedford, New York, in 1779, displays whimsical figures of sailors on the decks and in the rigging of wonderful sailing ships. The Henry Turner horn (No. 114), dated 1781, also features ships, as well as animals, trees, vines, birds, and a hunting scene. Turner, who lived near Pittsfield, Massachusetts, served from 1778 until the end of the war; he was commissioned a lieutenant in 1781. All the above-mentioned designs on the Forman and Turner horns represent peculiar blends of the Lake George and Siege of Boston formulas.

The necessity of confining this exhibition to a manageable number of horns has made necessary the elimination of many fine examples. Those included were chosen for their quality of workmanship, representation of schools of carvers, and availability. Although a number of significant horns were carved after the Revolution, they have not been included here because they fall outside the period when production was at its highest and most accomplished point.

To sum up the thesis of this exhibition: connoisseurship should be the most important aspect of the study of American carved powder horns, whether the object is to choose examples for inclusion in an exhibition or in a public or private collection. If carved powder horns are to be looked upon as art objects, then the basis for inclusion should be principally the quality of the engraving, secondly the content of the decoration, thirdly the disposition of the decoration, and lastly historical factors like provenance and events recorded. In other words, a horn can be an important historical document and still not be a work of art. A horn can also be masterfully carved but unimaginative or poorly planned, and so again not be a work of art.

98

Eighteenth-Century American Forts

THERE IS no chapter in American frontier history that does not include fortifications of some sort, because a protective structure was absolutely necessary to both individuals and armies facing an unknown wilderness. The French and the English competed in the construction of North American forts at strategic locations along waterways, portages, or high above critical routes, just as they competed for the rich timberlands and fur trade.

The standard system of European fortification was not practical in the North American wilderness and a simplified version evolved. With a few exceptions, the resultant design was fairly consistent, being composed of a walled square or pentagon with a two-story blockhouse located at every angle. A ditch seven to eight feet deep encircled the wall and loopholes in the cellar walls of the blockhouses, six feet above the ground, allowed firearms to be aimed directly into or over the ditch.

During the eighteenth century, forts changed hands between the French and English, were destroyed and rebuilt, and were often renamed. Some were named for important personages or for the commanders of the campaigns during which they were built. Some were the centers around which towns later grew up and in many instances all that remains of a fort is the name of the town that took shape around it.

For soldiers living in those frontier forts, carving powder horns may have been an enjoyable and satisfying outlet or merely a way to pass time between battles. It also seems evident that as a means of earning their livings professional carvers followed the troops to the various forts and offered to engrave powder horns. Because no records of powder-horn carvers are known today, however, all we can do is speculate about their activities during the French and Indian War.

Later, in the beginning stages of the Revolution, most horns were carved in the camps around Boston, which were actually small fortifications. Some French and Indian War forts were used again during the Revolution, but for the most part horn carving of this period was accomplished in the smaller camps during the Siege of Boston.

The subject of fortifications is immense and is introduced here in a very abbreviated way merely to acquaint the reader with the names and locations of forts where powder horns were carved and with some of the terms used to describe them. It also enables the reader to envision the surroundings in which horn carvers worked and to understand the enormous amounts of physical labor that were required—and that soldiers supplied—to build the frontier forts.

Drums A'beating, Trumpets Sounding

View of Fort William Henry (detail, Thomas Smith Diamond horn, No. 17).

EIGHTEENTH-CENTURY AMERICAN FORTS

FORTIFICATION TERMS

Abatis: A barrier of felled trees placed with their trunks facing toward the place to be defended and their branches, often sharpened, facing toward advancing enemy troops.

Bastion: A projecting part of a fortification, often taking the form of a blockhouse.

Battery: An emplacement where artillery is mounted.

Berm: A small space four or five feet wide between the foot of the rampart and the side of the moat.

Curtain: The part of a bastioned front that connects two neighboring bastions.

Demilune or Ravelin: A triangular construction placed outside the fort consisting of two embankments that form a salient angle in front of the curtain of the fortified position.

Emplacement: A prepared position for weapons from which they can operate effectively.

Glacis: That part of a fortification beyond the covered way, to which it serves as a parapet and terminates toward the field in an easy slope.

Machicolation: The opening between extended projections in a wall or floor through which to shoot at the attacking enemy.

Palisade: Stakes made of strong, split wood about nine feet long, fixed three feet deep in the ground and set closely together to make a strong fence.

Parapet: Part of the rampart of a work, eighteen or twenty feet broad, and raised six or seven feet above the rest of the rampart.

Platform: A floor made of strong planks, laid upon joists, on a battery.

Rampart: A broad mound of earth raised as a fortification around a place and usually surmounted by a parapet.

Redout: A kind of work placed beyond the glacis, or a square work without any bastions placed at some distance from a fortification to guard a pass or to prevent an enemy from approaching that way.

Talus: A slope made both inside and outside a fortification to prevent the earth from rolling down and to buttress the fortification.

Work: A fortified structure.

THE FORTS

Fort Edward

BUILT IN 1755, sixty-six miles north of Albany at the beginning of the twelve-mile portage between the Hudson River and Lake George, Fort Edward stood at the junction of the Hudson River and Fort Edward Creek. It protected the south end of the portage or Great Carry and the storehouses and magazines that had been built there as a base of operations for attacks on Ticonderoga, Crown Point, and Canada.

Originally called Fort Lyman by the English and Fort Lydius by the French, this spot was dubbed "The Great Carrying Place" by the Indians and early explorers. In 1709 Colonel Francis Nicholson, who headed an expedition against the French in Canada, built the earliest fortification erected here. Because of illness and failure to receive supplies, however, Nicholson's expedition proceeded only as far as the lower end of Lake Champlain. In 1731 John Henry Lydius, a white Indian trader of Dutch descent, erected a trading post here, and he or his agents ran it until the outbreak of the French and Indian War. Lydius had considerable influence among the Indians but a bad reputation among both the English and the French—and eventually, among the Indians, too.

In the mid-1750s, Sir William Johnson planned an expedition whose objective was the French fort at Crown Point. To prepare for the attack, Johnson ordered his second-in-command, Major General Phineas Lyman of Connecticut, to direct the construction of a fort at the site of Lydius's post. The work, begun in August of 1755 under the supervision of British Chief Engineer William Eyre, took longer than anticipated, and Johnson began his expedition before Fort Edward was completed. He defeated the French at Lake George (called Lac St. Sacrement by the French), but did not attempt to capture Crown Point.

The Sir William Johnson papers tell us that:

> The walls of the fort, which consisted of 2 bastions and 2 half-bastions, were made of timber and earth. It had a broad rampart, a bomb proof [ammunition dump], a deep fosse, with drawbridge, a covered way, and a glacis. The fort was located directly on the bank of the river, and several block houses stood near.*

The main part of the fort contained a barracks, magazine, blacksmith shop, guard house, and storehouses, and was surrounded by palisades and a moat, parapet, rampart, and blockhouses. On an adjacent island in the river now called Rogers Island were additional barracks, a brickyard, and a hospital. Johnson ordered the fort renamed "Fort Edward" on September 21, 1755, for Edward Augustus, Duke of York.

Fort Crown Point

Crown Point, designed by Gaspard Chaussegros de Lery, chief engineer of New France, was built in 1731. Strategically located halfway between Montreal and Albany, at a point where Lake Champlain narrows to less than one-quarter of a mile, it was first named Fort St. Frédéric after Frédéric Maurepas, the French secretary of state (the prefix "St." was added as a result of the Canadian custom of naming places after saints).†

The fort was a quadrangle with high, thick stone walls. Its entrance had a dry ditch and a draw bridge, and twenty cannon and swivels were mounted on its ramparts and bastions. There was a four-story citadel, or tower, with supposedly bombproof stone walls guarded by twenty cannon, the largest of which were six-pounders. There were also an underground passage to the lake and an armed windmill.

In 1759, the French destroyed most of the fort before evacuating it. Sir Jeffrey Amherst claimed it for the British and rebuilt it. Under his direction, the fort became a pentagon with twenty-five-foot-high stone walls. There were bastions at every angle and a fortified redout 250 yards ahead of each bastion. The fort, which mounted 104 cannon along its walls and

---

* *Sir William Johnson Papers*, Vol. 10, p. 100, ftnt.

† *Documents Relating to the Colonial History of New York*, Vol. X, p. 193, ftnt.

accommodated four thousand men, was called Fort Amherst for a short time, and then Crown Point. It was finally destroyed by fire in 1779.

Fort LaGallette

LaGallette, an Indian settlement on the north bank of the St. Lawrence River a little below the modern town of Prescott, Ontario, was founded by the French for rebels from the Iroquois Confederacy whom they persuaded to move to Canada under their protection. Fort LaGallette was a post on the south bank of the river; it was later called Fort La Présentation and is now Ogdensburgh, New York.

Fort No. 4

In 1736, the General Court of Massachusetts ordered the laying out of a range of townships from the Merrimack River to the Connecticut River and along both sides of the Connecticut. The townships were numbered from one to nine and No. 4, on the Connecticut River, eventually became the site of Charlestown, New Hampshire.

When it was established, No. 4 was the most advanced frontier settlement in the region. In 1743 a sawmill was erected there and in November of that year the proprietors petitioned for a meeting to consider building a fortification. At the meeting a committee selected by the proprietors voted to assess a levy of 300 pounds for the purpose of building a fort, which was begun in 1744.

John Maynard of Sudbury, Massachusetts, who was stationed at No. 4, drew the plans for the fort. Completed by 1746, it stood on the west side of present-day Main Street in Charlestown. Some houses may have been moved so that they and the fort formed a rectangle. Walls of eight-inch squared timbers were built between the houses and additional houses were built against these. There was a "great chamber" with a look-out tower over a gate in the south wall and a stockade of twelve-foot posts, each a foot thick, sharpened at the top and spaced four or five inches apart, was erected on the other three walls.

Fort No. 4 experienced numerous attacks until the end of the French and Indian War. Major Robert Rogers rendezvoused there in 1759 after the raid on the St. Francis Indian settlement. By 1763, however, it was probably no longer standing.

Fort Ticonderoga

Under the direction of the French engineer Michel Alain de Lotbinière, construction of the fort began in October of 1755. The French first named it Fort Vaudreuil, then Fort Carillon, and it became Fort Ticonderoga under the British. Its design was based on plans by the great French military engineer Sébastien Le Prestre de Vauban. Its site was a promontory the Indians called Cheonderoga, which was within hearing of the falls where Lac St. Sacrement (Lake George) emptied into Lake Champlain. When completed, the fort was about 530 feet across, from point to point, and consisted of four bastions containing the magazine, the bakery, and two storehouses; two demilunes; and a platform, counterscarp, ditch, and three stone barracks. A wide covered way, never completed, was to extend through the rock on the east side to a battery located at the extreme end of the promontory.

French traders lived in a village on the southern end under the protection of a high wall that guarded the entrance to Lake George. A sixty-foot watchtower formed the end of the east barracks. The most important barracks was situated on the west; its ground floor was divided into a mess room with a kitchen at one end and a scullery at the other and its second story was approached by an exterior wooden staircase.

The fort was still under construction when British General James Abercrombie attacked in 1758. Under General Montcalm, the French chose to meet the enemy on cleared ground west of the fort, where they dug trenches and constructed an abatis. They drove the British back, but the next year they burned the magazine and barracks and abandoned Fort Ticonderoga for Fort St. Frédéric. General Jeffrey Amherst then occupied it before moving on to Fort St. Frédéric, which he rebuilt as Crown Point. For the rest of the French and Indian War, the British left only a token garrison at Ticonderoga.

During the Revolution, the British had only a half company stationed at the fort. When Ethan Allen and Benedict Arnold captured it in 1775, they moved the stores and cannon from Ticonderoga to Boston, using them to sustain the troops during the siege that forced the British to evacuate the city in 1776.

The Americans who remained at Ticonderoga during the winter of 1775–1776 suffered severe hardships, and many froze to death. Benedict Arnold returned to Ticonderoga after the defeat of his forces at Quebec and began to build a fleet of whaleboats and ships. Although this fleet was defeated at Valcour Island in the fall of 1776, it did prevent the British from advancing on upstate New York until 1777, when General Burgoyne forced General St. Clair to evacuate the fort. Burgoyne then moved on to Saratoga, where he was defeated. The Americans attempted to recapture Fort Ticonderoga during the summer of 1777, but failed, and the British continued to occupy it until 1782. The military history of Ticonderoga ended after General Washington's 1783 visit in the company of Alexander Hamilton and General Clinton.

Fort Oswego

A French report of 1756 on the forts of Oswego, which were situated on Lake Ontario and the Oswego River, stated:

> Fort Ontario on the right of the river in the center of a high plateau consisted of a square of 30 toises [sixty yards] on each side, the faces of which, broken in the center, were flanked by a redout placed at the point of the brisure. Constructed of pickets 18″ in diameter, hewn on both sides, joined to one another, 8 to 9 feet out of the ground. A ditch 18 feet wide and 8 feet deep surrounded the fort. The excavated dirt was thrown up as a glacis on the counterscarp and a deep talus on the berm. Loopholes and embrasures were cut on the pickets on a level with the pickets and the earth thrown up on the berm and a wooden scaffolding ran all around to admit firing over the top. There were 8 guns and 4 mortars for double grenades.

Old Fort Choueguen (Oswego), situate on the left bank of the river consisting of a house with machicoulis [fort], and perforated on the ground and first story, the walls of which were 3 feet thick and surrounded at a distance of 6 yards by another wall 4 feet thick and 10 feet high, perforated and flanked by two large square towers. There was likewise a raised work which protected the fort on the land side. Here the enemy [the British] had placed 18 pieces of cannon and 15 mortars and howitzers.

Fort George situated 600 yards beyond Fort Choueguen on a height by which the latter was commanded. It was of pickets and very poorly entrenched with earth on two faces. It was principally by means of the advantages this establishment [all three forts] afforded the English that they flatter themselves to invade Canada.*

In 1727, New York began building a stone fortification at the location of Fort Oswego, a two-story building with musket loopholes surrounded by a parapet. A surrounding U-shaped stone wall with two integral blockhouses was added between 1741 and 1743. In 1755 Governor Shirley of Massachusetts received a major general's commission and became second-in-command of British forces in North America. He constructed outer earthworks to protect the stone walls from land-directed artillery and began building an eight-pointed, star-shaped log palisade fortification known as the original Fort Ontario, which occupied a tactically important bluff on the east side of the harbor's mouth. Shirley started another fortification, to be named Fort George, to protect a ridge to the west and initiated work on a squadron of seven vessels. The French destroyed most of these works in August, 1756. Sir William Johnson built a temporary log and dirt fortification there in the spring of 1759.

Fort William Henry

Royal Engineer Harry Gordon described Fort William Henry thus:

Fort William Henry is situated at the South End of Lake George—It is a Work that consists of 4 Bastions with intermediate Curtains—and a Ditch eight feet deep and about thirty wide from the North-West Bastion to the South East one. The Work of the Ramparts and Parapets is faced up with Large Logs of Timber bound together with smaller ones. The Rampart is in most Places fifteen foot broad on the Curtains— the Bastions are filled up—The Parapets are, in the Faces of the Bastions, most exposed, from fifteen to eighteen Foot thick, and on the Curtains from twelve to fifteen—The Rampart is between ten and eleven Foot high, and the Parapets from five to five and a half—There are Barracks for between three and four hundred Men—A Casement under the left Flank of the South East Bastion, and another under the East Curtain. Likewise a Magazine under the N. E. Bastion towards the Lake another smaller under the N. W. Bastion.

This Fort stands upon a high sandy Bank twenty Foot above the Lake which covers one front—A Morass [covers] another [front] which winds within fifty yds of the third—so that an Attack cannot be well carried against any but the Western Front.

---

* Documents Relating to the Colonial History of the State of New York, Vol. 10, pp. 457–458.

There is a rising Ground about 300 Yards distant before the South West Bastion which rises to between sixteen and eighteen Foot higher than the Ground the Fort stands upon—likewise the rising ground across the morass is higher.*

Sir William Johnson ordered the construction of Fort William Henry late in 1755 after his victory over the French at the Battle of Lake George. Constructed of wood and built on limestone, it was destroyed by the French after their victory in 1757 and never rebuilt. Fort George was subsequently built upon higher ground about a mile southeast of Fort William Henry, but was never involved in much action.

Fort Niagara

According to John Martin Hammon in *Quaint & Historic Forts of North America*, The Castle, the main building of old Fort Niagara, may be the oldest masonry structure in New York State, having been built by the French in 1726. Situated on a triangular tip of land at the mouth of the Niagara River where it empties into Lake Ontario, the fort covered about eight acres.

The fort's location was strategic, for through a system of waterways and a difficult portage at the falls it guarded the entrance, via the Niagara River, to the Great Lakes from the Atlantic Ocean. In 1679, the French had built a small stockaded post called Fort Conti on a high bluff at the mouth of the river; this accidentally burned within a year. In 1687 the French built another stockaded fort, Denonville, but it was under siege and succumbed to the Iroquois, who controlled the territory. For almost forty more years the Iroquois remained in control of the passage to the Great Lakes.

Then in 1726 the French gained permission from the Seneca to construct a "trading post" at the mouth of the Niagara River. In order not to offend the Indians by constructing a typical fort, the French built a machicolated stone structure that didn't resemble a fort but that had all the facilities of a fort within a wooden stockade (which was typical of trading posts of that period). The building, known in later years as "The Castle," was designed by the chief engineer of New France, Gaspard Chaussegros de Lery, who also designed the stone redoubt at Fort St. Frédéric (Crown Point).

Local tribes were not offended because the structure did avoid the appearance of a normal fort, even though it was much stronger than a typical trading post and its size made it resemble a small city. During the French and Indian War it served as a base of operations against the English until they captured it in 1759.

In 1755, at the beginning of the French and Indian War, The Castle with its wooden stockade and other wooden structures, which had provided adequate defense against Indian attack, was obviously inadequate against British cannon. Threat of attack in 1755 prompted the French to send a battalion of soldiers under the command of Captain Pierre Pouchot to convert the old fort into a modern fortification. During the winter of 1755–1756 Pouchot directed the construction of earthwork walls that were almost eight times larger in area than the old wooden stockade, which was then torn down. A dry ditch and outworks were con-

---

* *Military Affairs in North America, 1748–1765*, ed. by Stanley Pargellis, pp. 177–178.

structed and buildings erected within the walls including barracks, bakehouse, hospital, guardhouse, forge, and gates. Pouchet felt it unnecessary to construct heavy walls along the river and lake sides of the triangle. A drawbridge allowed access to the fort which now housed almost 1,000 men. In 1757, additional buildings were constructed, including a stone powder magazine.

The fort fell to the British in 1759 when General John Prideaux, commanding a force of about 2,000 men and 1,500 Iroquois warriors, arrived. The force landed on July 5 and began to dig trenches and bombard the fort; the French capitulated on July 14 and formally surrendered on July 25.

During the Revolution Fort Niagara played an important role as headquarters for Butler's Rangers and as a supply center for the crown's Iroquois allies, enabling the British to maintain control of the Great Lakes. The fort also played an important role during the War of 1812 and remained a United States border garrison well into the twentieth century.

View of Fort Edward
(detail, Edward Courtney horn, No. 20).

27

Catalogue

Note to the Catalogue

This is the format followed in cataloguing each object:

SCHOOL OF CARVERS
>Name of carver or attribution
>(w. = known working dates)

00 *Name of object*

Place and date of carving
Materials
Dimensions, given in inches (followed by centimeters). OL = overall length, w = width; in. = inches, cm = centimeters
Inscriptions
Name of lender

History of original owner, observations and comparisons, and physical description.

BIBLIOGRAPHY (for research on this particular object)

PLATE 1   SURGEON'S MATE WILLIAM WILLIAMS'S HORN CARVED BY JOHN BUSH, 1755 (NO. 11).

PLATE 2 CAPTAIN DAVID BALDWIN'S HORN CARVED BY JOHN BUSH, 1756 (NO. 14).

PLATE 3 THADDEUS BENNETT'S LAKE GEORGE SCHOOL HORN, 1757 (NO. 19).

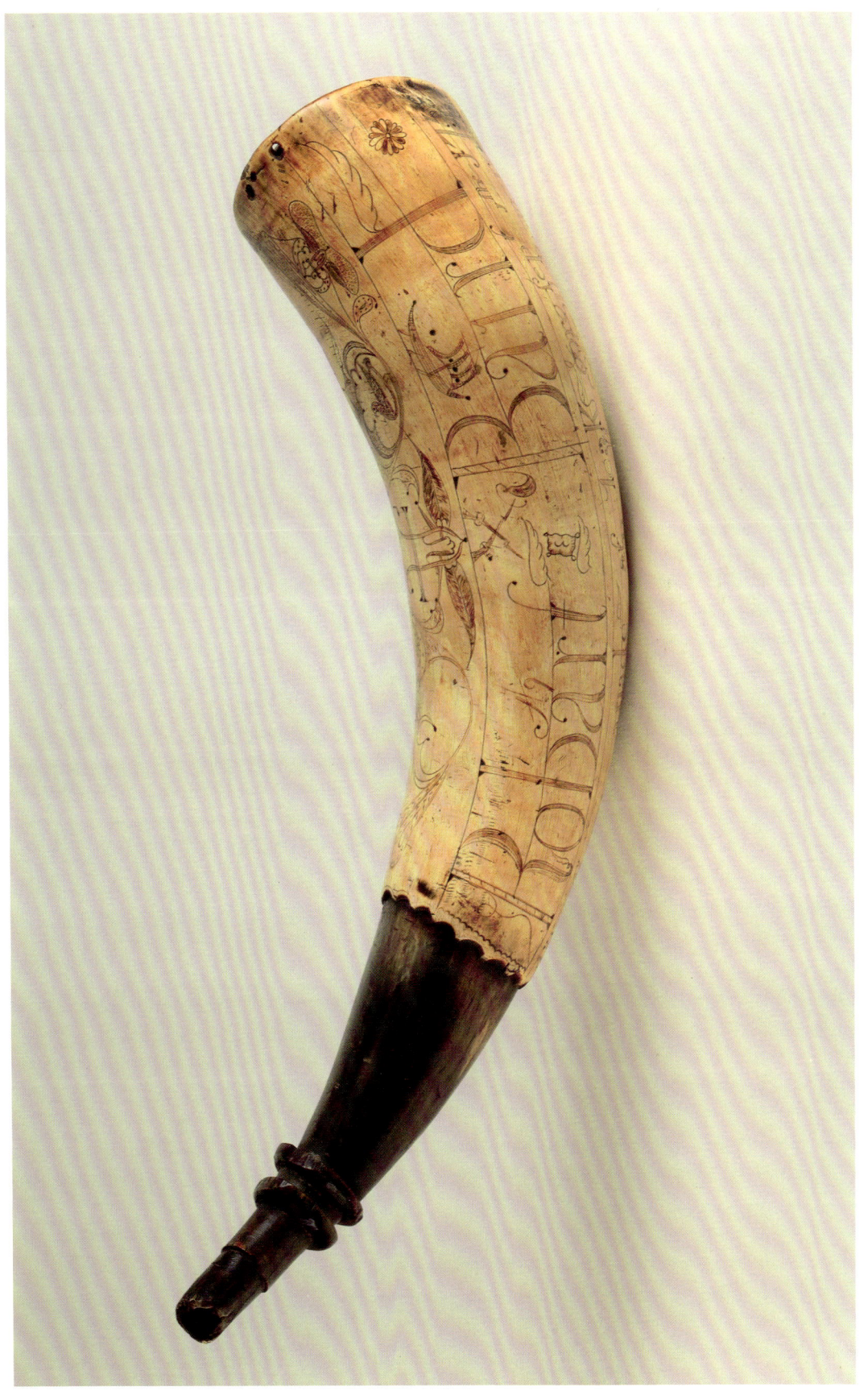

PLATE 4 ROBERT BAIRD'S HORN CARVED BY J.W., 1758 (NO. 32).

PLATE 5 ENOCH COOPER'S HORN CARVED BY J. W., 1758 (NO. 35).

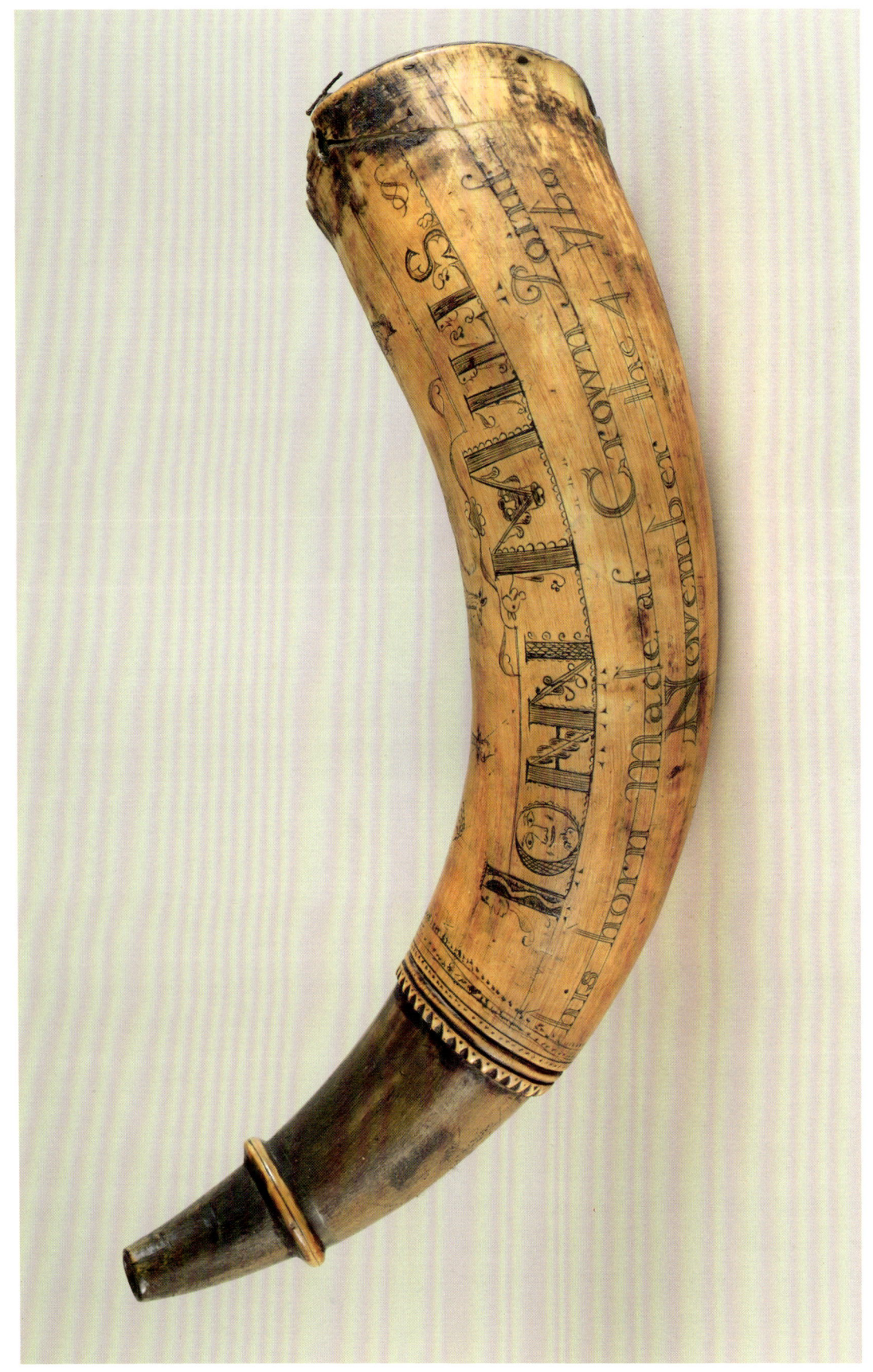

PLATE 6 JOHN MILLS'S HORN CARVED BY JACOB GAY, 1760 (NO. 43).

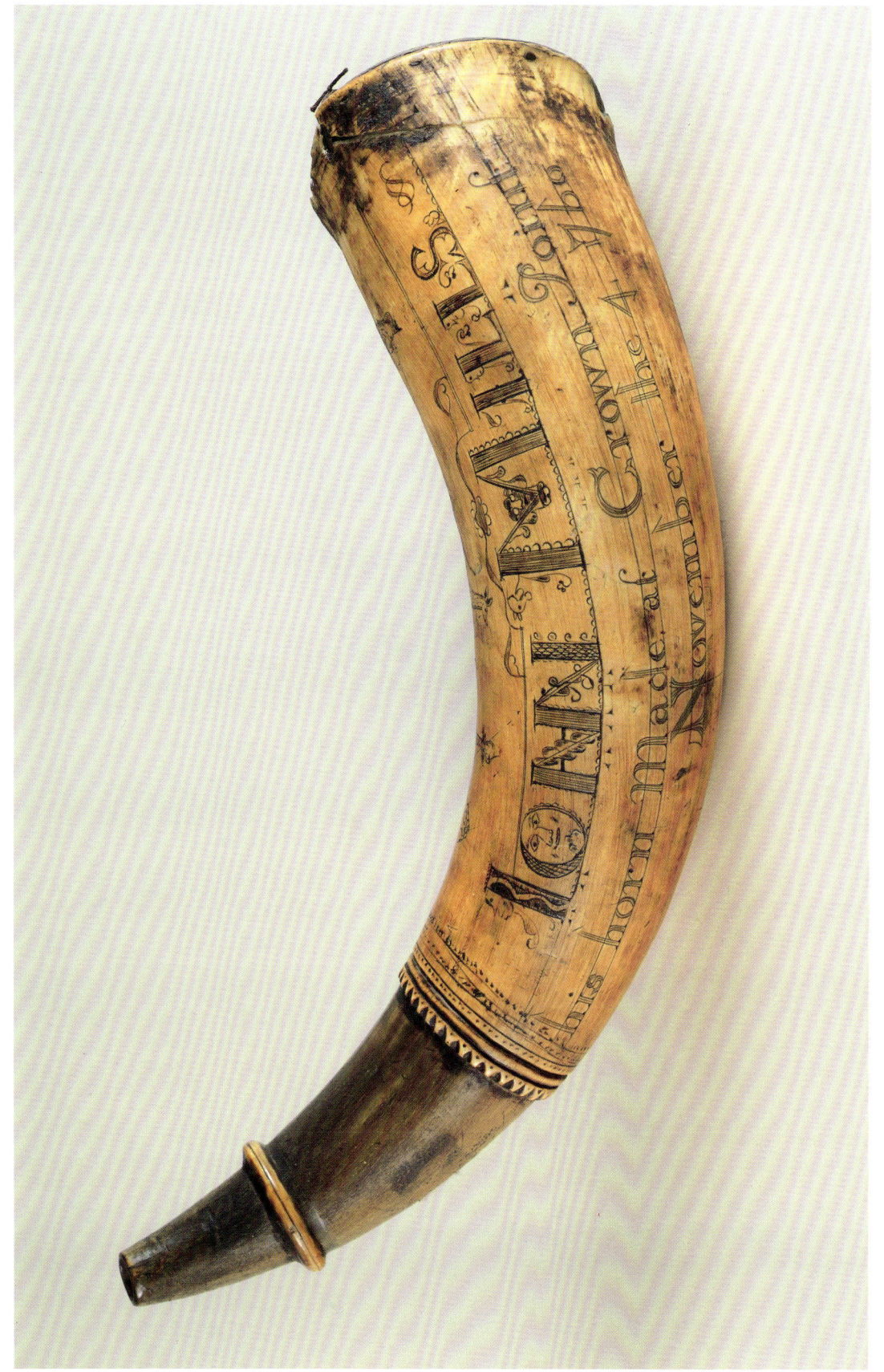

PLATE 6 JOHN MILLS'S HORN CARVED BY JACOB GAY, 1760 (NO. 43).

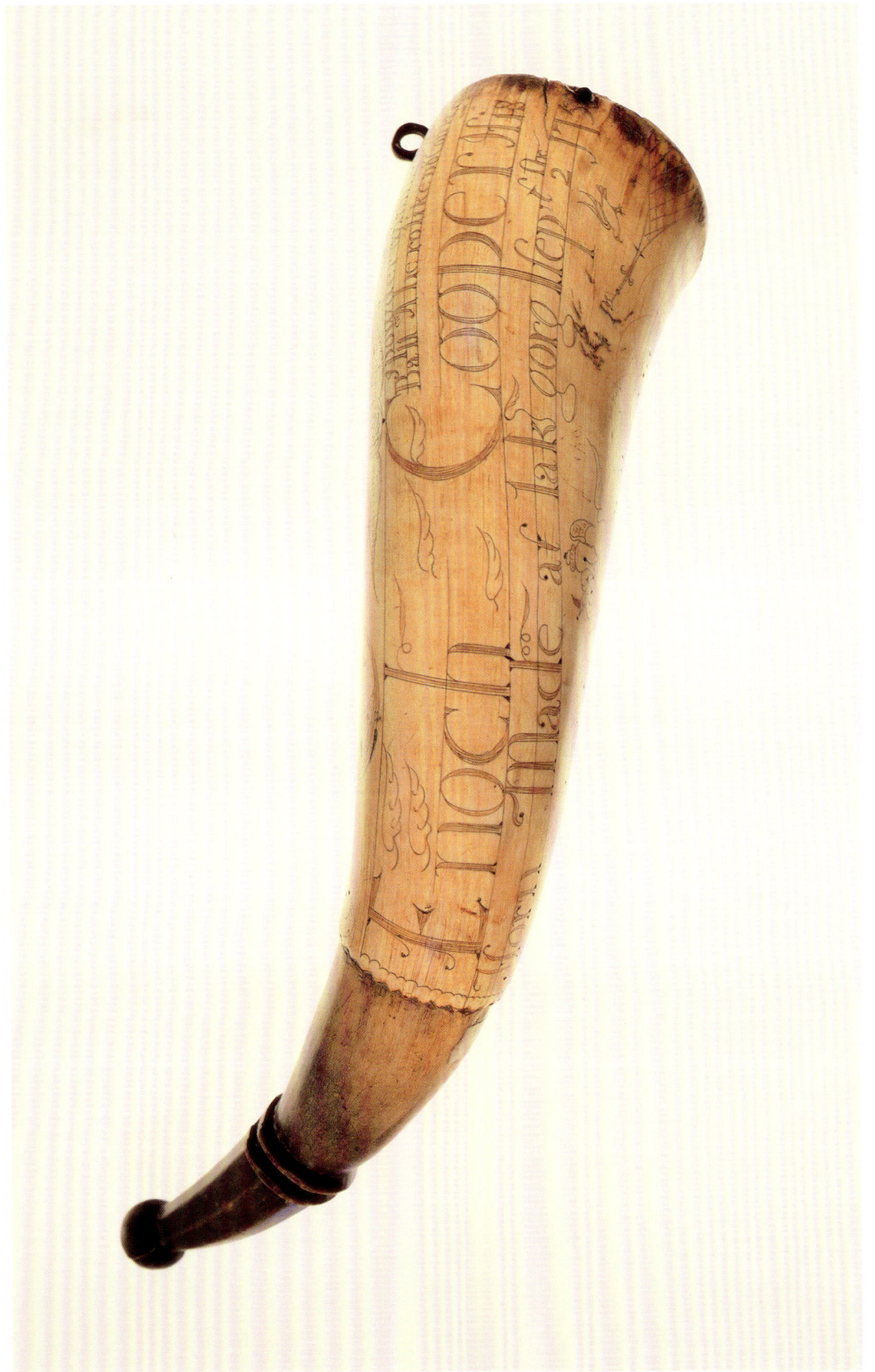

PLATE 5 ENOCH COOPER'S HORN CARVED BY J. W., 1758 (NO. 35).

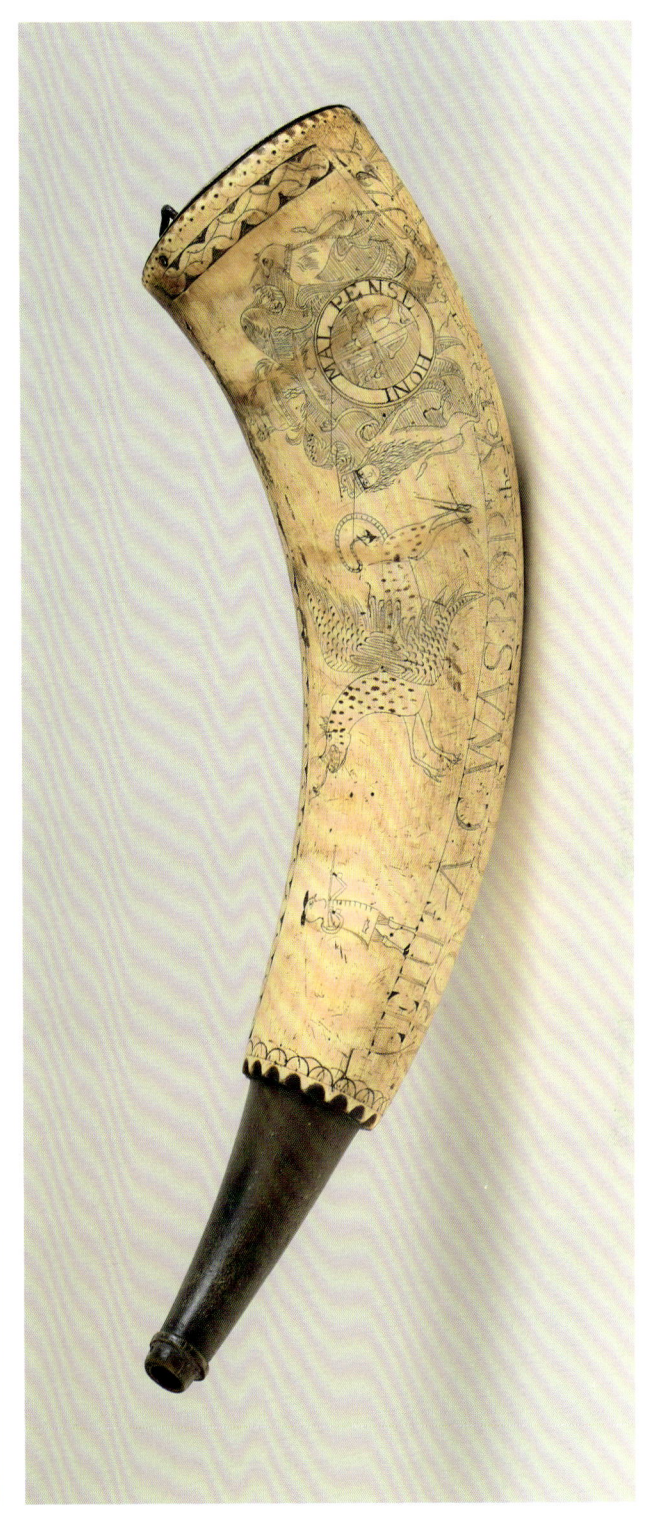

PLATE 7 ABOVE: JOHN TRIBBLE'S HORN CARVED BY THE MILLER-TRIBBLE CARVER, 1758 (NO. 45).
BELOW: AARON CLEAVELAND'S HORN CARVED BY THE CLEAVELAND-CARRIL CARVER, 1758 (NO. 49).

PLATE 8 RICHARDSON MINOR CARVED THE TOP HORN FOR STEPHEN TAMBLING, 1761, AND THE BOTTOM HORN FOR HIMSELF, 1760 (NOS. 52 & 53).

PLATE 9 TITUS CAVE'S HORN CARVED BY HUGH TOLFORD, 1772 (NO. 67).

PLATE 10 MINUTEMAN JOHN PARKER'S PRE-SIEGE OF BOSTON SCHOOL HORN, 1775, SHOWN WITH THE NEW ZEALAND WAR CANOE PUBLISHED IN THE "GENTLEMAN'S MAGAZINE," 1773, THAT INSPIRED THE CANOE ON THE HORN (NO. 71).

PLATE 11 ELIJAH CASE'S HORN CARVED BY THE SIMSBURY CARVER, 1775 (NO. 79).

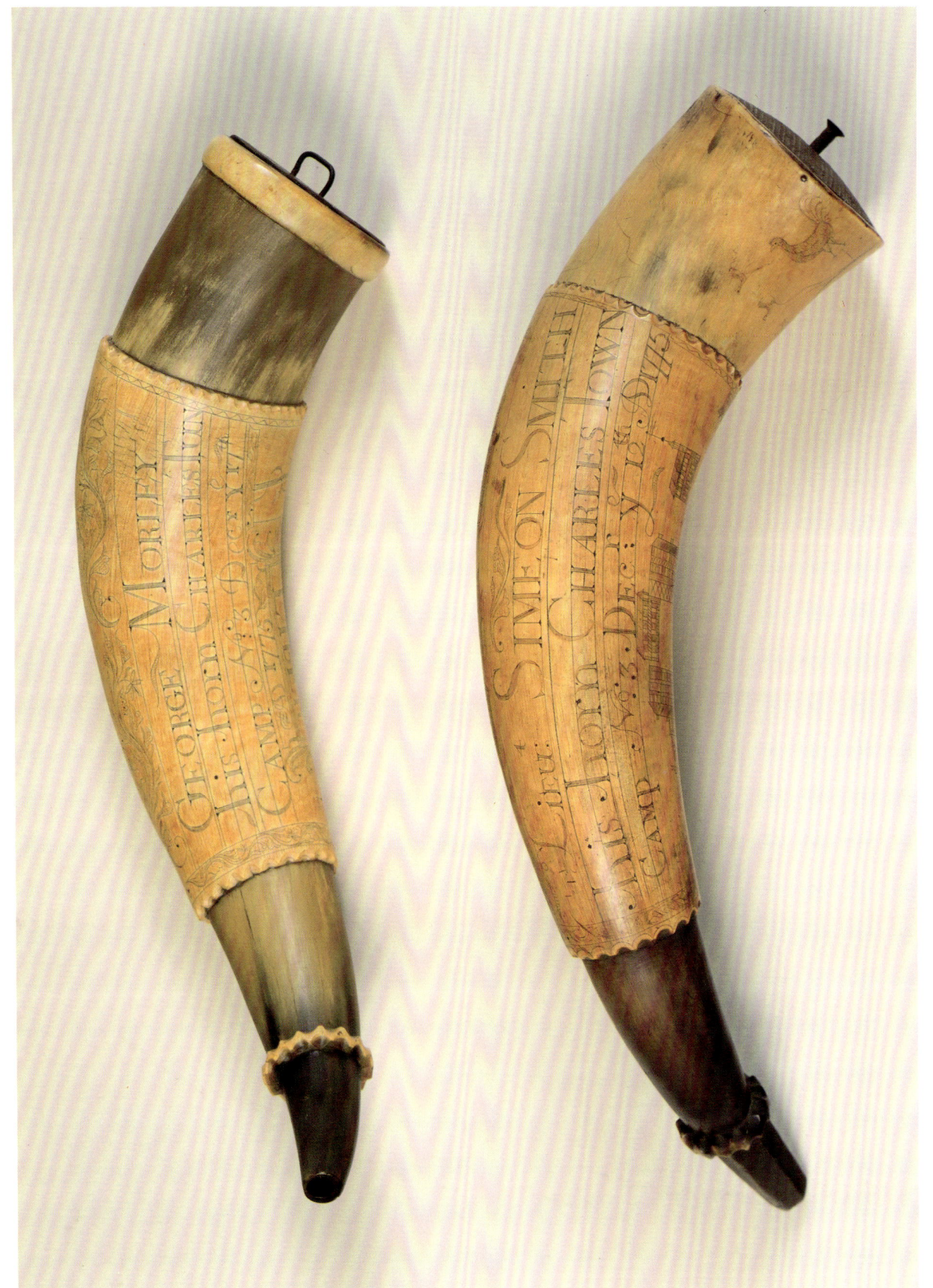

PLATE 12 GEORGE MORLEY'S HORN, TOP, AND SIMEON SMITH'S HORN, BOTTOM, BOTH SIEGE OF BOSTON SCHOOL, 1775 (NOS. 81 & 82).

PLATE 13 FREDERICK ROBBINS'S SIEGE OF BOSTON SCHOOL HORN, 1775 (NO. 83).

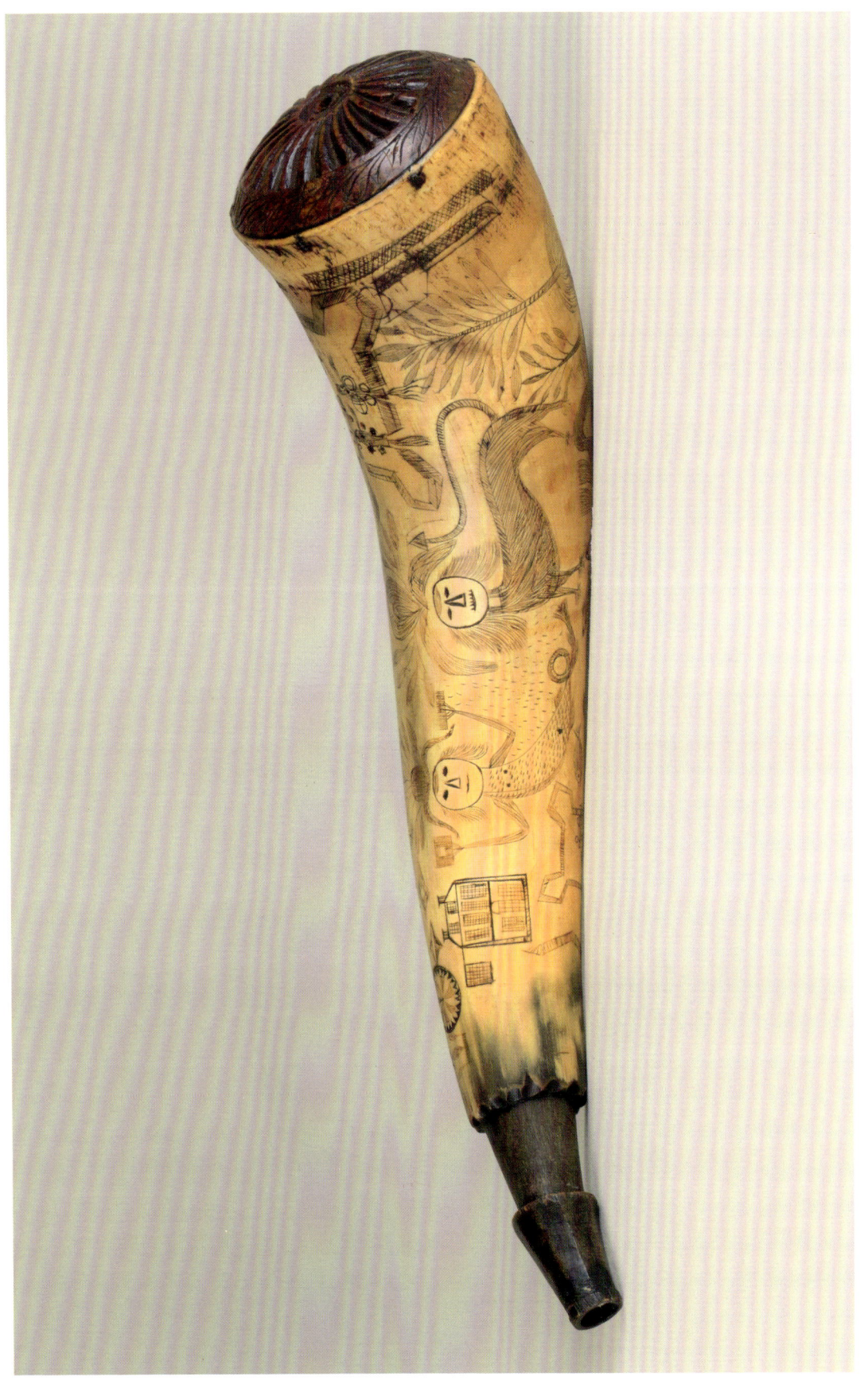

PLATE 14 NATHANIEL SUNSIMON'S HORN CARVED BY JAMES GREENFIELD, 1777 (NO. 87).

PLATE 15 EDWARD SHERBURNE'S HORN CARVED BY JACOB GAY, 1776 (NO. 91).

PLATE 16 JABEZ GOODDEL'S SIEGE OF BOSTON SCHOOL HORN, 1776 (NO. 15).

Catalogue

1

KING GEORGE'S WAR SCHOOL
Attributed to the Smith-Fletcher Carver
(w. 1746–1747)

1 *William Smith horn*

New England, 1746
Horn, pine, iron, pigment
OL: 15¼ in. (38.7 cm), w (of plug): 2⅞ in. (7.3 cm)
Inscriptions: WILLIAM: SMITH: 1746 / THE / DUKE . OF / CUM / BERLUND / PEACOCK, FOX, COCK, BUCK, BUCK, BUCK

The Connecticut Historical Society, Museum Purchase Fund

Undoubtedly, William Smith was a New England soldier during King George's War. His horn is extremely important because it exhibits features characteristic of the earliest form of carved powder horn in North America as well as features that remained popular on horns up through the second quarter of the nineteenth century. It and the David Fletcher horn (No. 2) were carved by the same engraver.

This horn's earliest feature is the highly stylized lettering consisting of rigid, double-line letters filled in with dashes between the double lines. A sharp V-shaped joint is formed wherever branches of the letters are joined and at the point where the date numerals "1" and "7" terminate. The "W" has the ornate serifs favored by John Bush, the Selkrig-Page Carver, and others of the Lake George School of the 1750s and 1760s.

The name is enclosed within a double-line rectangular cartouche that is decorated with more short double lines set at an angle and cross-hatched with smaller double lines. A similar rectangular cartouche

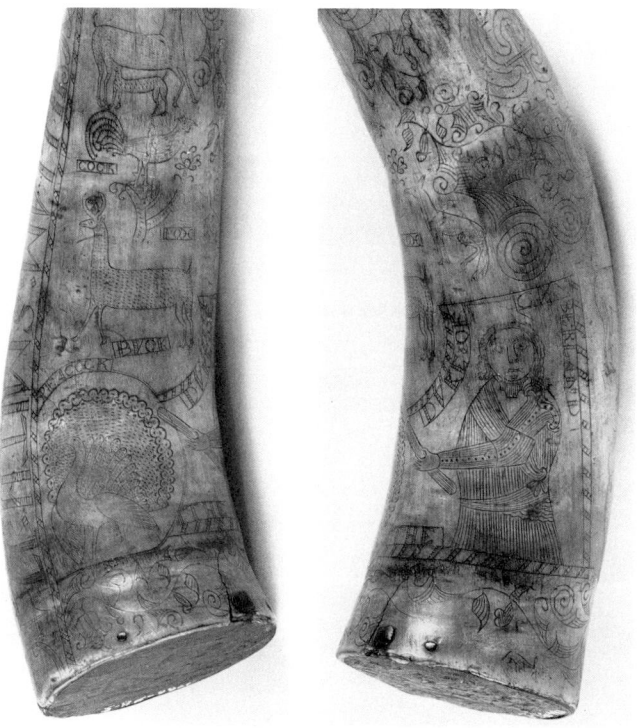

1 Left: *animals and birds borrowed from traditional European decoration.*
Right: *Duke of Cumberland in a rigid, regal pose characteristic of King George's War School horns.*

spaced ¼ inch below the name cartouche extends the length of the horn and encloses a floral and vine design.

An irregular 3-inch cartouche at the plug end of the horn contains a portrait of the Duke of Cumberland. The rest of the space is filled with graceful, detailed birds and animals whose names are incised in rectangular cartouches set above, below, or adjacent to them, with the exception of a doe nursing a fawn, which is not labeled. Many of these features appear in the work of Jacob Gay.

A crude, deeply incised line is ¼ inch from the beginning of the recessed portion about 5¼ inches from the spout. A double raised ring demarcates the octagonal tip of the spout 1¾ inch from the tip, which has an uneven, raised lip. The flat pine plug, painted brownish black, is secured by four wooden pegs and a rosehead nail. An extension of the horn that formed a lobe with a hole has disappeared.

The decoration of this horn, one of the finest of the early examples, departs from the characteristic severity of King George's War horns. Its capable carver disposed his ornament and figures carefully throughout the space and obviously influenced later carvers.

2 Left: *Duke of Cumberland in a majestic pose characteristic of the European tradition.* Right: *animals and Admiral Warren in traditional European style.*

KING GEORGE'S WAR SCHOOL
Attributed to the Smith-Fletcher Carver
(w. 1746–1747)

2 *David Fletcher horn*

New England, 1747
Horn, pine, iron, pigment
OL: 14¾ in. (37.5 cm), W (of plug): 2⅞ in. (7.3 cm)

Inscriptions: HIS.HORN / DAVID:FLETCHER: 1747 / THE DUKE ADMIRAL WARREN / OF CUMBERLAND / BUCK, DOE AND FAWN, FOX, COCK, LION / PEACOCK, UNICORN
Maine Historical Society

Catalogue

2. Animals and birds taken from traditional European decoration.

Probably, David Fletcher was a New England soldier during King George's War. His horn is a virtual duplicate of the William Smith horn (No. 1) but has some additional details, like the figure of Admiral Warren and assorted animals. Both horns were carved by the same engraver, who also made an example inscribed *Meshach Taylor, 1747* (No. 3).

There is no incised line at the point where the recessed portion of the horn begins, but a border of double-curve type, similar to Penobscot Indian designs, is 5 inches from the tip of the spout. The horn is octagonal in shape from the beginning of the recessed portion to the raised ring 1¾ inches from the tip, and the tip has three raised rings. The flat brown-painted pine plug is secured by four wooden pegs. A vine-and-geometric border outlines the plug end; a rosehead nail and notches indicate the location of an original carrying tab.

BIBLIOGRAPHY: Grider, NYHS; Grancsay, No. 333, 52.

KING GEORGE'S WAR SCHOOL
Attributed to the Smith-Fletcher Carver
(w. 1746–1747)

3 *Meshach Taylor horn*

New England, 1747
Horn, pine, iron
OL: 12¼ in. (31.1 cm), W (of plug): 2⅞ in. (7.3 cm)
Inscriptions: MESHACH TAYLOR HIS HORN AD YE 1747 / AM
The Colonial Williamsburg Foundation

No military records for Taylor exist, but he probably served during the Louisburg campaign. His horn is engraved by the same hand as the William Smith and David Fletcher horns (Nos. 1 and 2). The foliated scroll border above the inscription is almost identical to the borders on the other two. A peacock, doe, nursing fawn, and other animals and fish are also nearly identical to those on the other examples. The figure of a woman is in the same style as the Duke of Cumberland figures on the related examples. Also typical of the carver's work is the double-line lettering with vertical dashes between the double lines and sharp, V-shaped terminals of the letters and numerals.

Just one letter, the "M" in Meshach, has the ornately foliated, scrolled, and shaded appendage extending from a single upright. However, above and parallel to the inscription the carver engraved a series of the same foliated C-scroll designs; these were bordered on the bottom by a diamond-shaped design and above by a zigzag series of shaded V's. Ornate floral and vine designs appear after the words at the end of each line. Extremely fine elongated notching demarcates the point where the recessed portion begins 5

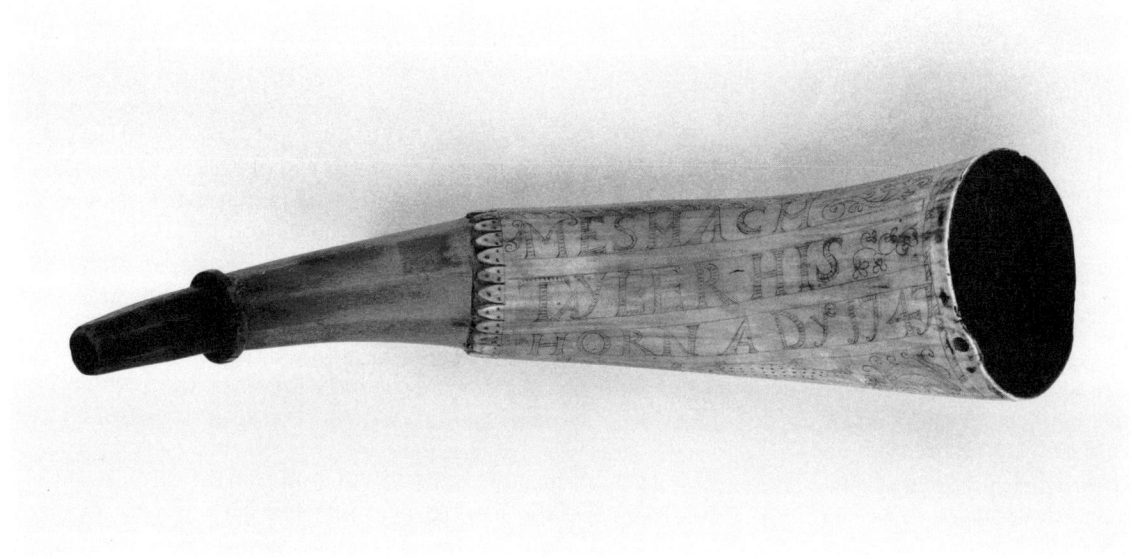

KING GEORGE'S WAR SCHOOL

4 *David Willson horn*

Hollis, New Hampshire, 1747
Horn, pine
OL: 13 in. (33 cm), W (of plug): 2⅝ in. (6.8 cm)
Inscriptions: DAVID WILLSON / HIS HORN 1747 / HOLLES BF / E:P / JEREMIAH:BURCH / 1754
Private Collection

3. The figure of a woman and a nursing fawn pictured in traditional European fashion.

David Willson and Jeremiah Burch have not been identified. However, Stephen Grancsay's checklist in *American Engraved Powder Horns* includes another horn (No. 956) which is inscribed *His Horn David Willson Derefeld the 18th 1747 February / David Willson of Holles*. Grancsay states that he obtained this listing from the 1925 Bannerman sale catalogue; Bannerman, in turn, bought the horn from the A. E. Brooks collection in Hartford (Brooks published a catalogue of his collection in 1899). Bannerman further described the horn as being ornamented with deer, snakes, turtles, and other animals and gave its length as 13 inches. The price was $100.00.

The Nathaniel Willson horn (No. 5) has the identical form of calligraphy and was carved by the same hand as this example. The highly stylized lettering consists of rigid double-line letters that are filled between the lines with dots or dashes. Wherever branches of the same letter meet, a sharp V-shaped joint is formed, and the date numerals often end in V-shaped points. The letters have fancy geometric appendages that are either a part of the letter or an addition to it. The appendages are usually straight multilined geometric forms that widen as they move away from the letter and end in a foliated scroll. Smaller geometric designs are mixed in with the lettering or serve as fillers in areas not utilized for the inscription.

The borders appear to have been made with a red-hot file, and they converge as they move from the wider plug end toward the narrower spout; they end at a raised ring 1¾ inches from the tip. A small ship is added to the borders between lines of the inscriptions.

Unlike other King George's War horns, this example does not have a raised ring at the tip of the spout to secure one end of the carrying strap, which appears to have been placed behind rings farther down the body of the horn. The tip of the spout is encircled with a series of incised V-shaped notches. A carved wooden stopper remains. The plug end is decorated with a ¼-inch border, burned in the same manner as the other border lines. The flat pine plug is incised with the date 1754 and secured by eight wooden pegs.

13/16 inches from the tip of the spout. The recessed portion is faceted in octagon shape up to the raised ring 2⅛ inches from the tip. From that point the tip is faceted, but not in as pronounced a fashion. The flat pine plug is secured by six square iron pegs. A hole at the edge of the horn extends through the plug for securing the carrying strap.

Catalogue

4

4. *Rigid geometric design characteristic of the King George's War School.*

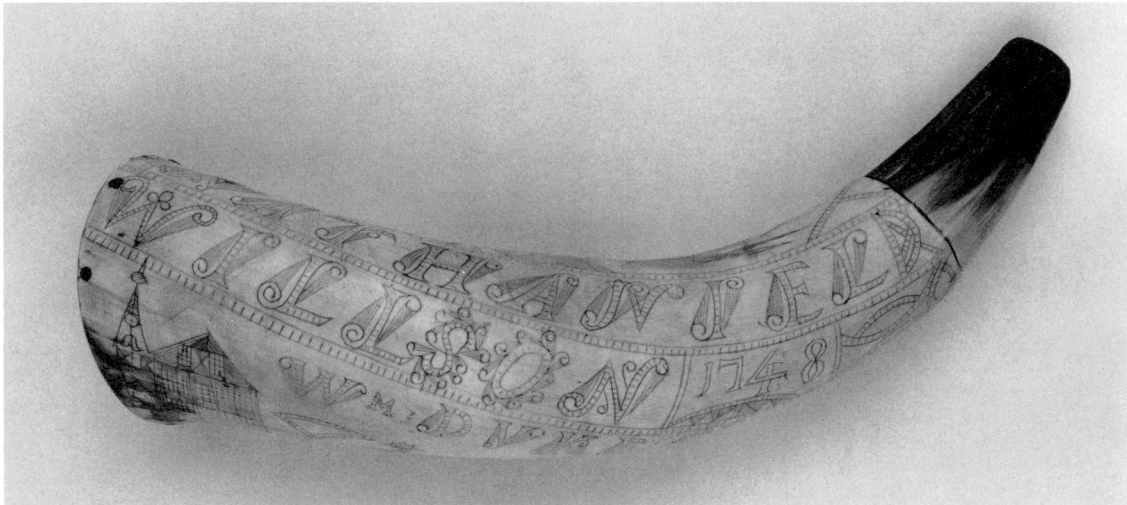

5

5. A horseman, animals, and stylized lettering characteristic of the King George's War School.

KING GEORGE'S WAR SCHOOL

5 *Nathaniel Willson horn*

New England, 1748
Horn, pine
OL: 11⅝ in. (29.5 cm), W (of plug): 2 1/16 in. (5.2 cm)
Inscriptions: NATHANIEL WILLSON 1748 / Wm DUKE OF CUMBERLAND
The Colonial Williamsburg Foundation

Carved by the same hand as the David Willson horn (No. 4), this horn has excellent ornate calligraphy. Each letter consists of double lines with short dashes filling the space between them and the branches of the letters have sharp V-shaped ornaments where they meet. Each letter also has double-shaded C-scrolls extending from a curled finial at the top of a branch to a similar curled finial at the bottom of the same letter, with, in some cases, additional small curls extending from the uprights or the curves of the branches of the letters. Each line of the inscription is bordered by a double line, with upright short dashes filling in the space. Motifs include two sailing ships, a building, a horse and rider, two dogs holding a deer at bay, and a fish.

The recessed area, which is worked into nine facets, begins without a border 2¼ inches from the tip. The slightly rounded pine plug is secured by six wooden pegs, but it is probably a replacement because the plug end of the horn has been trimmed slightly.

Catalogue

6

FORERUNNER OF THE LAKE GEORGE SCHOOL

6 *Jonathan Cook horn*

Possibly Fort Dummer, Massachusetts (now Vernon, Vermont), 1747
Horn, pine, brass, pewter, pigment, twine
OL: 15¼ in. (38.7 cm), W (of plug): 2⅞ in. (7.3 cm)
Inscriptions: JONATHAN COOK***1747 / JM / JM / JAMES MELVEN 1746
William H. Guthman

Massachusetts records list a Jonathan Cook serving at Fort Dummer in 1749 as a private in Captain Josiah Willard's Company. Fort Dummer, built at the site of present-day Vernon, Vermont, near Brattleboro, was one of a line of forts constructed by Massachusetts in 1724. Obviously, Cook may well have been serving there in 1747. James Melven was almost certainly a later owner of the horn.

A mixture of mediocre Gothic and block lettering, the calligraphy on the Cook horn displays both double-line letters with dots and single-line letters. The inscription is contained by a single-line border with dots at either end. The initials JM are carved in two places in a more accomplished hand, while the second inscription JAMES MELVEN 1746 is a badly engraved obvious addition.

The horn is profusely engraved with animals, birds, and flowers whose deep incising shows remarkable skill. The figures are both realistic and fantastic, with amusing and animated expressions. Placed helter-skelter over the surface of the horn, these animals, probably derived both from picture books and from direct observation, represent the beginning of a tradition that lasted throughout the period covered by this exhibition and on into the nineteenth century.

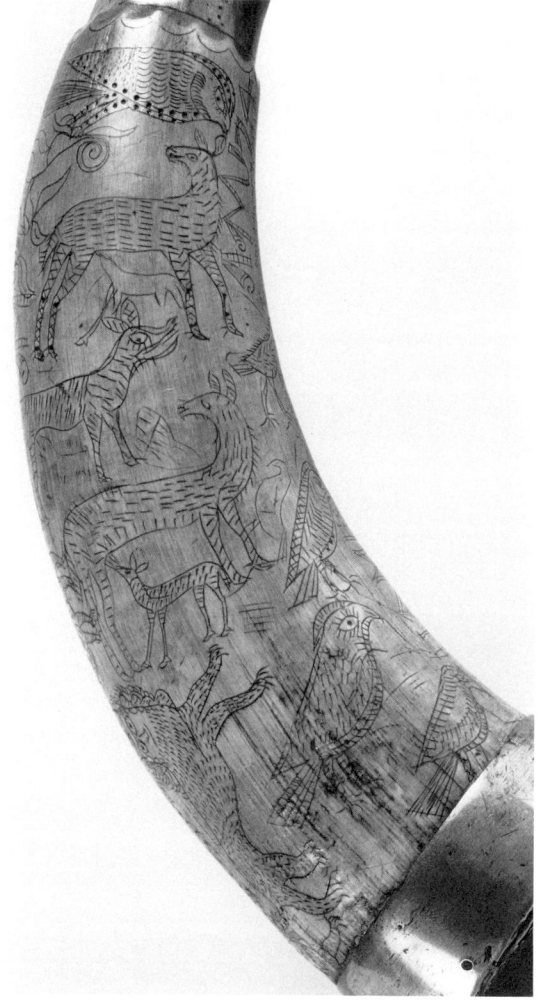

6. Animals, birds, and nursing fawn depicted with a new freedom of spirit that signifies the arrival of the American provincial manner.

6. Animals from picture books translated by the new American frontier spirit.

FORERUNNER OF THE LAKE GEORGE SCHOOL

7 *Samuel Crosby horn*

Shrewsbury, Massachusetts, May 13, 1748
Horn, pine, iron
OL: 15 in. (38 cm), W (of plug): 2⅞ in. (7.3 cm)
Inscriptions: SHREWSBURY ∗∗ MAY ∗ Y 13 ∗ 1748 / SAMUEL ∗∗ CROSBY ∗∗ JUNER ∗ HIS : HORN ∗∗
William H. Guthman

Although a number of Massachusetts soldiers with the name Crosby are listed in the rosters for 1744 through 1748, no Samuel Crosby is listed. A Dr. Samuel Crosby is, however, listed in the 1757 alarm list of Colonel Artemus Ward's Company. Ward, later a prominent Revolutionary general, was from Shrewsbury, and most of his company was recruited in the area.

The recent discovery of this horn is a stroke of sheer luck, for it supplies the key to the origins of the Lake George School. The horn is a composite of many features found on Lake George School horns, and although the quality of the carving is not as high as in later horns, it is still fine work.

The neat double-line uppercase block letters with diagonal line shading combine the Gothic lettering typical of King George's War horns and the more complex style of the Lake George School. The geometric devices used to separate the letters and the scrolled border with stylized shaded and dotted tulips underneath the inscription are motifs commonly found on Lake George School horns.

Below the scroll border is another consisting of a row of shaded diamonds similar to the incised chevron device found on horns by John Bush and the Selkrig-Page Carver. The sailing ships with animated figureheads and the cartoonlike characters in uniform and with tricorn hats resemble those seen on horns by the Selkrig-Page Carver and Jacob Gay. Graceful tulips and birds resemble motifs employed by the J. W. Carver. The circle of lunettes at both the throat and the plug end are found widely in the later school. Because this horn has so many attributes of horns dating a decade later, it is tempting to think that John Bush or the man who trained him carved it.

The recessed portion of the horn begins 6½ inches from the tip of the spout at a border of lunettes and triangles. A neat raised ring ¼ of an inch wide is located 2¼ inches from the spout. The section near the tip has nine facets and flares outward at the tip, much like the muzzle of a fowling piece of the period. A simpler lunette border encircles the plug end. The flat pine plug is secured by four wooden pegs. An iron

The free choice and placement of motifs and animated renditions are not usually seen in European work.

A neat geometric border runs the entire length of the horn and ends at the border of the recessed portion, which is preceded by a neatly scalloped border that is outlined by a single-line border following the pattern of the scalloping. The recessed portion is 5 inches from the tip of the horn. A raised ring 1¼ inches from the tip of the spout has broken off and is fragmentary. A pewter cover was added to protect the broken spout, and it is now fragmentary, as well.

A heavy brass band 1 inch wide encircles the butt end of the horn and is secured with five tiny brass nails. This is undoubtedly a later addition, for it covers part of the engraving. The rounded plug, painted a reddish brown, extends ¼ inch beyond the horn. Two holes 1 inch apart at the outer edge of the brass band have fragments of twine in them, undoubtedly from a later carrying strap.

CATALOGUE

nail is driven into the center of the plug, but vestiges of two holes at the probable location of a protruding tab remain.

BIBLIOGRAPHY: Richard Withington auction brochure for January 3, 1987; Auction report, March, 1987, *Maine Antique Digest*.

7

7. Colonial figure with stylized tulips embodies a total departure from European tradition.

7. The cock weathervane atop a ship's mast represents the imagination and humor characteristic of American frontier art.

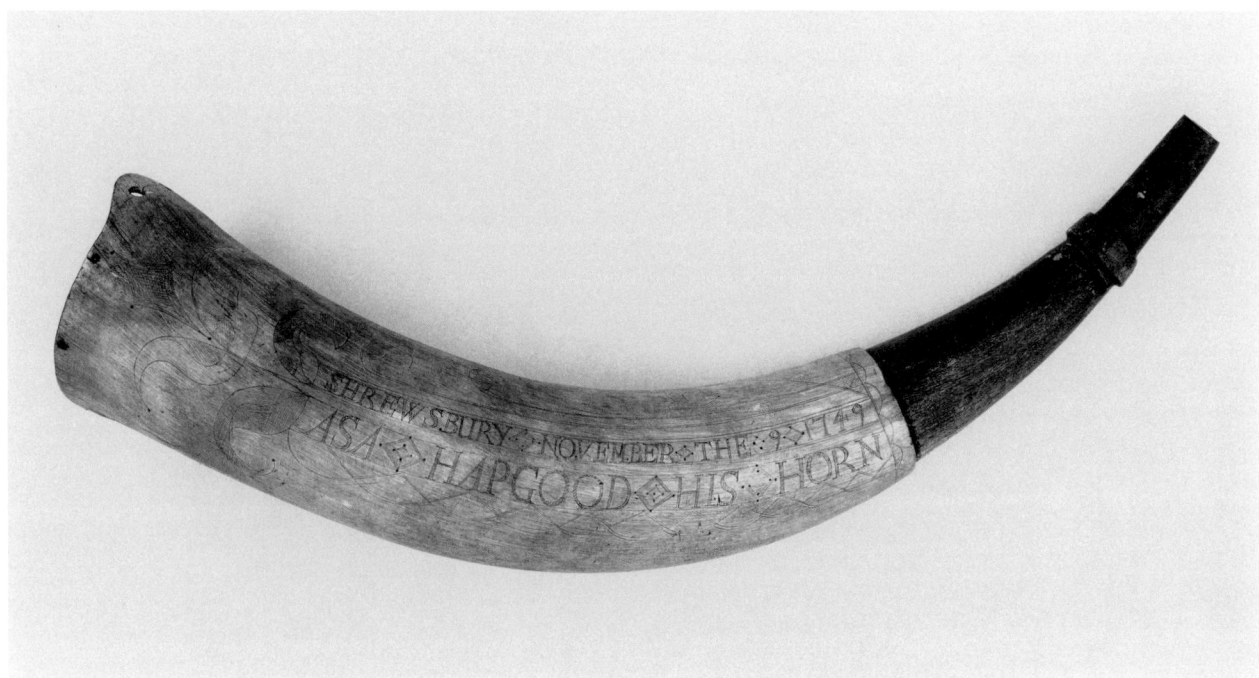

8

Catalogue only
FORERUNNER OF THE LAKE GEORGE SCHOOL

8 *Asa Hapgood horn*

Shrewsbury, Massachusetts, November 9, 1749
Horn, pine, hardwood pegs
OL: 16½ in. (41.9 cm), w (of plug): 2¾ in. (7 cm)

Inscriptions: SHREWSBURY*NOVEMBER*THE*9*1749 / ASA *HAPGOOD*HIS*HORN; *initials applied at a later date are* JAL *in a different hand and* H4TH *on the plug in yet a different hand.*

The Metropolitan Museum of Art, J. H. Grenville Gilbert Collection

This horn is illustrated because of the importance of examples carved in Shrewsbury during the period just prior to the beginning of the French and Indian War. This one is plain, with mediocre calligraphy underlined by a simple border and within a cartouche embellished with stylized tulips and a few leaves.

There is a simple incised demilune border at the beginning of the recessed portion which is 5½ inches from the tip of the spout, 3¼ inches from the beginning of the recessed portion. An extension lobe with two holes to secure the carrying strap extends ½ inch beyond the end of the horn, where there is a flat pine plug secured by five wooden pegs.

BIBLIOGRAPHY: Grancsay, No. 415, 55, illus. pl. XVII, cat. no. 1, and pl. VII, cat. no. 1.

FORERUNNER OF THE LAKE GEORGE SCHOOL

9 *Levi Whitney horn*

Shrewsbury, Massachusetts, January 14, 1750
Horn, pine, brass
OL: 14½ in. (36.8 cm), w (of plug): 2⅞ in. (7.3 cm)

Inscriptions: SHREWSBURY * JANUARY * 14 ADOMINI / LEVI * WHITNEY * HIS * HORN * 1750 / DIVEL / BRAZON SERPENT / LF

William H. Guthman

Family tradition relates that this horn belonged to Samuel Whitney of Stratford, Connecticut, but it is more plausible to place the horn in or near Shrewsbury, Massachusetts. It has been suggested that Whitney was born in Westminster, Massachusetts, and served in the Massachusetts militia.

The double-line uppercase block lettering with diagonal shading is nearly identical to that on the Crosby horn (No. 7), but is not so finely engraved. It appears to be by the same hand, but is executed as if the carver was pressed for time. Each word is separated from the next by a diamond-shaped device, or lozenge, that is simpler than those on the Crosby horn.

Instead of the sailing ships seen on the Crosby horn, the carver used a large snake that runs the length of the horn and is labeled BRAZON SERPENT. Brazen Serpent is a Biblical term from John 3:14–15 that

Catalogue

9

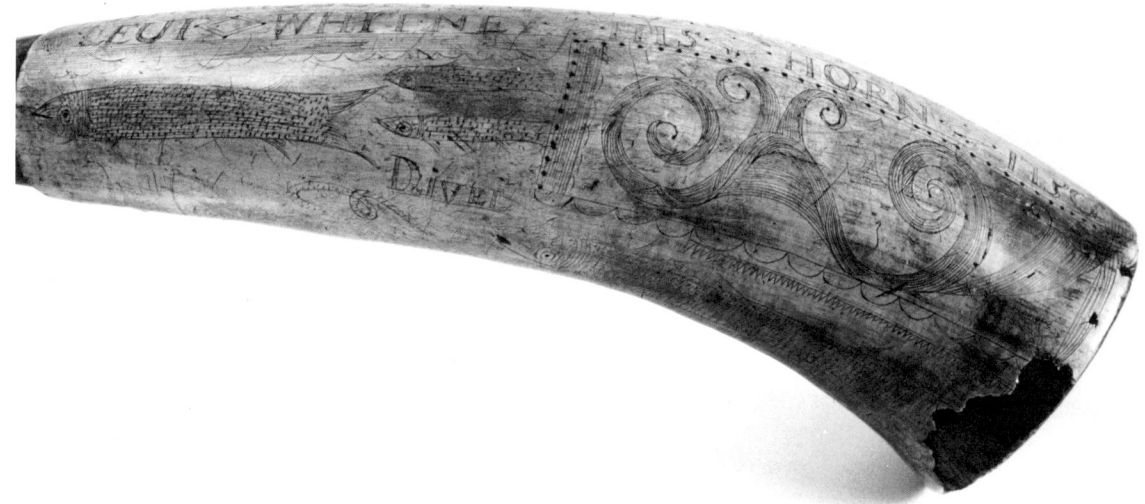

9. These imaginative depictions of the octopuslike Devil (middle) and the Brazen Serpent (below) show the importance of religion to provincial soldiers.

symbolizes the redemptive death of Christ. A weird, octopuslike design within a dotted rectangular cartouche is captioned DIVEL. Schools of freshwater fish and rows of sprawling plants with slender shaded leaves extend the aquatic motif. There are also three sunbursts and occasional lightly engraved scrolls and lunettes.

A crudely incised border precedes the recessed portion 4½ inches from the tip of the spout. A raised ring to secure a carrying strap is 1¼ inches from the tip of the spout. The plug end has no border. The lightly rounded pine plug, secured by eight pegs and extending ¹⁄₁₆ of an inch beyond the end of the horn, has a small brass screw in the center for securing the carrying strap. Four Maltese cross devices are stamped into the plug.

LAKE GEORGE SCHOOL
    Attributed to the Hill-Tyler Carver (w. 1755)

10 *Rufus Hill horn*

Lake George, New York, just after September 8, 1755
Horn, wood, paper
OL: 14¼ in. (36.2 cm), W (of plug): 3¼ in. (8.3 cm)
Inscriptions: RUFUS HILL / Lake George / September Ye 5 1755 / July 1755 Ye 3rd /Allbone [Albany] / Sept Ye 8th 1755 / Our Battle At The / Lake St George / Was Fought With 2300 / French / RUFUS HILL: / Lake George Call'd by / The French Saucreman / Sepr ye 7th A:D:
William H. Guthman

During the 1755 campaign, Rufus Hill served in Major Robert Denison's 3rd Company of Major General Phineas Lyman's 1st Connecticut Regiment. Solomon Tyler, whose horn (not in the exhibition), made at Carrying Place (Fort Edward) on August 24, has almost identical calligraphy and is by the same hand, also served in that company. Hill's horn traces the company's itinerary: Albany in July, Fort Edward in August, Lake George in September. Hill is listed again as serving in 1757 at Fort William Henry. Captain John Baldwin's Company, which he served in for that campaign, was one of the militia companies raised in the alarm for the fort's relief; the service was sixteen days. Over fifty Connecticut militia companies were called out in the alarm, but the fort was destroyed by the French during the campaign. These militia companies were not part of the 1,400-man regiment raised by the Colony of Connecticut for the 1757 campaign; those troops were raised during the early spring. Hill's name does not appear in rosters for later campaigns.

This profusely carved horn incorporates both accomplished and mediocre elements, as well as a battle chronology. The owner's name is elaborately engraved in large flowing uppercase letters with multiple C-scrolls used as extensions of serifs, crossbars, and curved elements of letters. At the forward end of the inscription are similar scrolls. The legend about the battle is neatly inscribed within an elaborate cartouche. Floral and geometric designs of stylized flowers, including tulips, are found all over the horn. Around the plug is a 2-inch border of deeply incised flowers, birds, and a bell with the numeral "7" and the initials "R.H."

The plug is one of the most elaborate known. It is flat, with a beautifully inscribed ink-on-paper legend giving the owner's name. The paper is covered with

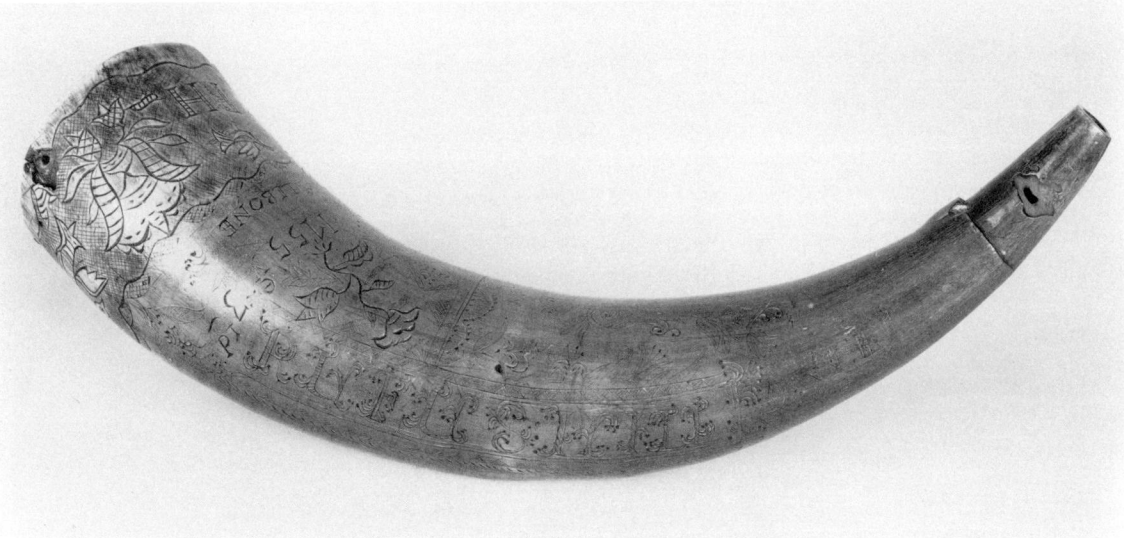
10

Catalogue

10. *Inscription commemorating the Battle of Lake George framed into a carved plug.*

LAKE GEORGE SCHOOL
Attributed to John Bush (w. 1755–1756)

11 *William Williams horn*

Probably Lake George, New York, about September, 1755
Horn, pine, iron, pigment
OL: 16 in. (40.6 cm), W (of plug): 3¼ in. (7.3 cm)
Inscription: WILLIAM WILLIAMS
William H. Guthman

thin horn and the edges of the horn are enclosed within an elaborately carved wooden frame that fits into the end of the horn. The plug itself is recessed ¹⁄₁₆ of an inch to accommodate the frame. The inside border of the frame is a deep, two-line molding, while the outer edges are decorated with incised leaves. The frame is made of four parts, each of which is secured to the plug with two wooden pegs. The plug itself is secured to the horn by seven pegs. A lobe extension for a carrying strap is missing. A slight recessing is 1½ inches from the tip of the horn, and a large, deep hole in the center of the spout was probably created when a staple for a carrying strap broke away. Both this and the Sol Tyler horn are a dark, slightly greenish-brown color, and the engraving does not show well against the horn. Both were carved by the same hand.

This horn is virtually identical to the Thomas Williams horn (No. A). The decorative devices and calligraphy are of the same high quality. William Williams served as surgeon's mate during the 1755 campaign, assisting his uncle, Surgeon Thomas Williams. John Bush carved the horns of both men just after the Battle of Lake George on September 8, 1755.

William's father, Colonel William "Billy" Williams (1713–1788), had studied medicine but became a military leader and land speculator instead of a doctor. He received a captain's commission in Sir William Pepperell's 51st American Provincial Regiment, helped in the construction of the string of Massachusetts forts west of the Connecticut River during King George's War, and was commander of several of the forts during the Louisburg campaign. He was one of the settlers of Pittsfield, Massachusetts, and held the status of a "River God" for most of his life (that is, he was a member of elite Connecticut River Valley society).

The elder Williams received a colonel's commission for the 1758 and 1759 campaigns and requested that his son receive a commission in his regiment. William did receive a commission as surgeon's mate and served in that capacity in 1759 and 1760. Massa-

11

85

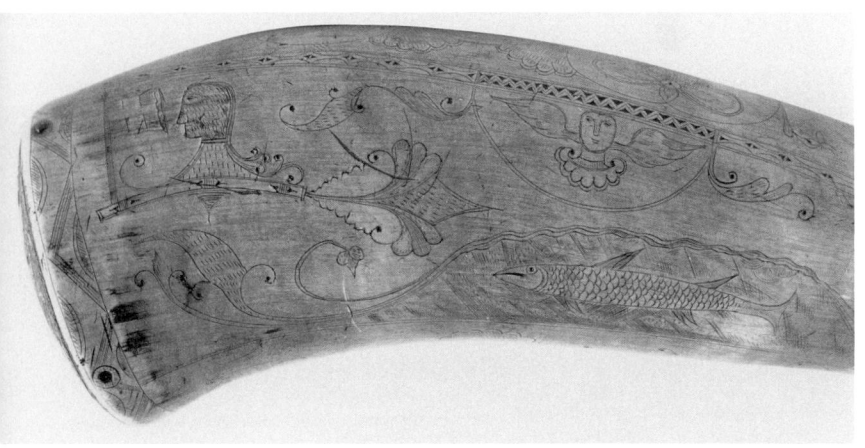

11. Motifs illustrating the masterful ability of carver John Bush.

chusetts records list his residence as Sheffield and show that he was surgeon in Brigadier General Timothy Ruggles's regiment for the Crown Point campaign of 1759. In 1760, his residence was cited as Salisbury, Connecticut, and he is listed as serving again in Ruggles's regiment. He died of smallpox soon after this date.

The importance of the two Williams horns to John Bush's small surviving oeuvre is enhanced by their high quality. All known Bush horns have some highly stylized capital letters, especially the letter "W," and his lowercase letters are fine and precise. Each of the letters on the William Williams horn has two graceful C-scrolls save for the "M" in Williams and the "s," which appears to have been squeezed in at the end. The borders and much of the other decoration consists of featherlike shells and flowers. A fine incised zigzag and incised geometric devices surround or separate words of the inscription. There are also a death's-head or winged angel, an Indian head facing a British flag, and a fish within a cartouche.

A floral-scroll border precedes the recessed throat 4½ inches from the tip, a feature found only on the Thomas Williams horn, and there is a raised ring 1¾ inches from the tip. There is no scalloping. The rounded pine plug retains original red paint and is secured by three wooden pegs and the prongs of an iron staple.

BIBLIOGRAPHY: *Antiques*, 1978, 322; Guthman, *Guns and Other Arms*, 144; Bulletin, ASAC, March, 1981, 50.

LAKE GEORGE SCHOOL
Signed by John Bush (w. 1755–1756)

A *Thomas Williams horn* (Grider drawing)

Lake George, New York, September 8, 1755
Materials and dimensions not known
Inscriptions: THOMAS WILLIAMS / This Horn Was made at Lake George The Battle 8th of Sepr 1755 / I Powder with my brother ball Im hero Like I Conquer all / John Bush: Fecit / When Bows and weighty Spears were usd in Fight / Twere nervous Limbs Declard men of might / But Now Gun Powder Scorns such Strength to own / And Heroes not by Limbs But Souls are Shown
Rufus Grider Collection, New-York Historical Society

This horn, known only through this Grider drawing, is the only signed horn by John Bush and, with its death's-head, shell-like scrolls, and wonderful calligraphy, forms the basis for the attribution to him of twelve other horns.

Thomas Williams (1718–1775), son of Ephraim Williams (1691–1754) of Hatfield, was born in Newton, Massachusetts. In 1755, he served in the regiment of his brother Colonel Ephraim Williams (1715–1755) who was killed leading his men on September 8 at the beginning of the Battle of Lake George. Thomas graduated from Yale College in 1737 and studied medicine in Boston under Dr. Wheat, settling in Deerfield, Massachusetts, to practice. During King George's War, he served at the string of forts along the Massachusetts border west of the Connecticut River, including Fort Dummer and Fort Massachusetts on the Hoosick River near present-day Williamstown, Massachusetts. Thomas's brother Ephraim commanded at those forts during this period.

Thomas Williams's cousin, Esther Williams (1726–1800), became his second wife in 1746. She was the daughter of the Reverend William Williams (1688–1760) of Weston and the sister of Colonel William Williams (1713–1788). Thomas Williams trained his nephew William Williams (No. 11) in medicine and took him along as surgeon's mate on the Crown Point expedition of 1755.

In that engagement, Thomas Williams saw his brother Colonel Ephraim killed and ministered to the captured French leader Baron Dieskau, who was wounded in the battle and died a few months later. Dieskau was shot by accident when a Provincial soldier mistook his gesture of offering his sword in surrender as an attempt to reach for a pistol. Ephraim Williams's will gave land to establish an educational institution that eventually became Williams College.

BIBLIOGRAPHY: Grider, NYHS; Grancsay, No. 955, 74; *Antiques*, 1978, 317; Guthman, *Guns and Other Arms*, 139.

Catalogue

A

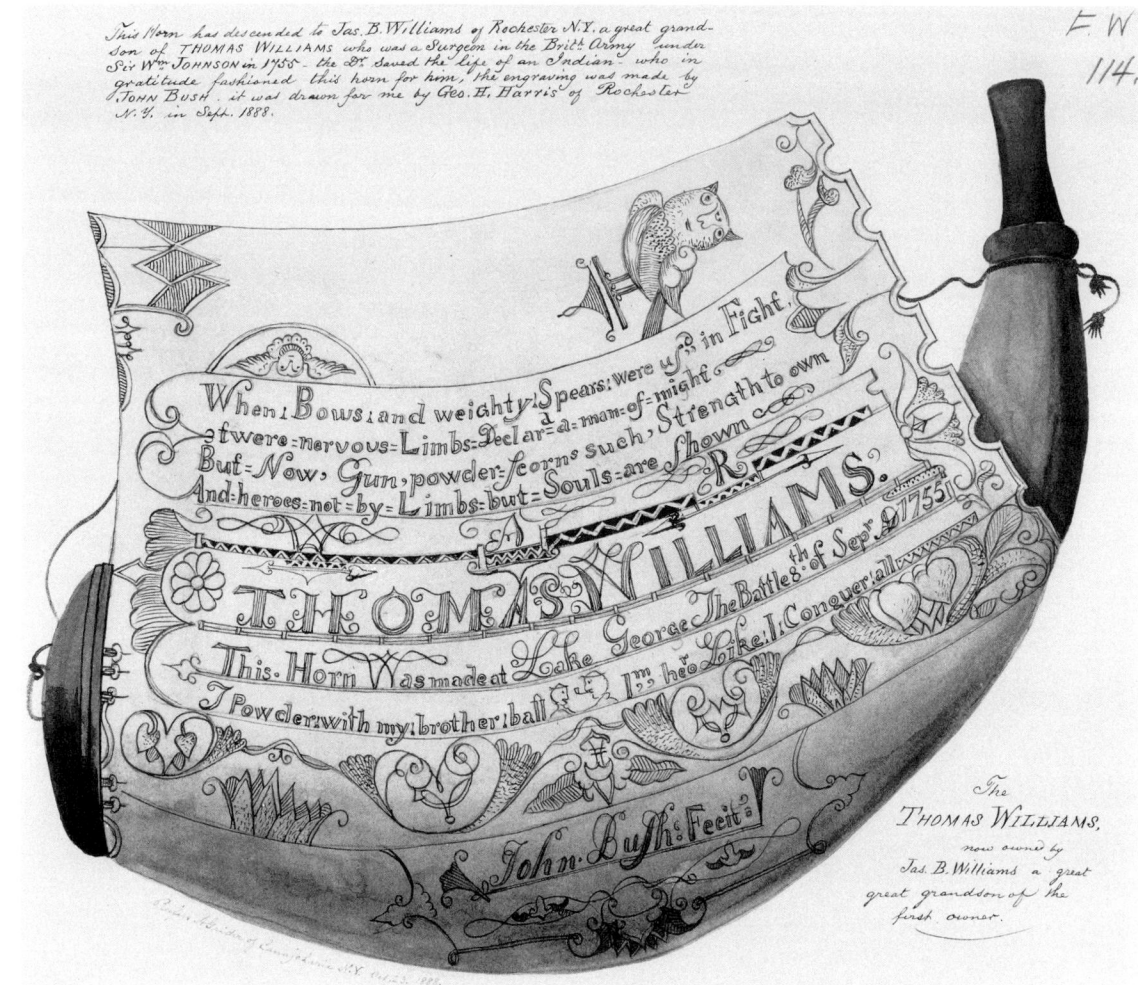

LAKE GEORGE SCHOOL
Attributed to John Bush (w. 1755–1756)

B *Robert Rogers horn* (Grider drawing)

Fort William Henry, New York, June 3, 1756

Horn, pine, red varnish, iron

OL: 15 in. (38 cm), w (of plug): 2¾ in. (7 cm). Measurements taken from original horn when it was in the Cuneo collection.

Inscriptions: WM WHITE / ROBERT ∗ RODGERS / his. horn. Fort. Wm henry. June. Yᶜ. 3. 1756; *other initials and date are later additions.*

Rufus Grider Collection, New-York Historical Society

Robert Rogers and Israel Putnam were probably the most famous figures of the later colonial wars. Rogers remained a renowned hero until Pontiac's Rebellion of 1763, when he was relieved of command at Fort Michilmackinac because of power struggles and jealousy among the commanding officers. The modern novelist Kenneth Roberts based *Northwest Passage* on the story of Rogers and the New Hampshire Rangers whom he commanded. Rogers was feared by the French and their Indian allies and applauded by both Provincial troops and British regulars. During the wars English and American newspapers were filled with his exploits. William White's identity has never been established.

Rogers's horn was carved by John Bush and reflects his mastery. Although it is now very worn, with a great deal of the carving barely visible, enough remains to indicate that the horn was a masterpiece. The calligraphy has the typical Bush "W," as well as beautiful block lettering with C-scrolls and geometric designs extending from the uprights of the letters and between words. The border below the inscription is in Bush's style of fine C-scrolls and shell motifs with cloud scrolls. The same border precedes the recessed portion, which is 3½ inches from the tip. A similar border

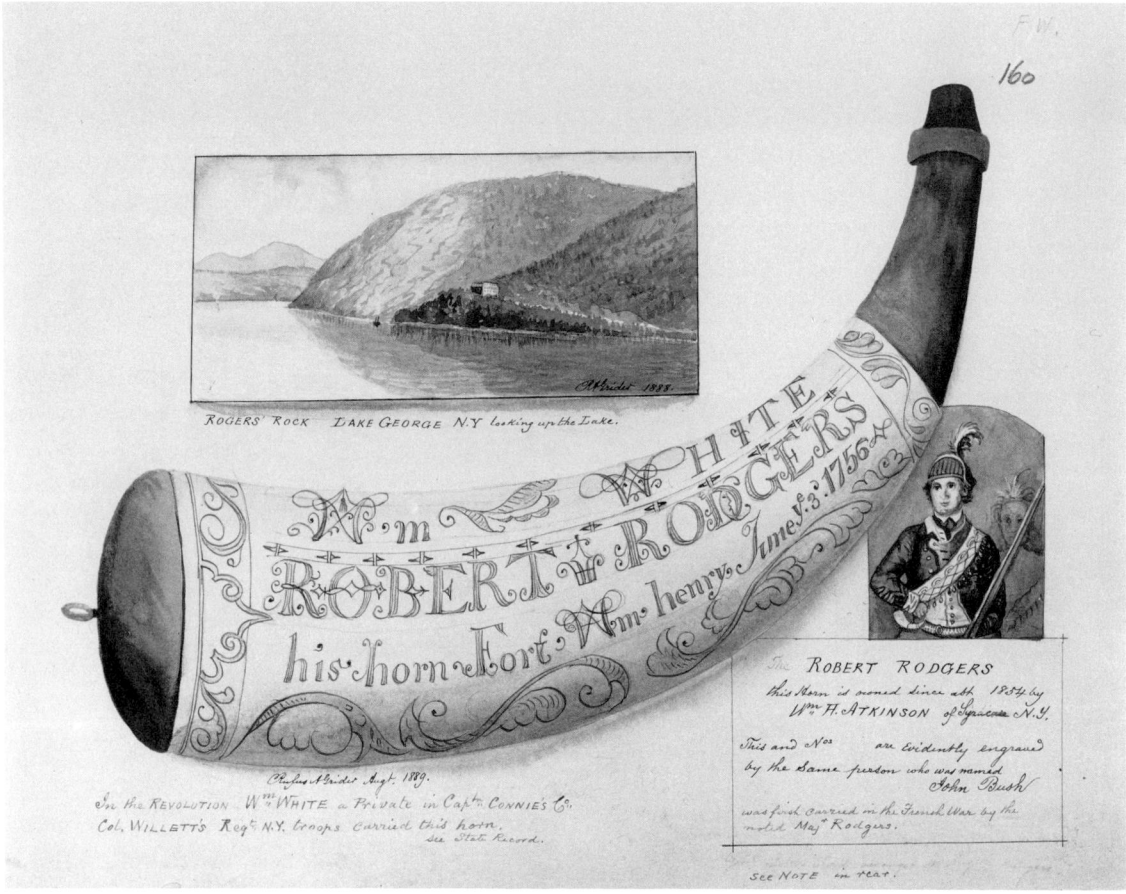

B

1¼ inches wide encircles the plug end. The rounded, raised, red-varnished plug extends ½ inch beyond the edge of the horn and is secured to it by six round wooden pegs and an iron staple ¾ of an inch long. A hole in the center of the plug probably held a metal carrying-strap attachment.

BIBLIOGRAPHY: Grider, NYHS; Grancsay, No. 726, 66; American Art Association auction catalogue, January 31, 1921, checklist No. 991; *Antiques*, 1978, 316 and 318; Guthman, *Guns and Other Arms*, 138 and 140; The Old Print Shop catalogue, Vol. XVIII, No. 9 (May, 1959), 211.

LAKE GEORGE SCHOOL
Attributed to John Bush (w. 1755–1756)

12 *Colonel Nathan Whiting horn*

Fort William Henry, October 11, 1756

Horn, pine, wood, iron

OL: 16 in. (40.6 cm), W (of plug): 3 in. (7.6 cm)

Inscriptions: Colº Nathan Whiting Esqʳ / His Horn made at Fort Wᵐ Henry / The 11ᵗʰ of Octᵇʳ, AD 1756 / When Bows and weighty Spears were us'd. in / Fight. twere nervous Limbs Declrᵈ. a man of might / But Now Gun powder Scorns such strength to Own / And heroes not by Limbs but Souls are Shown / War

The Connecticut Historical Society, Bequest of Clarence Everett Bacon

Lieutenant Colonel Nathan Whiting of New Haven, Connecticut, was commissary at Fort Edward during the winter of 1755–1756. From March, 1755, until December, 1755, he was lieutenant colonel and captain of the 2nd Company in the 2nd Connecticut Regiment under Colonel Elizur Goodrich of Wethersfield, Connecticut. From December 8, 1755, through May 30, 1756, Lieutenant Colonel Whiting

Catalogue

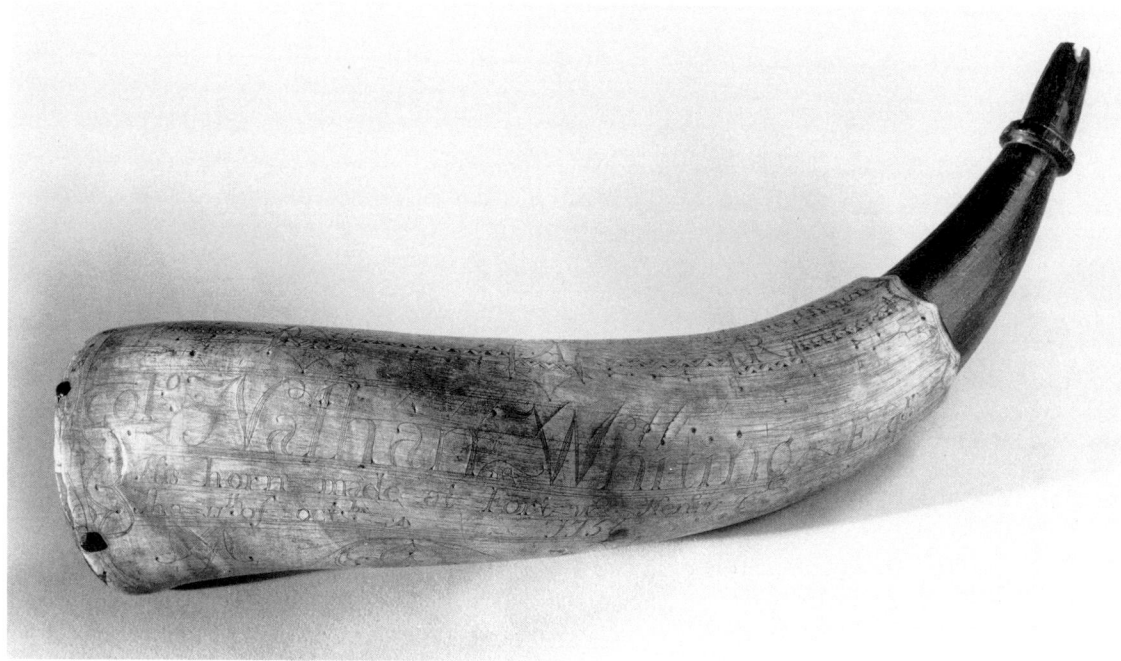

12. *John Bush's typical fine engraving.*

served under Colonel Jonathan Bagley of Massachusetts in a regiment consisting of officers and men from Massachusetts, Connecticut, New York, New Hampshire, and Rhode Island. This regiment manned Forts Edward and William Henry and Whiting was in command of 430 men at Fort Edward. During the 1756 campaign Whiting was appointed colonel of the 3rd Connecticut Regiment, in command of the 1st Company. He served from March until December 2, 1756. During the 1757 campaign he again served as lieutenant colonel, in command of the 2nd Company of the Connecticut Regiment raised for that campaign and commanded by Colonel Phineas Lyman of Suffield, Connecticut. He served from February 21 to November 24, 1757. In 1758 he was appointed colonel in command of the 2nd Connecticut Regiment, commanding the first company from March 10 through November 20. During the 1759 campaign he was again appointed colonel in command of the 2nd Connecticut Regiment, heading the 1st Company from

12. *John Bush's fluid style and imaginative repertory of decorative motifs may be seen here.*

March until December of that year. He was again appointed colonel of the 2nd Connecticut Regiment in the 1760 campaign, commanding the 1st Company from March 24 to November 27 of that year. He was appointed colonel of the 2nd Connecticut Regiment in the 1761 campaign, again commanding the 1st Company and serving from April 1 through December 4 of that year. The records are not clear about the 1762 and 1763 campaigns, but he did serve as colonel in both of them.

This previously unpublished horn is of a fine quality that is typical of Bush's better work. Bush inscribed the illuminated lettering for the word WAR using the incised zigzag design to separate the letters as well as to underline them. The calligraphy of the rhyme is typical, fairly neat and studied, while the flourishes in the name and rank make this calligraphy quite beautiful. An unusual feature is the use of the hyphen to separate the words in the rhyme.

The throat is scalloped 4⅜ inches from the tip, where the recessed portion begins. The scalloping is wide and is accented by a double incised line that follows the line of the scalloping and serves as a border. There is a single raised ring 1⅜ inches from the tip. The pine plug is flat, secured by two small original round wooden pegs and four iron rosehead nails that replace four missing wooden pegs (these nails are a period replacement). A unique removable turned wooden dowel is held in a drilled hole in the plug's center by friction.

LAKE GEORGE SCHOOL
Attributed to John Bush (w. 1755–1756)

13 *Jesse Austin horn*

Fort Edward, New York, September 14, 1756
Horn, pine, iron, rosewood, black pigment
OL: 17 in. (43.2 cm), W (of plug): 2¾ in. (7 cm)
Inscriptions: When Bows and weighty Spears were us^d in / Fight twere Nervous Limbs Declar^d a man of might / But Now Gunpowder Scorns Such Strength to / Own and heroes not by Limbs but Souls are Shown / Jesse Austin / his horn Was made at Fort Edward / Sept^{br} the 14th AD 1756 / W   A   R
William H. Guthman

Jesse Austin is listed as serving in only one campaign, that of 1756. From March 29 to November 12 he served as a private in General Phineas Lyman's 1st Company, General Phineas Lyman's 1st Connecticut Regiment (in the Provincial forces the commanding officer of a regiment was also the commander of the 1st company of that regiment).

The Austin horn is virtually identical in format and style of carving to the David Baldwin horn (No. 14). The quality of the carving, the calligraphy with featherlike devices between words, and the geometric decorative devices all appear similar to the work of the carver J. W. This four-line rhyme is also seen on the Baldwin horn, the Putnam horn (No. 15), and the Thomas Williams horn (No. A). The deep incised zigzag carving colored with pigment and underlining the stylized word WAR is typical of Bush's work. The copperplate calligraphy demonstrates the versatility of Bush's engraving skills.

Unlike the Baldwin horn, the Austin horn has a

Catalogue

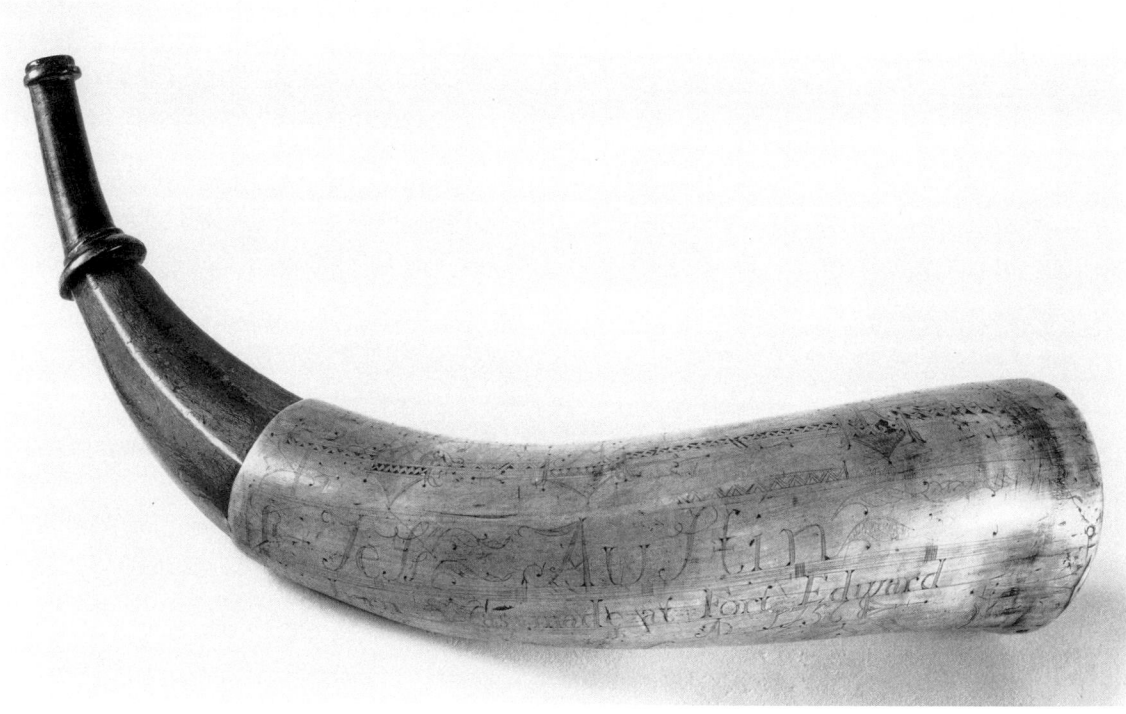

13

simple incised border preceding the beginning of the recessed portion 7 inches from the tip. The recessed portion is faceted octagonally, with a large raised ring 2¼ inches from the tip. The remainder of the throat is smooth, ending in a double raised ring at the spout. The large shaped stopper is a replacement.

There is no border at the butt end and the wooden plug, which is flat and has traces of reddish-brown paint, is secured by six small wooden pegs. A large L-shaped iron nail ⁹⁄₁₆ of an inch long and ³⁄₁₆ of an inch wide is at the center of the plug. Another round-headed nail is driven into the plug alongside the large nail.

LAKE GEORGE SCHOOL
Attributed to John Bush (w. 1755–1756)

14 *David Baldwin horn*

Fort William Henry, New York, October 18, 1756
Horn, pine, brass, black pigment
OL: 16½ in. (41.9 cm), w (of plug): 3¼ in. (8.3 cm)
Inscriptions: When Bows and Weighty Spears were Usd In / Fight Twere Nervous Limbs Declrd a man of might / But now Gunpowder Scorns Such Strength / to own and heroes not by Limbs but Souls are Shown / W   A   R / Cpt David Baldwin Esqr / his horn made at Fort wm Henery Octbr ye 18th 1756
William H. Guthman

David Baldwin of Milford, Connecticut, served in the 1755 campaign as a first lieutenant in a New York regiment. New York enlisted three companies of Connecticut men in that campaign, commanded by their own Connecticut officers. During the 1756 campaign David Baldwin was captain of the 7th Company, Major General Phineas Lyman's 1st Connecticut Regiment. In 1757 he served as a private in a Connecticut militia company commanded by Captain Isaiah Brown of Stratford. The company was enlisted for sixteen days, from August 7 to August 23, during the alarm for the relief of Fort William Henry. During the 1758 campaign Baldwin served as captain of the 4th Company, Colonel Nathan Whiting's (No 12) 2nd Connecticut Regiment. During the 1759 campaign, he served as major in Colonel Nathan Whiting's 2nd Connecticut Regiment and as captain of the 3rd Company, and continued in that rank during the campaigns of 1760–1762.

The horn descended in Baldwin's family until it was sold at auction in 1987. In 1891 a descendant, the Reverend DeWitt Clark of Salem, Massachusetts, corresponded with Rufus Grider, and Grider's six-page letter to Clark has survived along with the horn. Clark sent Grider a sketch of the horn, which Grider copied for his collection, and the drawing is now part of the Grider Collection at the New-York Historical Society.

In his letter, Grider refers to Clark as Baldwin's

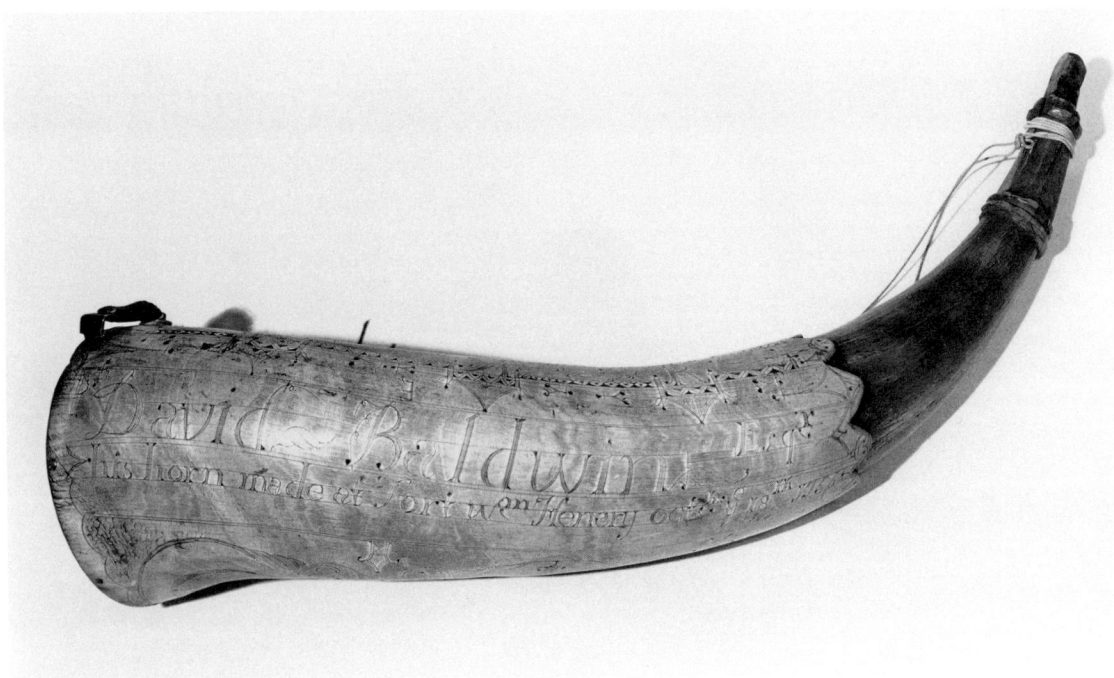

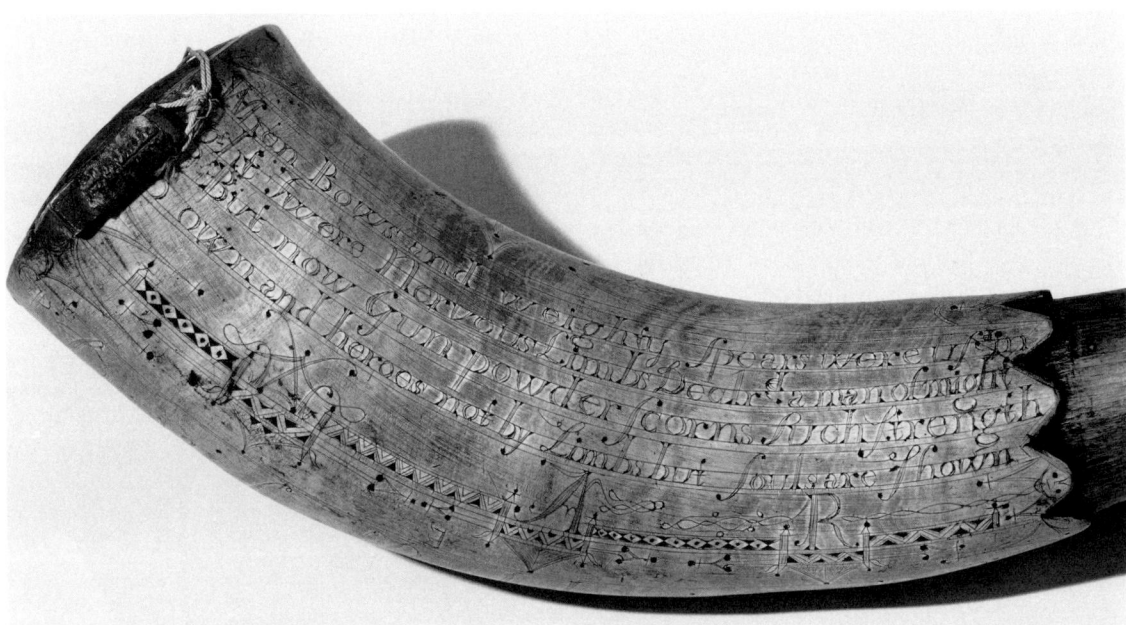

14. *Precise engraving and incised decoration were trademarks of carver John Bush.*

great-grandson, and thanks him profusely for the several drawings Clark had made of the horn: "I rec'd your welcome reply today & rejoice in the result—it is of much value to the Collection.... The drawings you made are of more value to me than a Photo. They give the Size—the Photo would not."

Grider's letter to the Reverend Clark also indicates that Grider was aware of Bush's importance as a horn carver. He cites the Israel Putnam horn (No. 15), the Thomas Williams horn (No. A), and the Robert Rogers horn (No. B) as related examples. He also recognized that professional carvers were responsible for the best horns, for he wrote, "Your horn was made by a professional horn decorator—his name was John

Bush. He had a certain sett of patterns—which he used partly on every horn he decorated & altho he may not have subscribed himself, once acquainted with his work, one can know it . . . ."

The Reverend Clark appears to have cleaned the horn of surface dirt and varnish when he made the sketches. Then before the 1987 sale, someone artificially darkened the horn. Happily, when the darkening agent was removed, the surface was found to retain a satisfactory amount of its original color and patina.

This is one of Bush's most beautiful horns. It features all of his standard motifs and several that do not exist on other examples. The copperplate calligraphy of the name, as well as the featherlike devices between words and the geometric decorative device centered between two baroque floral designs strongly resemble decorative devices of the carver J. W. The four-line rhyme is identical to that on the Israel Putnam horn (No. 15), and the deeply incised zigzag accenting the illuminated letters of the motto WAR are colored with dark pigment and are superior to treatments on other Bush horns. The geometric and floral designs are expertly executed, and the winged cherubs are on a par with the cherub on the William Williams horn (No. 11). The large brass staple at the edge of the base is identical to that on the Putnam horn, and the sure free-flowing movement of Bush's burin is emphasized by the exquisite quality of the engraving, as it is on the Austin horn (No. 13).

An unusual feature is the treatment of the border preceding the recessed portion of the horn 6 inches from the tip of the spout. The sawtooth scalloped edge consists of nine large deeply cut chevron devices that average ¼ inch in length. Each of these is decorated with a winged death's-head effigy separated by devices that might be interpreted as crosses; these are contained within a festoon border. The recessed portion is plain up to within 2 inches of the tip, where a double raised ring leads to a hexagonal portion and another raised ring. The wooden stopper in the spout appears to be original.

A series of geometric devices identical to those at the throat surround the plug end. The slightly rounded pine plug extends ¼ inch beyond the edge of the horn and is secured to it by five T-headed brass nails and a large brass staple. The large staple has a fragment of leather attached to it which may be part of the original carrying strap.

BIBLIOGRAPHY: Grider, NYHS; Grancsay, 39, 42; *Antiques*, 1978, 317; Guthman, *Guns and Other Arms*, 139.

LAKE GEORGE SCHOOL
Attributed to John Bush (w. 1755–1756)

15 *Israel Putnam horn*

Fort William Henry, November 10, 1756
Horn, pine, brass, black pigment, red paint
OL: 19¼ in. (48.9 cm), W (of plug): 3¼ in. (8.3 cm)
Inscriptions: When bows and weighty Spears were us'd in Fight / Twere nervous Limbs Declr'd a man of might / But now Gun powder Scorns Such Strength to own / And Heros not by Limbs but Souls are shown W A R / Capt Israel putnam's Horn made at / Fort w^m, Henry Nov^r, the 10th. AD: 1756 / a plan of the Stations / From albony to / Lake George / The River, The Road
William H. Guthman

Israel Putnam is one of the legendary heroes of colonial American history. Born in Massachusetts in 1718, he moved to Pomfret, Connecticut, in 1740, where he prospered, becoming a substantial member of the community. Folk legends, such as that telling of his entering a wolf's den and killing a large wolf during the winter of 1742–1743, became part of his biographical image so that he, like Robert Rogers, remained an American hero in history books, historical-society displays, and at historical sites and markers. Also like Rogers, Putnam's ability began to fade during the American Revolution, even though the legends grew stronger.

Israel Putnam is listed as a captain in Major General Phineas Lyman's 1st Connecticut Regiment during September of the 1755 campaign. This regiment was to garrison during the winter at Fort William Henry and Fort Edward. Putnam continued to serve in the Connecticut forces throughout the war, commanding a company of Rangers under Robert Rogers and being promoted to the rank of major in 1758. He was captured on August 8, 1758, taken to Canada, and remained a prisoner for three months until he was exchanged. Legend has it that he was tied to a stake and was about to be burned alive by his Indian captors but was rescued at the last minute. He continued to serve until the end of the war, achieving the rank of lieutenant colonel during the 1759 campaign.

In 1762, Putnam commanded a company in the Havana Expedition; he was one of the few survivors of his company when their ship was wrecked off the coast of Cuba during a hurricane. Struggling to stay alive on the island, most of the men were annihilated by fever, exposure to the elements, and starvation.

In 1764 Putnam first served as a major under Bradstreet in the expedition against Pontiac at Detroit, and in May he was promoted to lieutenant colonel, commanding five Connecticut companies in Pontiac's War. After the Pontiac expedition Putnam

remarried (his first wife, who bore him ten children, died in 1765), thus improving his fortune and social position. He opened the General Wolfe Tavern in Pomfret, which was frequented by ex-soldiers and patriots, and became involved in every major patriotic effort against Great Britain in his community. He was prominent in the Sons of Liberty and was a representative of the General Assembly. Legend relates that when the Port of Boston was closed in August of 1774 Putnam drove a herd of 125 sheep to the stricken town to aid the hungry people.

With General Phineas Lyman, Putnam undertook an arduous trip in 1772–1774 to study land granted to veterans of the Havana Expedition and to consider the possibility of land speculation. The exploratory expedition took the men to the West Indies, the Gulf of Mexico, and up the Mississippi to Natchez. The grants were never given, however.

Another Putnam legend, which became famous in the nineteenth century as a result of a print depicting the event, has it that Putnam heard the news about the Battle of Lexington while plowing his field in Pomfret. He unhitched one of the horses, abandoned the plow, and rode the 100 miles to Cambridge in eighteen hours, leaving word for the Connecticut militia to follow. He was indeed a lieutenant colonel in the 11th Regiment of the Connecticut militia, and he did rush to Cambridge, although he was soon recalled to Connecticut to become a brigadier general. He returned to Boston and became actively involved in the councils of war. Another legend states that at Bunker Hill it was he who commanded, "Don't shoot until you see the whites of their eyes."

Many other legends evolved about Israel Putnam. He became a major general in the Continental army, serving in positions of responsibility from Boston to New York and then on up the Hudson. However, Washington eventually realized that the French and Indian War hero was not qualified for high command. Putnam, who served so bravely as a field commander during the French and Indian War and who inspired the men under him at Bunker Hill (he was one of those in high command at that battle) was not qualified to command large operations or to map out military procedures, as Washington learned through experience.

David Humphreys perpetuated the Putnam legends in the biography he wrote after the war. Although Putnam had little education and was barely literate, his ability as a company commander and ranger cannot be questioned. The legends that evolved from his exploits undoubtedly began during his career in the French and Indian War. There is no question that he inspired men and was a fearless warrior, but his ability was in fact limited.

The fact that this horn was carried by an impor-

15. *John Bush's use of the motto* WAR, *of four-line rhymes, of copperplate calligraphy, and of flowing floral and geometric designs was the basis for the Lake George School of powder-horn carving.*

tant—even legendary—leader during the French and Indian War made it an object in constant demand for display during the nineteenth century. It was exhibited at the celebration of the opening of the Bunker Hill Monument, at the Centennial exhibition in Philadelphia, and at the Putnam Cottage in Greenwich, Connecticut. This is one of the few legitimate horns that belonged to a legendary hero (in spite of the fact that much of the legend was fiction).

The horn might have accompanied Putnam to Bunker Hill during the battle of June 17, 1775, where, indeed, he was a leading force. It was the frontispiece for Volume II of *The American Pioneer*, published in Cincinnati in 1843; it was included to accompany an article about the Putnam horn entitled "An Ancient Relic." This short-lived periodical was the organ of the Logan Historical Society. The article related that the horn descended in the Putnam family and was at that time owned by William Pitt Putnam of Belpre, Washington County, Ohio. The rendition of the horn, in two different engravings, is remarkably accurate. The artist, John H. Lovejoy, a Cincinnati wood engraver, captured the essence of Bush's style and produced an artist's interpretation of Bush's motifs and incised designs. This artistic duplication did not occur again until Rufus Grider copied several Bush horns. Even Grider was not so completely faithful to the original as Lovejoy was. Bush's artistic ability was not noted again until the 1970s, in the article that appeared in *The Magazine Antiques*.

This horn incorporates both the perpetuation of the myth of a man who was a living legend and the artistic ability of a carver who was a master during the period when the most beautiful horns were carved. Letters to Putnam-family members who possessed the horn in the nineteenth century requesting permission to borrow it for various exhibitions, along with the printed history of the horn, render this and the David Baldwin example the finest existing powder horns with totally genuine provenances. Rufus Grider must also have felt the importance of this horn, since it is number one in his series of over 500 horns.

The Putnam horn compares favorably with those made for Jesse Austin (No. 13) and David Baldwin (No. 14). With some additions, its format is the same, and the style of its calligraphy is almost identical. The engraving on the Baldwin horn is probably a shade better than on this and the Austin horns. The difference is slight, however, and the quality is about equal.

As on the other horns, the calligraphy of the name and the featherlike devices are similar to the work of J. W. The border at the throat end of the horn is the same as those on the William Williams and Robert Rogers horns. The four-line rhyme, the baroque floral design, and the deeply incised dark-colored zigzag accents of the illuminated lettering W A R are compatible with decorations on other Bush horns.

A unique feature not seen on other Bush horns is the *plan of the stations from albony to Lake George*. There are two diagrams of forts, probably William Henry (nearest the plug) and Edward (toward the throat), each over 2 inches square, along a numerically marked road that has five other smaller blockhouses placed along it. According to several accounts this road, originally cut through the woods during the 1709 expedition against Canada, is the oldest military road in

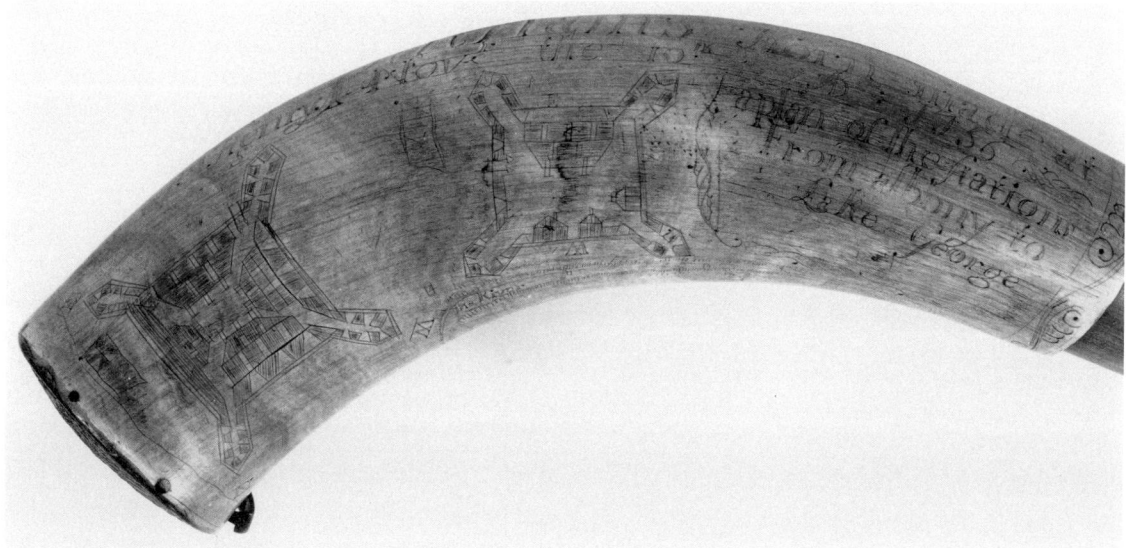

15. *Detail of a map showing two forts and the Hudson River on the route from Albany to Lake George.*

North America. There are also the course of *the River* marked on the plan, a British flag on pole flying above each fort, and N, S, E, W directionals on the two large forts.

The border preceding the recessed portion 6½ inches from the tip of the spout is almost identical to the William Williams border. There is no scalloped edge and the floral-scroll border preceding the recessed edge is almost identical to the borders on several other Bush horns. There are two raised rings ¼ inch apart; the ring closest to the tip is 2 inches from the spout. The slightly rounded pine plug, which still retains about twenty percent of a red paint coating, extends about ⅛ inch from the end of the horn. It is secured by five small round wooden pegs and a 2¼-inch long brass staple with crosshatch engraving that is at the edge of the upper portion of the horn. It is almost identical to the staple on the Baldwin horn.

BIBLIOGRAPHY: *American Pioneer*, Vol. II, Cincinnati, 1843, frontispiece and pages 3–16; Grider, NYHS; Grancsay, No. 699, 65; *Antiques*, 1978, 317, 319; Guthman, *Guns and Other Arms*, 139, 141; *Sketch*, by Lyman C. Draper, Wisconsin Historical Society; "Powder Horns of Men Who Made Our History," Stewart Culin, *The Philadelphia Press*, Feb. 20, 1898, 33; "Rufus Grider, A Memorial," Thomas A. Dickenson, *Proceedings*, Worcester Society of Antiquity, Vol. XVII (1900), 110–113; "Historical Military Powder Horns," J. L. Sticht, *St. Nicholas Magazine*, Vol. XXIII (1896), 993–997; "Historical Military Powder Horns," Gilbert Thompson, The Society of Colonial Wars of the District of Columbia, *Historical Papers*, No. 3, Washington, D. C., 1901; duMont, 16 (caption).

LAKE GEORGE SCHOOL
Attributed to the Goldthwait-Smith-Diamond Carver (w. 1756)

16 *Michael B. Goldthwait horn*

Fort William Henry, New York, October 2, 1756
Horn, pine, brass, iron
OL: 16½ in. (41.9 cm), W (of plug): 3¼ in. (8.3 cm)
Inscriptions: Michael : B Goldthwait^s horn . 1756 / At Fort W^m henry Oct^{br} 2 AD / AP
Maine Historical Society

Michael B. Goldthwait is listed in 1755–1756 as a private in Colonel Jonathan Bagley's Massachusetts Regiment, Colonel Jonathan Bagley's 1st Company at Fort William Henry.

What this horn lacks in profuse ornament it makes up for in a revealing vignette that extends from the plug to the recessed portion: there are a lake, a single-masted cutter or galley anchored offshore, and a rowboat with five oarsmen leaving the ship and heading for a wooded bank. On the opposite shore is a profile view of Fort William Henry with the Union Jack flying high overhead.

The calligraphy is typical of the Lake George School's neat, ornate style, but is not so elaborate as some other work. Four of the capital letters in Goldthwait's name have flourishes extending from the uprights of the letters. The letter "O" in October has a diamond-shaped device as a central motif. Underneath the inscription is a handsome S-scroll border. This was executed by the engraver who also decorated the Thomas Smith Diamond horn (No. 17), although the fort engraving differs.

No border precedes the recessed portion, which begins 5⅞ inches from the tip. A raised ring is 2⅝ inches from the tip, and the spout has a wooden stopper. A ¾ inch brass ferule takes the place of a border at the plug end. The ferule is neatly engraved with floral designs and is carefully scalloped on the spout side. The slightly rounded pine plug extends ⅜ inch beyond the edge of the horn and is secured to it by five small round iron tacks driven through the ferule. The initials AP are engraved at the bottom of the cap.

BIBLIOGRAPHY: Grider, NYHS; Grancsay, No. 385, 54; New York Public Library, "Calendar of the Emmet Collection of Manuscripts, Etc. relating to American History," New York, 1900.

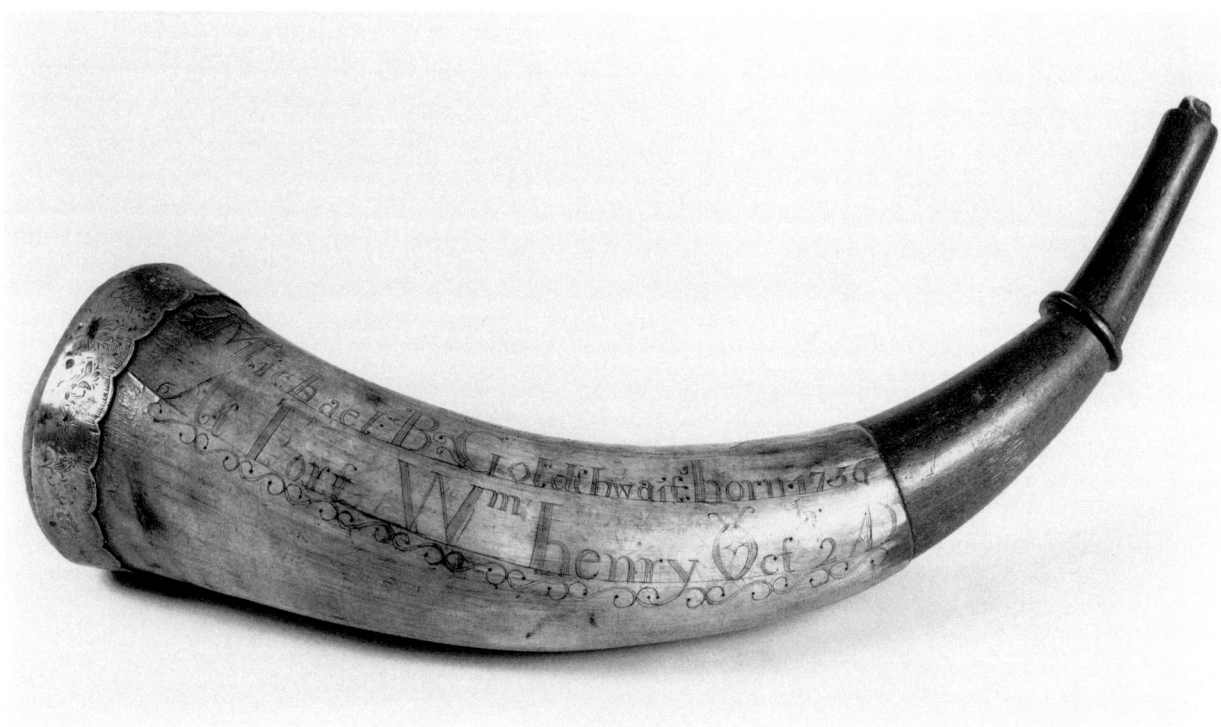

16

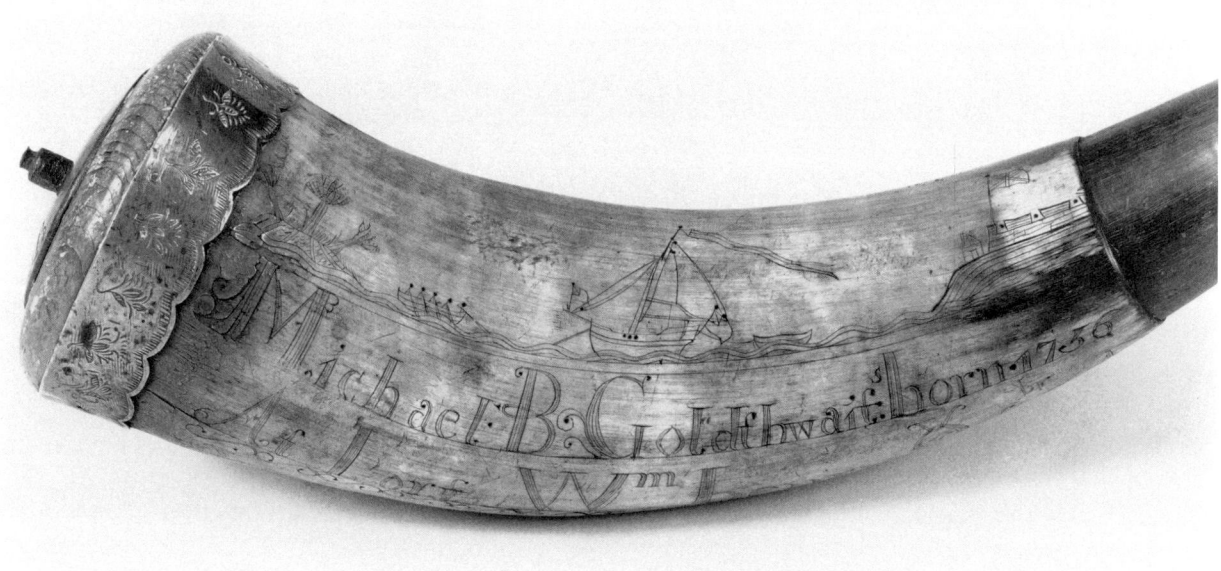

16. *Profile view of Lake George and Fort William Henry.*

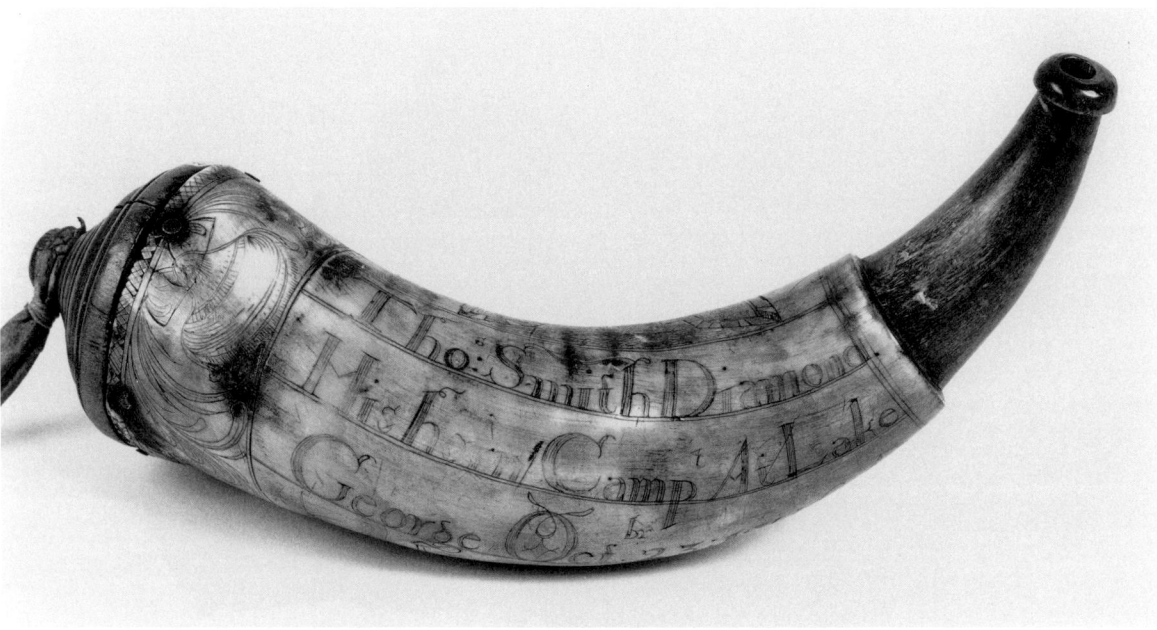

17

LAKE GEORGE SCHOOL
    Attributed to the Goldthwait-Smith-Diamond Carver (w. 1756)

17 *Thomas Smith Diamond horn*

Lake George, New York, October 23, 1756
Horn, iron, cherry, leather
OL: 12 in. (30.5 cm); W (of plug): 2½ in. (6.4 cm)
Inscriptions: Tho: Smith Diamond / His horn Camp At Lake / George Oct^{br} 23: 1756 AD / Fort / William / Henry / Gate / The Road
William H. Guthman

Thomas Smith Diamond's service records have not been found, but his descendants claim that he was from Massachusetts.

The Diamond and Goldthwait (No. 16) horns were carved by the same engraver. The shaded thick-and-thin, or copperplate, lettering is almost a carbon copy of J. W.'s lettering. The lower case "g" is extremely close, and the serifs are almost identical except that those with flourishes at the end have underdeveloped wings. The "S" and "h" in the name Smith both have these curling serifs, almost like a pig's tail, but the end of the serif on the "S" continues in a double streamer connected to the end of the curled serif on the "h." The "O" in October forms a round bowknot with the bow at the top, and it has a diamond-shaped geometric design in the center, something like those of Jacob Gay's letters. The lower loop of the lower-case "G" in George twists gracefully, as if it were a winding river, and is strikingly similar to J. W.'s letters. The engraving is almost identical to that of the Goldthwait horn.

The most remarkable feature of this horn is the highly detailed and expertly carved plan of Fort Wil-

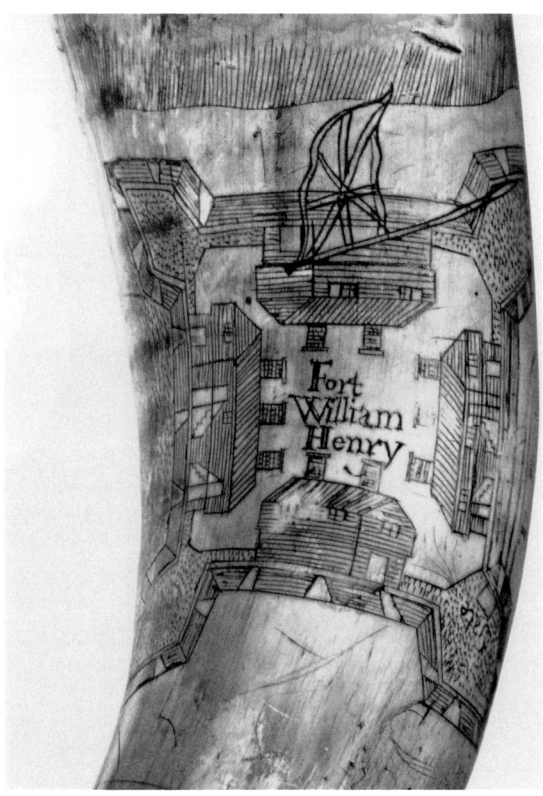

17. Accurate, detailed plan of Fort William Henry.

liam Henry. Such is the detail and depth of this engraving that it looks as if it had been executed by a military engineer. The three-inch-square plan shows a large British flag flying from a tall staff.

The plug end of the Smith horn has a beautifully engraved floral border connected with graceful C-scrolls. This 1½-inch-wide border is similar to those in J. W.'s work, and one of the devices on the border is like more developed geometric devices on later horns by J. W. The cherry plug is ornately carved in a spiral and extends 2 inches beyond the end of the horn, where it ends in a knob to which a portion of leather thong from the carrying strap is still attached. Seven iron nails securing the plug to the horn are believed to be replacements for seven wooden pegs. The only border at the spout end consists of two incised lines ¼ inch from the point where the recessing begins. The recessing extends for 2⅞ inches, and the tip of the horn is a raised lip ⅛ inch thick for securing the strap.

LAKE GEORGE SCHOOL

18 *George Willson horn*

Probably Fort Edward, Fort William Henry, or Crown Point, New York, about 1756
Horn, maple, iron, dark pigment
OL: 15¾ in. (40 cm), W (of plug): 3 in. (7.6 cm)
Inscriptions: GEORGE WILLSON / NA . NOME . IMPUNE . LASESSET / LION UNICORN / I'm Powder With / my Brother ball / I'm Hero like / I conquor All / 1756
Carol and the late Thomas J. Segal

George Willson is listed as serving in Captain Edmund Moore's Company, Colonel Ichabod Plaisted's Massachusetts Regiment, raised for the Crown Point campaign.

This beautifully carved horn incorporates features of several recognized carvers. The illuminated lettering of the "G" in George and the "W" in Willson are taken from calligraphy styles of the sixteenth century that remained popular throughout the seventeenth and eighteenth centuries. John Bush, the Selkrig-Page Carver, and Jacob Gay all shared some of the calligraphic features seen here. The "G" in George incorporates a new moon in profile. The neatly executed floral and animal designs are typical of the better carvers and a vignette of a bird with a large fish in its beak partakes of their whimsical aesthetic.

A finely scalloped and incised border precedes the recessed portion 5¼ inches from the tip. The sawtooth scalloping becomes raised as it extends into the recessed portion. The deeply incised center is accented with dark pigment. A raised ring is 2 inches from the tip and a deeply incised border above the owner's name extends from the recessed portion to the plug end. It consists of twenty linked diamonds accented with dark pigment and resembles carved ornament in the work of the Cleaveland Carver (No. 49). The opposite side of the horn also has a lengthwise zigzag border. A wide floral border at the plug end precedes a sawtooth edge like that at the spout end. The flat maple plug, secured by four wooden pegs, is decorated with a relief-carved eight-pointed star. The remnant of an extension lobe remains on the plug end, indicating that the large wrought-iron nail in the plug is not the original attachment device.

BIBLIOGRAPHY: Exhibited in "Powder and Ball: The Colonial Soldier in the 18th Century," William L. Clements Library, University of Michigan, May 17–July 27, 1984.

18

LAKE GEORGE SCHOOL
Possibly by the J. W. Carver (w. 1758–1761)

19 *Thaddeus Bennett horn*

Fort No. 4 (Charlestown), New Hampshire, July 5, 1757
Horn, pine, iron, brass
OL: 15 in. (38 cm), W (of plug): 3½ in. (8.9 cm)
Inscriptions: Thadeus Bennitt : s Horn / made at No 4 July ye 3 AD 1757 / The Rowse is Red The Vilet Blue and A Fols Love / Can not Be Tru / D S 1759
William H. Guthman

Thaddeus Bennett served in Captain Samuel Hubbell's 5th Company (recruited in Fairfield), Colonel Phineas Lyman's Connecticut Regiment, during the campaign of 1757. David Wheeler (No. 34) served in the same company during this campaign, and although the two horns are by different hands, they are extremely similar. Bennett volunteered to serve in Captain Reuben Ferris's Company of Rangers stationed at Fort No. 4 during the winter of 1757–1758. This company was under the command of Lieutenant Colonel William Haviland of the 27th Regiment of British regulars. Captain Israel Putnam (No. 15) was in command of another company of Rangers under Haviland in that campaign. Other horn owners who served in Captain Ferris's company include Ensign Zebulon Butler, Sergeant Ichabod French (No. 21), Nathaniel Selkrig (No. 24), Sergeant David Hamilton (No. 26), and Sergeant Isaac Whelpley (No. 33).

The unidentified carver of the Bennett horn conformed to Lake George School calligraphic standards. The lettering is carefully measured and engraved with a burin. The thick-and-thin copperplate letters are precise and both the upper- and lowercase letters are of equal height. The thick portions are shaded and the thin portions usually end in decorative curls. Ornamental appendages sometimes extend from the upper-case letters and the serifs end in flowing designs like J. W.'s wing designs, also utilized here.

This carver depended on deeply incised geometric designs, graceful floating-leaf and feather devices, and stylized vegetation and flowers. Only one figural element, a fish, is present.

The horn is scalloped three inches from the tip, and the tip itself has a ¼-inch band. The border in front of the scalloping consists of three incised lines. The pine plug is secured with four iron tacks and two additional tacks are driven into the plug at the location of a missing extension lobe. The original hanging arrangement was later changed to a small brass furniture post with a brass ring, which was driven into the plug. The rounded plug extends ¼ inch beyond the horn.

BIBLIOGRAPHY: Crosby Milliman, "An exhibition of American Engraved Powder Horns of the Colonial and Revolutionary Periods," *Bulletin*, Fort Ticonderoga Museum, Vol. 12, No. 3 (October, 1967), 181–182; Swayze, 52–53; duMont, 17.

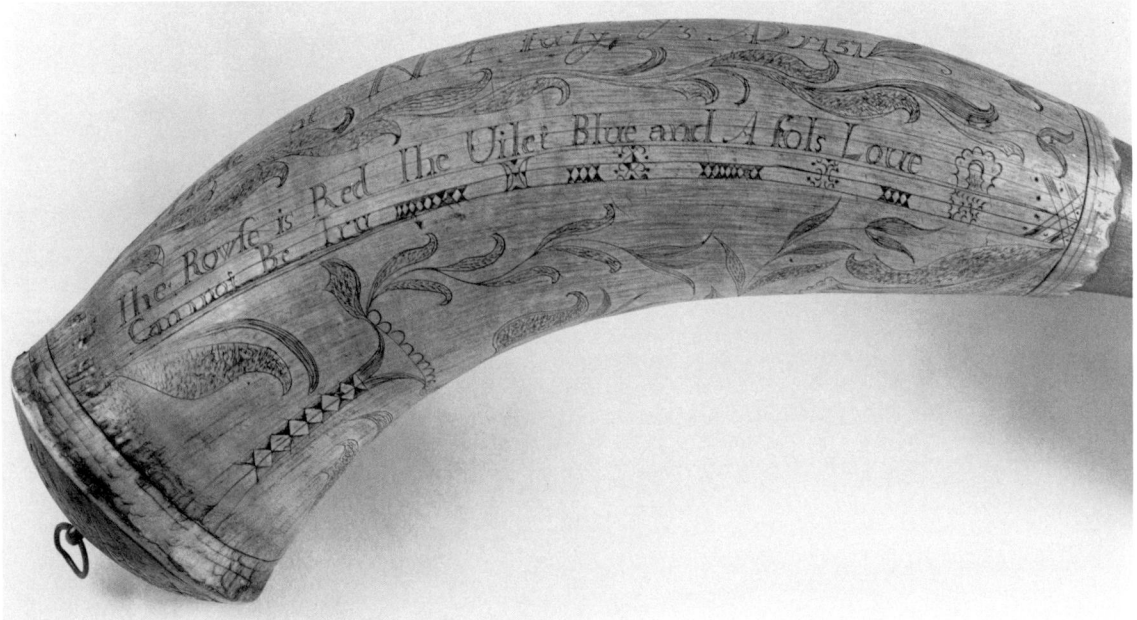

19

LAKE GEORGE SCHOOL

20 *Edward Courtney horn*

Fort Edward, New York, 1757
Horn, wood, brass, vermillion paint
OL: 13 in. (33 cm), W (of plug): 3 in. (7.6 cm)
Inscriptions: EDWARD COURTNEYS / POWDER×HORN×
FOTR×*[sic]* EDWARD / HEAR×AM I×POWDER G 1757 /
WITH×MY / FXR / EDWARD
William H. Guthman

No record of Courtney's service has been found. Although the engraving of his horn is not of the best Lake George School quality (some of the spelling is poor—"fotr" for fort—and the rhyme was left incomplete), it is competently executed, has graphic power, displays novel though naive calligraphy, and has accurate details of Fort Edward.

Especially compelling is the horn's intense, ideologically charged imagery. Two naked Indians with two-feather headdresses are shown shooting muskets at each other from behind trees. A spear and a tomahawk appear behind each of them. These typical depictions underscore the idea that soldiers did not view their foes with respect. Rather, they regarded Indians as unclean savages whom they hated and feared. It follows that all allegedly eighteenth-century American frontier artifacts with images that present Indians as glamorous warriors with full headdress, exotic clothing, and sophisticated weaponry are suspect.

"The Noble Savage" was not an American but a European and English concept and the literature and art of the Old World often portrayed Indians as being of royal blood or as gentle noblemen. An example is John Verelst's paintings, commissioned by Queen Anne in 1710 when the four Mohawk chiefs visited London, which show them as English noblemen standing in front of their manor houses. In contrast to this, as the Courtney horn illustrates, artists and writers on the frontier portrayed the Indians as almost naked fierce warriors, referring to them as "Savages." Colonel George Townshend, who assumed command of the British army in North America in 1759 upon General Wolfe's death on the Plains of Abraham, was an amateur artist whose sketches of Ottawa Indians (allies of the French) show them as blood-thirsty naked savages on the warpath.

Connecticut soldier Jonathan Hobart's powder horn, inscribed while he was at Fort No. 4 in 1757, displays the following message from a frightened and vengeful young conscript: I NOW AT NUMBER FOUR REMAIN / THO TIS AGIN MY WILL / I HOPE I SHALL NO ENEMY MEET / BUT WHAT I WOUND OR KILL. The rhyme reveals hate and fear of the American Indian enemy.

The coat-of-arms on Courtney's horn is extremely imaginative. A laughing rampant British lion with a "G" above his crown (for George II) holds a rose in his paw as he supports one side of the royal arms, while the rampant dragon on the other side holds a sprig of leaves in his outstretched claw. Immediately below the dragon is a miniature deer, or perhaps a mouse. Instead of an "R" above the dragon for "Rex," the carver substituted the date, 1757.

Underneath the coat-of-arms is a large and fairly accurate outline of Fort Edward with an unidentifiable flag flying from an ornate pole. A tall tree stands on either side of the fort, and a large deer and fox

20

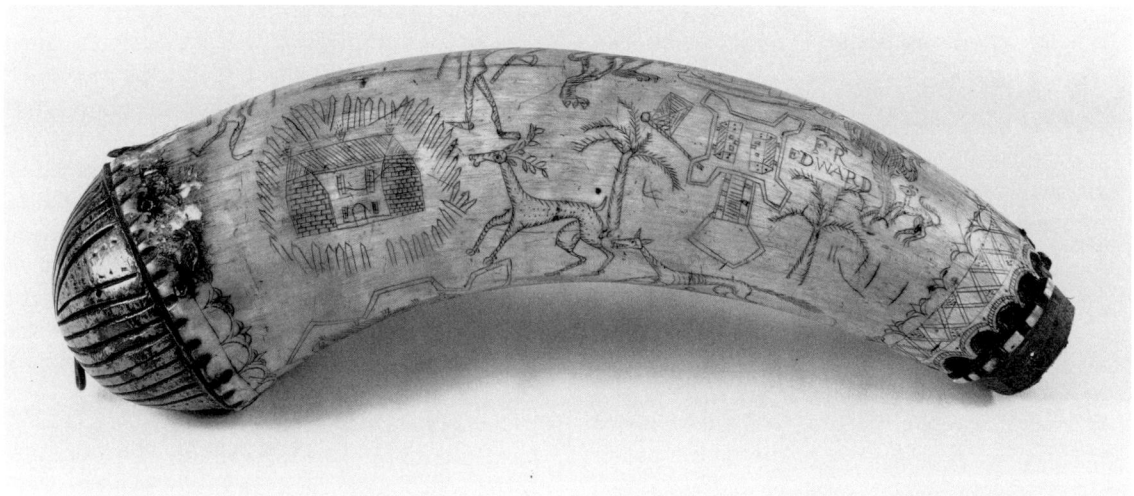

20. *Reliable plan of Fort Edward and view of the brick yard.*

20. *"Naked," as opposed to "noble," savages shown in Woodland battle scene.*

move forward underneath the vignette. Below and to the left of the animals is a drawing of several structures enclosed by a palisade, perhaps the island adjacent to Fort Edward. An unfinished dragon or sea monster menaces a single-masted sloop mounting six cannon that floats next to the island.

To the extreme left and above the island is a two-story brick building with three chimney stacks within a stockade (there was, in fact, a brick kiln at Fort Edward, so this might depict an actual brick barracks or brickyard). Just above Courtney's name are a naked Indian behind a tree with his tomahawk raised to strike and an Englishman behind another tree firing a musket.

All this imagery seems to be an uneducated man's reaction to disparate experiences, for the bulk of his images are derived from observations made on the spot. For example, he may have seen plans of the fort and he was clearly indulging in a satire on the British coat-of-arms, which he could have seen on official documents. An image like the dragon, however, may be a misunderstood interpretation of a map ornament. It may be that the brick barracks were reserved for British regulars and that the carver or Courtney resented them.

A neatly executed floral border encircles the plug end, which is deeply notched. The rounded plug, carved in a series of chevrons, extends 1¾ inches beyond the horn. Painted vermillion, it is secured to the horn by ten wooden pins and has a brass ring at the extreme peak. A handsome floral and geometric border, incised 1¼ inches from the beginning of the recessed portion of the horn, ends ½ inch from it. At this point a beautifully and deeply scalloped and engraved border begins and continues to the point of recessing, where the horn is deeply notched into fifteen separate rectangular sections about ⅛ inch apart and 1/16 inch wide. The tip was cut off ⅜ inch from the notching and a brass charger affixed, probably during the second quarter of the nineteenth century.

BIBLIOGRAPHY: *Antiques*, 1978; *Bulletin*, ASAC, No. 44 (March 1981), 22.

LAKE GEORGE SCHOOL

21 *Sergeant Ichabod French horn*

Probably Lake George, New York, about 1755–1757
Horn, pine, iron
OL: 19½ in. (49.5 cm), W (of plug): 3⅝ in. (9.3 cm)
Inscriptions: serjant Ichabod French's H[orn] / A MAN OF / WORDS / AND NOT OF / DEEDS IS LIKE / A GARDEN / FULL OF / WEEDS; Jan: 10:1748 *is in a later hand.*
William H. Guthman

Ichabod French (1730–1763) of Guilford, Connecticut, served as a private from April 10 to December 9, 1755, in Lieutenant Colonel John Pitkin's 2nd Company (Hartford), Major General Phineas Lyman's 1st Connecticut Regiment. He is not listed for the 1756 campaign, but served during the 1757 campaign as sergeant in Captain Andrew Ward's 14th Company, Colonel Phineas Lyman's Connecticut Regiment. French enlisted as a Ranger during the winter of 1757–1758 and then served as sergeant in Captain Reuben Ferris's Company of Rangers at Fort No. 4, from November 14, 1757, to May 14, 1758. Lieutenant Colonel William Haviland of the 27th Regiment of British regulars was in command of the Rangers at Fort Edward that winter and probably commanded those posted at Fort No. 4 as well. He was roundly disliked by both Provincial troops and Rangers, and his conduct may explain French's failure to enlist for the remainder of the war.

Other Ranger companies serving that winter were the New Hampshire companies under the command of Robert Rogers, who also hated Haviland, and Connecticut companies under the command of Israel Putnam and John Durkee. John Bush carved both the Rogers (No. B) and Putnam (No. 15) horns. Also serving in Ichabod French's company at this time were Thaddeus Bennett (No. 19); Nathaniel Selkrig (No. 24); Sergeant David Hamilton, whose horn (No. 26) was carved by the Selkrig-Page Carver in the style of Bush; and Sergeant Isaac Whelpley, whose horn (No. 33) was carved by J. W. Zebulon Butler, whose Lake George School horn (private collection) is dated 1756, served during this winter campaign as an ensign in Captain Ferris's Company of Rangers at Fort No. 4.

The French horn is neatly carved, but is not of the same high quality as the Bush, J. W., or Selkrig-Page examples. French's name is lettered in bold uppercase letters 1¼ inches high and lowercase letters ⅝

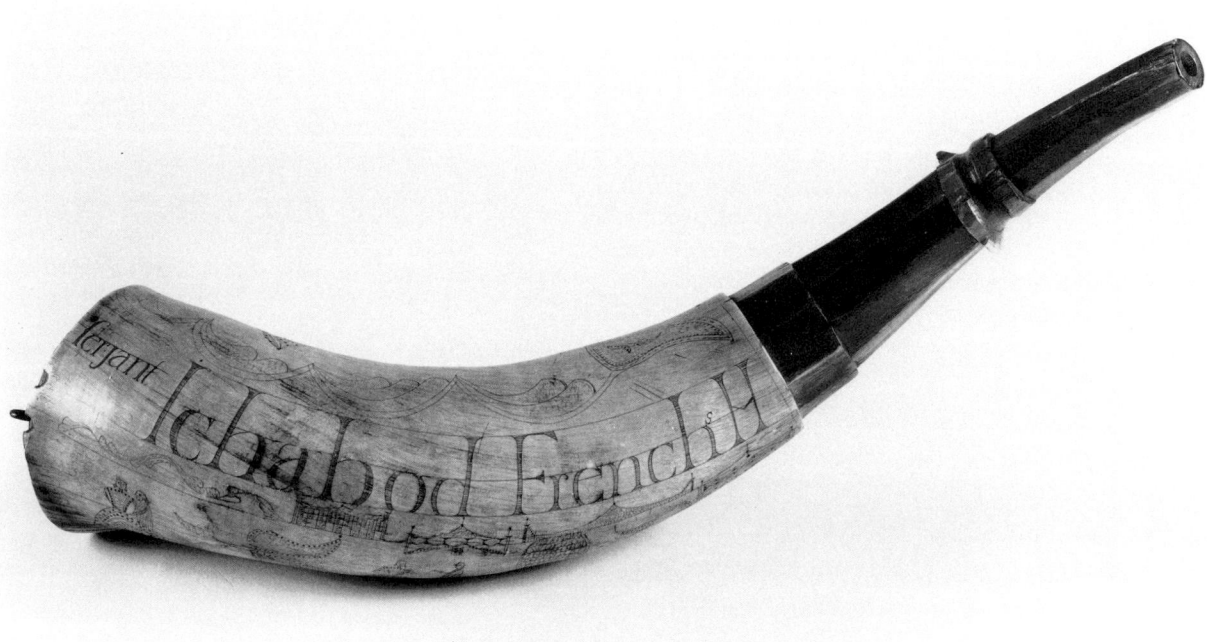

21

21. *Dancing couples and a Lorelei very likely indicate the Provincial soldier's fear that while he was away his wife was unfaithful.*

inch high. The rhyme, in letters that are less than ⅛ inch high, is squeezed into a round cartouche that measures 1⅞ inches in diameter. Of all the lettering, only that of French's name and rank approaches the thick-and-thin copperplate style.

Above the name and toward the spout is a large neatly carved tulip with a stem that evolves into partial C-scrolls and becomes a border for the name. Opposite the tulip is a sizable two-masted sailing ship under full sail with a ghostlike figure, possibly a mermaid or Lorelei, at the stern; she is rising on the breeze, as if she were a flag on a staff, with outstretched arms clutching a comb and a mirror. Below the name are scrolled and geometric designs, and below them, in succession, are a dancing couple dressed for a fancy ball, another large tulip, the same dancing couple doing a different step, the rhyme in a circle, yet another stylized tulip, and a group of three stylized tulips enclosed in a rectangular cartouche composed of sawtooth carving and incised dots. Finally, odd-looking animals fill the remainder of the space. All the motifs are accented with deeply-cut dots. The Lorelei and the dancing couple could represent a homesick soldier's imagined fear of an unfaithful spouse.

At one time a lobe extended beyond the plug end of the horn, as indicated by two notches ⅜ inch apart, which are the only broken portions of an otherwise smooth rim. An iron rivet was driven into the plug between the notches as a substitute. The rounded wooden plug has a flattened center and is secured by five wooden pegs. The plug extends ¾ inch beyond the horn, with the flattened area about 2 inches in diameter. The recessed portion begins 7¼ inches from the tip, with the first inch neatly faceted, then with deeper recessing and more faceting for 2¾ inches. Two large raised rings are carved 3¾ inches from the tip, and the remainder of the spout is faceted. There are no borders at plug or spout ends.

BIBLIOGRAPHY: *Antiques*, 1978, 10; *Bulletin*, Fort Ticonderoga Museum, Vol. 12, No. 3, 179–181; Grancsay, 53; Grider, NYHS; Guthman, *Guns and Other Arms*, 134–153; Beauchamp, W. M., "Rhymes from Old Powder Horns," *Journal of American Folklore*, vol. ii (1889), 117–122, vol. v (1892), 284–290; Swayze, 36–38; *Bulletin*, ASAC, March, 1981, 48; *Bulletin*, KRA, Fall, 1985, 10.

LAKE GEORGE SCHOOL

22 Asa Spaulding horn

Fort No. 4 (Charlestown), New Hampshire, 1757
Horn, pine
OL: 14½ in. (36.8 cm), w (of plug): 2¾ in. (7 cm)
Inscriptions: Asa Spalding's Horn No. 4. AD 1757 / In Billiards and Snuff while Beaus take Delight / With Powder and Ball how the brave Heros fight; *later initials* J D F *appear twice and* J D Frashr *once.*
William H. Myers

The Reverend Asa Spaulding (1729–1808?) of Fairfield, Connecticut, served as chaplain in Colonel Phineas Lyman's Connecticut Regiment from May to November of 1757. He was appointed chaplain for Fort No. 4 for two months during the 1758 campaign.

This horn has an interesting abstract cartouche containing the British coat-of-arms with lion and unicorn supports, handsome flowers behind the lion,

Catalogue

22

22. *Stylized flowers and a whimsical British coat-of-arms are characteristic of the Lake George School.*

and two ducks in flight being chased by arrows, as if they had just been fired upon. A cartoonlike hunter being pursued by two dogs fires at a graceful buck that looks back toward the hunter from among trees. A group of three soldiers is on either side of the rhyme. The first soldier in each group is kneeling and firing, while the second two are standing and firing.

A simple scalloped edge precedes the recessed portion 4½ inches from the tip. A raised ring is 1⅞ inches from the tip. A wide border of vines encircles the plug end. The slightly rounded pine plug is secured to the horn by friction only and extends ¼ inch beyond the horn. Four holes in the plug indicate four different devices used to attach the carrying strap at different times.

Catalogue only
LAKE GEORGE SCHOOL
    Attributed to the Selkrig-Page Carver (w. 1758)

23 *Samuel Lounsbury horn*

Fort No. 4 (Charlestown), New Hampshire, June 20, 1757
Horn, pine, iron
OL: 15 in. (38 cm), W (of plug): 2¾ in. (7 cm)
Inscriptions: Samuel Lounsbeary / his Horn Made at Charlestown / alias No. 4 June 20 1757 / W   A   R / I Powder with my brother ball / A Hero Like I Conquer all / c R His Horn
James E. Dresslar

Samuel Lounsbury is recorded as serving in Captain Israel Woodward's 6th Company, Colonel David

105

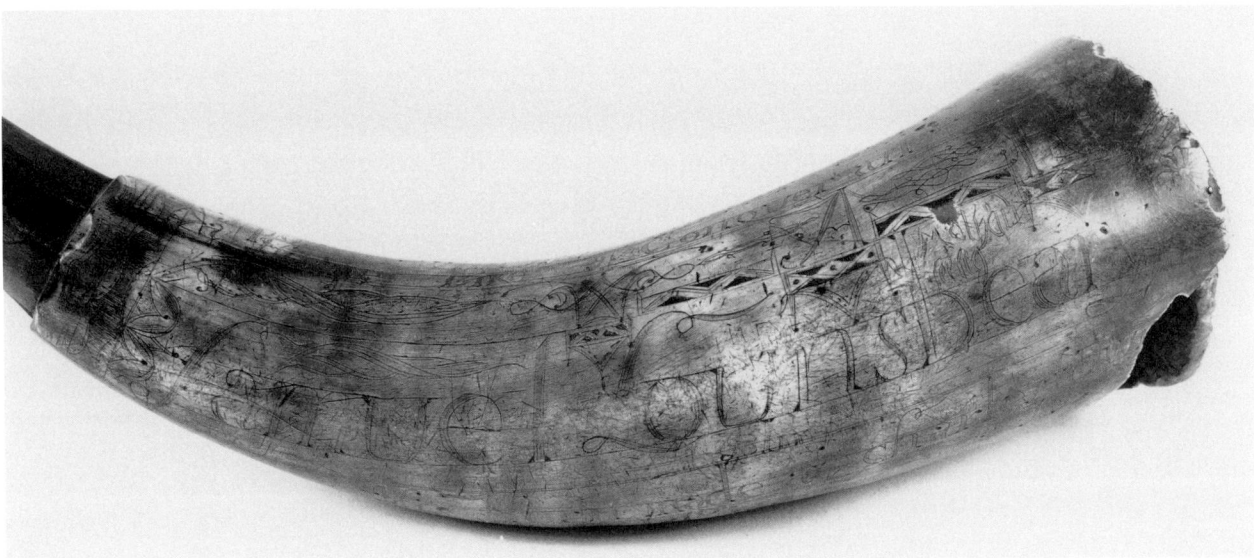

23

Wooster's 2nd Connecticut Regiment, from April 7 to October 18, 1756, in the Crown Point expedition. He served from March 28 to November 23, 1757, in Captain Ephraim Preston's 13th Company (Wallingford), Colonel Phineas Lyman's Connecticut Regiment, in the 1757 campaign. David Hamilton, whose horn was carved by the same hand (No. 26), also served in this company. During the 1759 campaign, Lounsbury served from March 28 to November 20 in Captain Samuel Gaylord's 6th Company (Middletown), Major General Phineas Lyman's 1st Connecticut Regiment, and in the 1760 campaign, he served in Captain Eldad Lewis's 7th Company (Southington), Colonel Nathan Whiting's (No. 12) 2nd Connecticut Regiment.

This fine horn is by the same carver who made horns for Nathaniel Selkrig (No. 24) and Aaron Page (No. 25), both of whom served in the same regiment in 1757. Other members of this regiment whose horns are in the exhibit include Ichabod French (No. 21), Isaac Whelpley (No. 33), and David Wheeler (No. 34).

This horn has the floral and vine designs and the highly stylized motto WAR with deeply incised sawtooth design also seen on the Selkrig and Page horns. These motifs all show a strong reliance on John Bush's work.

A scalloped edge with three-line border precedes the recessed portion 4⅛ inches from the tip. The recessed portion is octagonal for 1⅝ inches, then falls into a ⅜-inch-wide deeper recess bordered by a raised ring 2 inches from the tip and resumes its average recessing to the tip. A simple wooden stopper has a hole in the finial where a guard cord was attached to the tip. The slightly raised pine plug is recessed a fraction of an inch from the end of the horn and is secured by four surviving pegs of an original five. The portion of the horn beyond the plug is badly chipped and may have extended farther. An iron staple in the center of the plug secured the carrying strap.

LAKE GEORGE SCHOOL
Attributed to the Selkrig-Page Carver (w. 1758)

24 *Nathaniel Selkrig horn*

Fort No. 4 (Charlestown), New Hampshire, March 15, 1758
Horn, pine, iron
OL: 19 in. (48.3 cm), W (of plug): 3¼ in. (8.3 cm)
Inscriptions: Made / Nathaniel Selkrig At N° 4 March th 15 1758 / Street Firing / Make Ready Present / I Powder With My Brother Ball A Herow Like Do Concor all / W   A   R / Drums A beating Collers Flieing Trumpeths Sounding Men / A Dieing These are The Bloode Affects of Wars
William H. Guthman

Nathaniel Selkrig spelled his name on various muster rolls as "Selkrig," "Silkrag," "Silrag," "Silhrig," and "Silkreg." He is first listed as serving in Lieutenant Colonel Nathan Whiting's Company (No. 12; New Haven), Colonel Phineas Lyman's Connecticut Regiment, during the 1757 campaign. Serving in the same regiment was Sergeant David Hamilton (No. 26), whose horn is by the same carver. Selkrig volunteered to serve in Captain Reuben Ferris's Company of Rangers stationed at Fort No. 4 from November 14,

24

1757, to May 14, 1758. The company was under the command of Colonel Haviland of the 27th Regiment, British regulars. During the 1758 campaign, Selkrig again served in Colonel Nathan Whiting's 1st Company from May 14 to November 17. Aaron Page (No. 25) also served in this company. During the campaign Selkrig achieved the rank of corporal. He also served as corporal during the 1759 campaign in Major David Baldwin's (No. 14) 3rd Company (Milford), Colonel Nathan Whiting's 2nd Connecticut Regiment. Selkrig became a sergeant and served at that rank during the 1761 and 1762 campaigns in Colonel Nathan Whiting's 1st Company, Colonel Nathan Whiting's 2nd Connecticut Regiment. Aaron Page (No. 25) also served in this company.

The Selkrig-Page Carver copied the style of John Bush. The horns these two masters produced are of the highest quality and represent the high point of the Lake George School. The stylized lettering of the motto WAR was probably derived from illuminated lettering on colonial commissions. The calligraphy of the didactic verses, as well as the name, date, place, and captions of vignettes, are all in the same thick-and-thin copperplate style, carefully laid out and neatly shaded. Floral designs are accentuated with shading, dots, and fine lines, and geometric designs are highlighted by bold deeply incised carving that is accented with dark pigments.

An outstanding motif characteristic of this style is that showing whimsical cartoonlike soldiers marching in formation under the watchful eye of two officers, one holding a halberd, the other the regimental colors. Opposite them is a formation of mounted soldiers led by a trumpeter and under the supervision of a mounted officer with sword in hand. The troopers all have raised swords.

Another vignette depicts soldiers in firing formations, with captions taken from the military manual. Opposing groups are shown going through the correct procedure for "street firing," a tactic used to attack or defend a street, bridge, road, or pass. The commands are "Street Fire; Make Ready; Present; Fire." This was not a standard maneuver used in frontier warfare, but represented the artist's or owner's selection of an interesting subject. The use of mounted troops in the northern woodlands was equally fictitious. The uniforms are accurate, however, as is the use of trumpets by mounted troops (as opposed to drum signals for foot soldiers). The amusing faces are all the same, including those of the horses.

The horn is scalloped 5¾ inches from the tip. The recessing begins at the scalloping and proceeds for 3¼ inches to a ¼-inch raised ring and on to the slightly raised spout. The pine plug is secured by three ⅜-inch iron tacks with rectangular or T-shaped heads. Underneath two additional rosehead nails are fragments of the original carrying strap.

BIBLIOGRAPHY: *Bulletin*, Fort Ticonderoga Museum, 181–182; Swayze, 58–59; *Antiques*, 1978, 312, 314–315; duMont,

27; Guthman, *Guns and Other Arms*, 134–153; *Bulletin*, KRA, Fall, 1985, 9; *Bulletin*, ASAC, March, 1981, 45; Guthman, *U.S. Army Weapons*, 92.

LAKE GEORGE SCHOOL
Attributed to the Selkrig-Page Carver (w. 1758)

25 *Aaron Page horn*

Lake George, New York, July 8, 1758
Horn, wood, iron
OL: 14½ in. (36.8 cm), w (of plug): 3¼ in. (8.3 cm)
Inscriptions: Aaron Page His Horn / Made Att Lake Gorg july the 8 ano 1758 / W   A   R / Fire / Powder With My Brother Ball A herow Like / Do Conquer all
William H. Guthman

Page served in the same company as Nathaniel Selkrig (No. 24) during the campaigns of 1757, 1758, and 1762. His horn is almost identical to Selkrig's but has less verse, only one scene of foot soldiers, and more floral and vine designs. The missing wooden plug was secured by four wooden pegs, only one of which survives. The iron nails in the three other original peg holes are later replacements.

This horn is beautifully scalloped 6 inches from the tip. Its throat is splintered and remnants of a raised ring 1½ inches from the tip remain. Its spout is slightly flared and slightly faceted.

BIBLIOGRAPHY: *Antiques*, 1978, 314–315; *Bulletin*, ASAC, March, 1981, 44.

LAKE GEORGE SCHOOL
Attributed to the Selkrig-Page Carver (w. 1758)

26 *Sergeant David Hamilton horn*

Fort No. 4 (Charlestown), New Hampshire, about 1757
Horn, pewter, pine, iron
OL: 11½ in. (29 cm), w (of plug): 2½ in. (6.4 cm)
Inscriptions: AB / Sarjt. David Ha[milton] / Horn made at No: fou[r] / I Powder with my [brother ball] / A Hero Like do Con[quer all] / W   A   [R]
William H. Guthman

Hamilton served as sergeant in Captain Ephraim Preston's 13th Company (Wallingford), Colonel Phineas Lyman's Connecticut Regiment (Nathaniel Selkrig served in the same regiment during the 1757 campaign). Hamilton again served as sergeant from November 14, 1757, to May 14, 1758, in Captain Reuben Ferris's Company of Rangers, Colonel Haviland's Regiment. He then served as a second lieutenant during the 1758 campaign in Captain Archibald McNeal's 10th Company (New Haven), Colonel David Wooster's 4th Connecticut Regiment.

Although this horn is cut down, it is included because of the quality of its engraving and because of the importance of its carver. Its calligraphy is in the same style as that on the Selkrig and Page horns (Nos. 24 and 25), but the lettering of its verse differs slightly and the formations of soldiers are absent. The exquisite illumination of the motto WAR is identical, however, and beautiful floral and vine designs cover the horn's surface.

The scalloped edge preceding the recessed portion

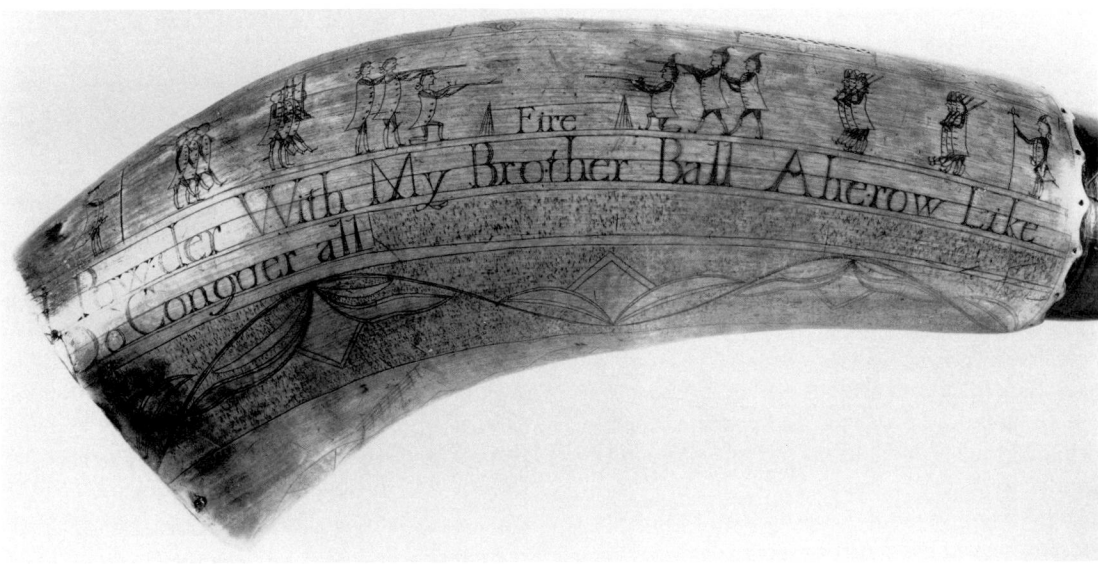

25

Catalogue

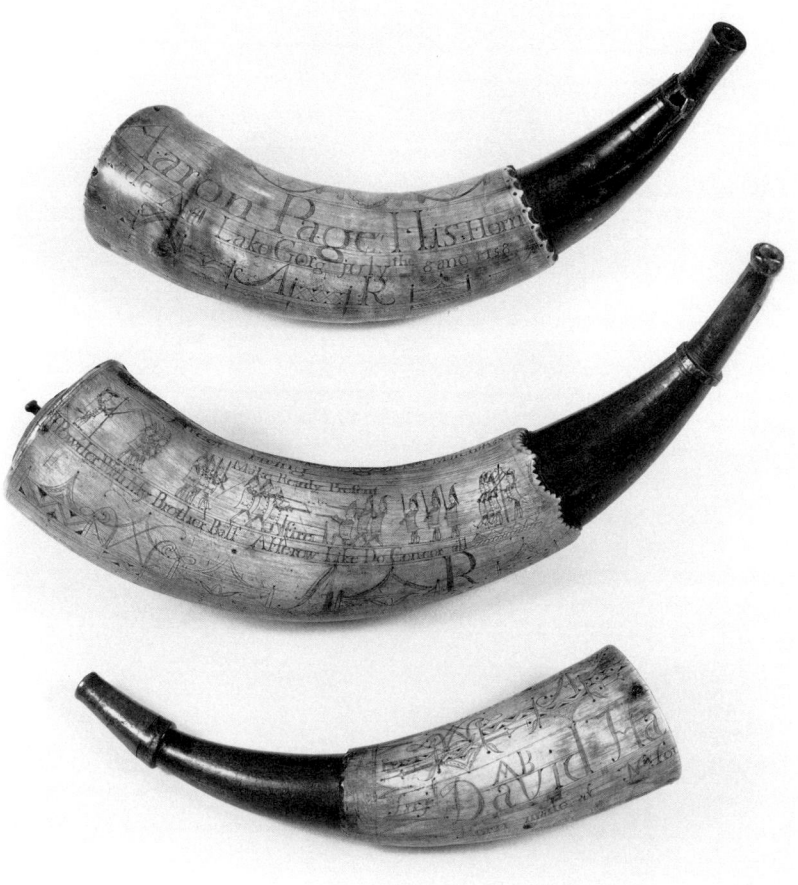

24 (Middle, Nathaniel Selkrig), 25 (top, Aaron Page), and 26 (bottom, David Hamilton): *these three horns illustrate the Lake George School's adherence to John Bush's stylized* WAR *motto and incised border decoration, and introduce the Siege of Boston School's formations of fighting soldiers.*

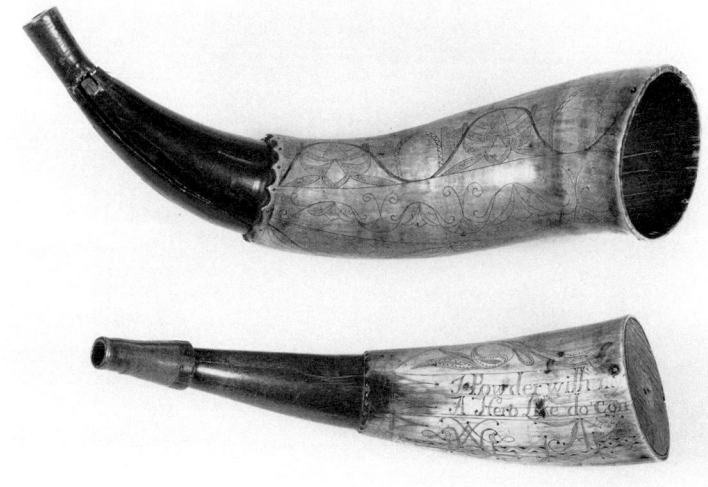

25 (Top, Aaron Page) and 26 (bottom, David Hamilton): *these horns illustrate the stylized calligraphy and floral and geometric decoration originated by John Bush and carried on by the Lake George School.*

is not as well executed as those on the other two horns, but it does have a high quality floral border. The scalloping is 5½ inches from the tip, the recessed portion continues 3⅝ inches to a raised ring, and a pewter cap or ferrule is affixed to the raised ring by three dovetails. The pine plug is secured by five iron tacks and two holes in the plug's center remain from a staple for the carrying strap. This may have been used as a priming horn after it was cut down.

BIBLIOGRAPHY: *Bulletin*, ASAC, March, 1981, 43; Guthman, *Guns and Other Arms*, 136; *Antiques*, 1978, 314–315.

109

27

27. The abstract coat-of-arms, the horseman, and the stylized flowers resemble the work of the Selkrig-Page Carver.

LAKE GEORGE SCHOOL
    Possibly by the Selkrig-Page Carver (w. 1758)

27 *Zebulon Waterman horn*

Lake George, New York, October 24, 1758
Horn, cherry, iron
OL: 19¼ in. (48.9 cm), w (of plug): 3 in. (7.6 cm)

Inscriptions: Steel Not This Horn For fear of Shame: For on it is The / Oners Name / Made at Lake George Zebulun Waterman His Horn October the 24 AD / 1758 / Street Fireing Make Ready Present / Fire; 1763 AE *is added in a later hand.*

William H. Guthman

Zebulon Waterman (baptized 1741–died 1794) of Colchester, Connecticut, served in Captain Noah Grant's 7th Company (Windsor) during the campaign of 1756. Serving in the same company were John Dodge (whose undecorated horn, not in the exhibition but sketched by Rufus Grider, displays calligraphy by J. W.) and John Tribble (No. 45). Waterman is also listed as having served in Captain Henry Champion's 12th Company (Colchester), Colonel Nathan Whiting's (No. 12) 2nd Connecticut Regiment, during the campaign of 1758. He also served in the Revolution.

The carver of this example could well be the same man who decorated the Selkrig, Page, and Hamilton horns (Nos. 24, 25, and 26). The formations of soldiers here are almost identical to those on the Selkrig and Page horns, and the calligraphy is in the same hand. Identical floral and geometric decoration, borders, and scalloped throat are also seen on the related horns, and the coat-of-arms, lion and unicorn, and mounted horseman are all comparable. However, the elaborate stylized coat-of-arms on the Waterman horn, as well as the sunflowerlike designs, are not found on the related horns, and the Waterman example also lacks the stylized motto WAR, which is a Selkrig-Page Carver signature.

An interesting abstract border precedes the scalloped edge at the beginning of the recessed portion of this horn. Beginning 5½ inches from the tip of the spout, it is broken by a raised ¼-inch ring placed 2¼ inches from the tip of the slightly turned spout. The replaced flat cherry plug is secured by two wrought-iron nails and another wrought-iron nail is driven into its center to serve as a carrying strap attachment. A nicely carved vine design surrounds the butt end.

All the horns by this carver appear to have been obtained from the same breed of cattle: they are exceptionally large and are all the same distinctive honey color. It may be pertinent that Henry Champion, the captain of Waterman's company, was a cattle drover and livestock speculator.

BIBLIOGRAPHY: *Bulletin*, KRA, Fall, 1985, 8.

ATTRIBUTED TO THE LAKE GEORGE SCHOOL

28 *Amasa Yale horn*

Lake George, October 18, 1758
Horn, pine, hardwood pins, iron
OL: 14½ in. (36.6 cm), W (of plug): 2¾ in. (7 cm)

Inscriptions: Amasa × Yale His Horn Made / At Lake George October the 18th / Ad 1758 I powder With My / Brother Ball, A Hero Like / do Conqur All Steel Notthis / Horn For Fear of Shame / For on it is the Oners name. / Make Redey / Present Fire

James E. Dresslar

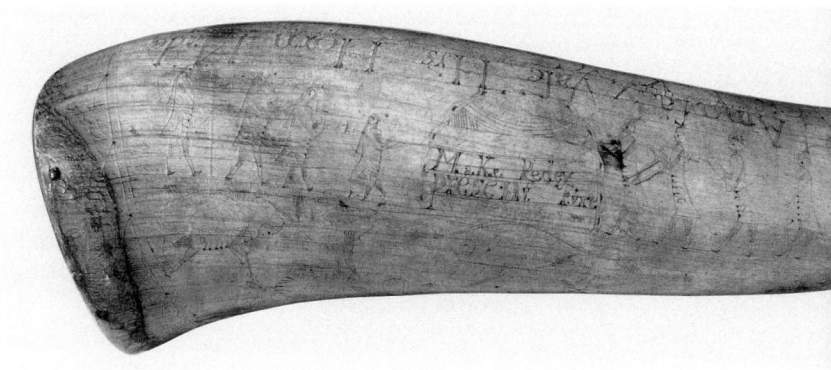

28. *Calligraphy and formations of soldiers by one of the untutored carvers of the Lake George School.*

Amasa Yale is not listed in the French and Indian War records but he is listed as having served as a minuteman from Wallingford, Connecticut, in 1775, in answer to the Lexington Alarm. He also served as a drummer from July 8 through December 20, 1775, in Captain Street Hall's company, Colonel Webb's 7th Connecticut Regiment, at Camp Winter Hall outside Boston. This may or may not be the same Amasa Yale who served at Lake George in 1758; he could have been a son of the man who served earlier. Since French and Indian War records are incomplete, however, the names of many men who served were lost with their records of service.

The calligraphy seen here is neatly executed in the style of block-letter copperplate with graceful serifs and uniform composition. A formation of four uniformed soldiers is firing at another formation of four uniformed soldiers who are returning the fire. The figures are crudely incised, but present an interesting group. The blank space on the horn is filled with

eight-petaled flowers. The horn's decorative style is typical of the Lake George School of carvers.

A neat, deeply scalloped edge precedes the recessed portion of this example, which begins 5¾ inches from the tip of the spout. There is a raised ring 2½ inches from the tip of the spout. A neat geometric border surrounds the scalloped edge.

The slightly rounded pine plug is secured to the horn by eight hardwood pins and the remains of three iron nails are visible in the center of the plug.

LAKE GEORGE SCHOOL
Attributed to the Memento Mori Carver
(w. 1756–1760)

29 *Colonel Nathan Payson horn*

Lake George, New York, October, 1756
Horn, wood, iron
OL: 16½ in. (42 cm), W (of plug): 2⅞ in. (7.3 cm)
Inscriptions: NATHAN.PAYSON Esq Col / LAKE.GEORGE. OCer 1756 / FORT. WM HENRY / FORT. EDWARD / MEMENTO.MORI
William H. Guthman

Nathan Payson of Hartford, Connecticut, was major and captain of the 4th Company of Major General Phineas Lyman's 1st Connecticut Regiment during the 1755 campaign. He was appointed commander of a company at Fort Edward in Colonel Jonathan Bagley's Massachusetts Regiment (consisting of recruits from New York, Connecticut, Rhode Island, Massachusetts, and New Hampshire) when his Connecticut enlistment expired, and he returned to Hartford at the end of December. He was appointed lieutenant colonel and commanded the 2nd Company in Colonel Nathan Whiting's (No. 12) 3rd Connecticut Regiment during the 1756 campaign. During the 1757 campaign he received only a major's commission and was captain of the 3rd Company in Colonel Phineas Lyman's Connecticut Regiment. He was appointed lieutenant colonel and captain of the 2nd Company, Colonel Phineas Lyman's 1st Connecticut Regiment, during the 1758 campaign. In the campaigns of 1759, 1760, and 1761, he again served as lieutenant colonel and captain of the 2nd Company in Major General Phineas Lyman's 1st Connecticut Regiment. Payson died in Hartford on April 17, 1761, without seeing active service that year.

This is an early horn by the Memento Mori Carver (see also Nos. 30 and 31). It does not have that artist's more elaborate elongated crosshatched curves decorating the cartouche, but it does have a shaded sawtooth design in that location. Similar borders consisting of large triangles edged with semicircles with dots in the center surround both plug end and throat. At the base the design has two opposed rows of triangles with the points directed at the intervals of the opposite row. Surmounting this is a row of diamonds, and engraved around the base of the horn above the band is the carver's signature motto, MEMENTO MORI. At the throat of the horn is a single row of triangles bordered on one side by a band with a wavy snakelike motif and on the other by a narrow shaded band with a wavy design. This latter band is bordered by a floral design and a human face in profile with a long-stemmed clay pipe in its mouth.

The horn is also decorated with a continuous map from Fort Edward to Fort William Henry that has detailed sketches of forts and fortifications in between. The central cartouche has flowers extending from the two corners facing the throat—another characteristic feature of this carver's work.

The rounded pine plug, extending ¼ inch beyond the horn, was held by six wooden pegs, four of which are missing. The initials I S are deeply carved in its center, and in between the initials is an iron screw that held the carrying strap. The recessing on the throat begins 4½ inches from the tip, and 2 inches from the tip a deep channel ¼-inch wide held the carrying strap. The channel is followed by two raised rings ⅛ inch apart; the remaining 1⅝ inch of the tip is faceted. The stopper contains a piece of antler.

LAKE GEORGE SCHOOL
Attributed to the Memento Mori Carver
(w. 1756–1760)

30 *Joshua Wolcott horn*

Probably Fort Edward, New York, April–November, 1758
Horn, wood, iron, brass, vermillion pigment
OL: 14 in. (36 cm), W (of plug): 3¼ in. (8.3 cm)
Inscription: JOSHUA. WOLCOT / MEMENTO . MORI.
William H. Guthman

Joshua Wolcott, probably the son of Joshua Wolcott of Wethersfield, Connecticut, who was born in 1730, served only in the campaign of 1758, in Captain Amos Hitchcock's 6th Company, Colonel Nathan Whiting's (No. 12) 2nd Connecticut Regiment. The diagram of a fort and the term of his enlistment are the evidence for the place and time suggested for this horn's manufacture.

Eight known horns that differ slightly but share many characteristics appear to have been carved by

Catalogue

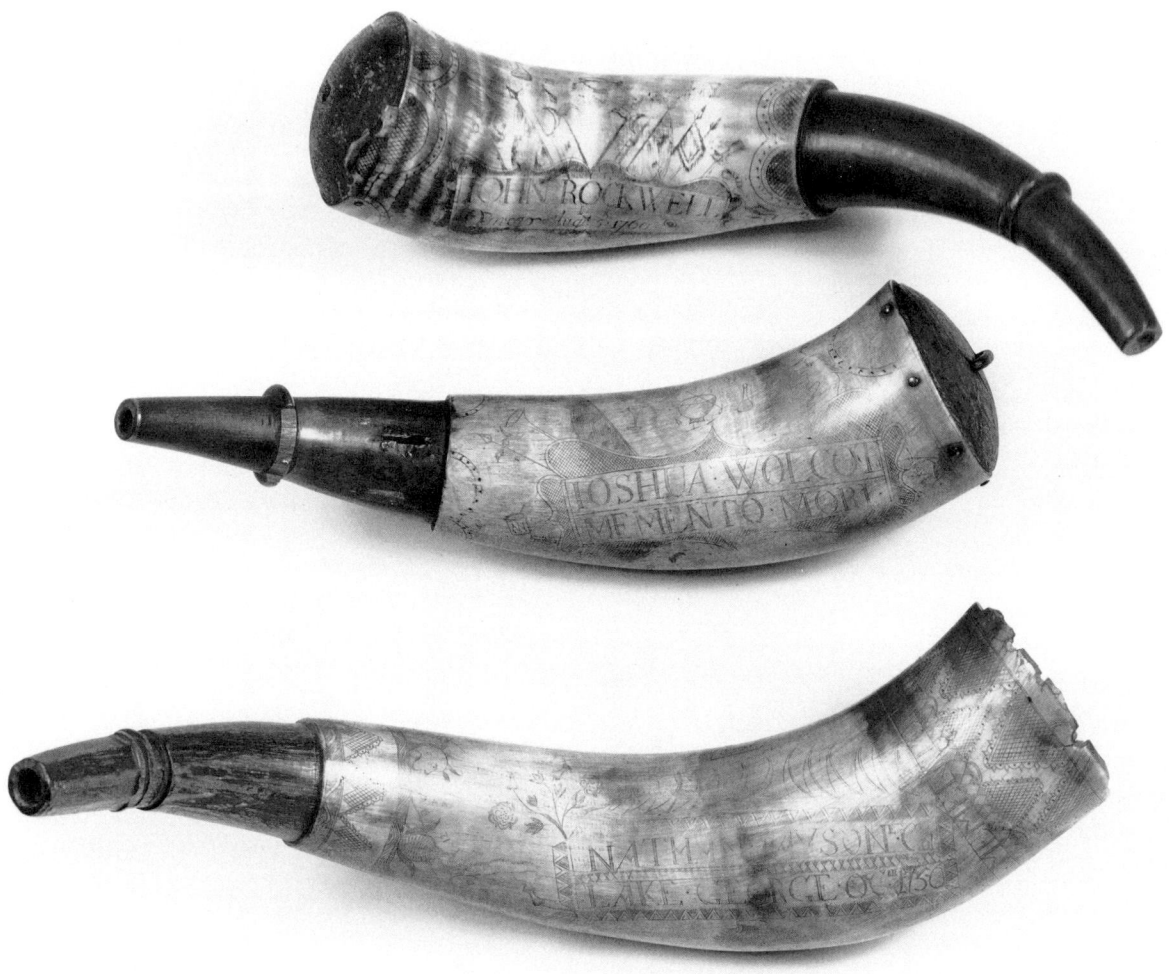

29 (Bottom, Nathan Payson), 30 (middle, Joshua Wolcott), and 31 (top, John Rockwell): *these horns exhibit the distinctive style of the Memento Mori Carver.*

this hand. All utilize crosshatching in border designs, some in a sawtooth fashion, others in lunette designs, and still others in elongated curves. On some examples the borders appear in one or two cartouches which can contain the owner's name, the place of manufacture, the date, a rhyme, or the motto MEMENTO MORI.

The Wolcott horn has a single cartouche containing Wolcott's name and the motto in two lines. The crosshatching is found within a graceful symmetrical framework of elongated curves that expand and recede in a wavelike manner. The Rockwell horn (No. 31) has a similar cartouche. Both of these horns have a border of semicircles around the butt end and at the border of the recessed portion. The semicircles are formed by two lines, roughly ⅛ inch apart, with a row of dots within the double lines. (An unsigned example has the same border without dots. The lack of an owner's name indicates that the carver engraved horns to be filled in after they were sold.)

Most of the Memento Mori horns have fort diagrams and the Wolcott example has a large diagram of what is almost certainly Fort Edward. The fort is shown at the junction of the Hudson River and Fort Edward Creek with the Union Jack flying from one of its bastions. Rogers' Island is also depicted in the river opposite Fort Edward. Halberds, flags, trumpets, and cannon protrude from points on the border of the cartouche where the curves recede and join the rectangular cartouche.

Vines and floral designs are also seen on all horns by this hand. A punch bowl with crossed long-stemmed clay pipes resting across the brim appears on the Wolcott and Rockwell horns. There is also a decanter, a wine glass, a drum and drumsticks, a musket, and a profile head with a pipe in its mouth. This profile

113

head appears on other Memento Mori horns with its own motto.

The calligraphy here is neat and plain copperplate engraving without flourishes or finials. The slightly rounded plug is painted vermillion and has an inch-long iron staple driven into its center for the carrying strap. It is secured by seven brass nails. The horn's recessed portion begins 4¾ inches from the tip. There is no decorative border here. (The example with no name, place, or date has carved sawtooth decoration at this point, while all the others have an engraved design.) A raised ring is 2¼ inches from the tip.

LAKE GEORGE SCHOOL
Attributed to the Memento Mori Carver
(w. 1756–1760)

31 *John Rockwell horn*

Oswego, New York, August 5, 1760
Horn, wood, vermillion pigment
OL: 16 in. (41 cm), W (of plug): 3¼ in. (8.3 cm)
Inscriptions: JOHN ROCKWELL / Oswego Augt 5$^{th.}$ 1760 / GARDEN / SEE WE FOOLS
William H. Guthman

John Rockwell (1733–1825) is recorded as serving during the 1757 campaign in Captain David Waterbury's 6th Company, Colonel Phineas Lyman's Connecticut Regiment. James Mead, another Memento Mori horn owner, served in this company, as did Isaac Whelpley, whose horn was carved by J. W. (No. 33). Rockwell served again in 1759 in Captain Timothy Hierlihy's 7th Company, Major General Phineas Lyman's 1st Connecticut Regiment. Roman Wetmore, who also owned a Memento Mori horn, was in this company. Rockwell is next listed in the 1762 campaign, serving in Captain Thomas Hobby's 5th Company, Colonel Nathan Whiting's (No. 12) 2nd Connecticut Regiment. Rockwell's service in the 1760 campaign is not listed.

This horn is similar to the Wolcott horn (No. 30), but the carver showed greater ability in this example's more profuse and slightly more detailed decorations and in the finely executed script of the place and date. The border decoration and the cartouche are almost identical on this and the Wolcott horns, but the fort here has far more detail and shows the encampment of tents, the order of the garrison, and the fort garden. The views of Lake Ontario and the Oswego River include boats, houses, and a flock of geese in flight. The panoply of arms around the border of the cartouche includes animals, a mounted horseman, an Indian with a bow approaching a bird in a tree, and a barking dog. The profile bust of a man is not smoking a pipe, but has a pair of horns on his forehead, a stubbly beard, and is uttering the phrase SEE WE FOOLS, suggesting that he has been cuckolded.

The rounded plug, which protrudes ⅜ inch beyond the horn, is secured by six wooden pegs and is painted vermillion like the plug on the Wolcott horn. A hole in its center indicates the prior location of a nail or lug. There is also the remnant of a lobe for securing the carrying strap. The recessed portion, which has a border like that of the Wolcott horn, begins 6¼ inches from the tip. A raised ring is 2½ inches from the tip.

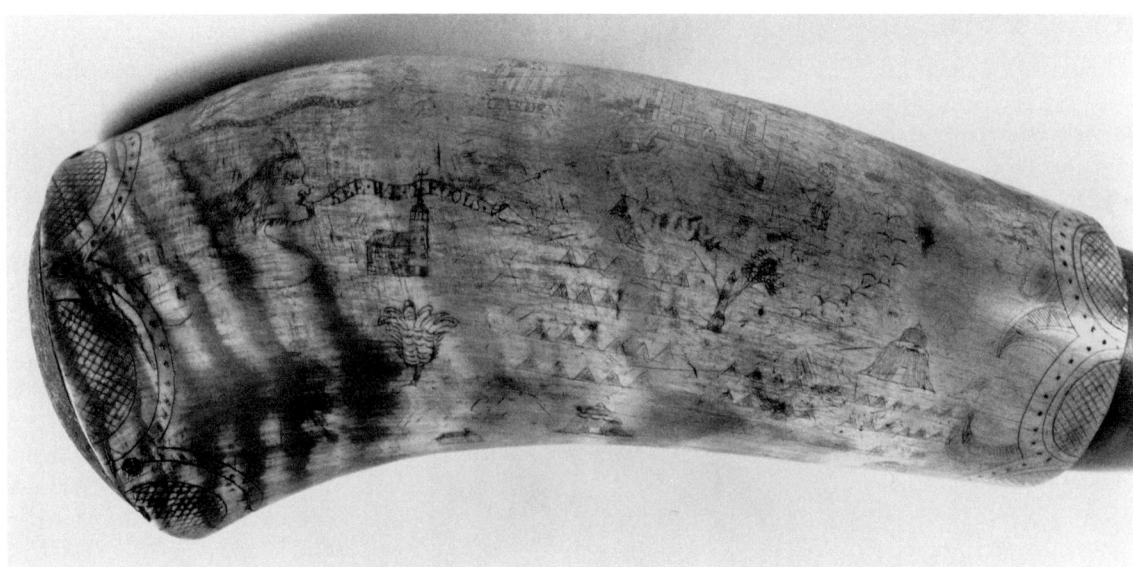

31. *A rare view on horn of the fortifications at Oswego.*

CATALOGUE

LAKE GEORGE SCHOOL
Signed by the J. W. Carver (w. 1758–1761)

32 *Robert Baird horn*

Lake George, New York, September 1, 1758
Horn, pine, iron
OL: 14½ in. (36.8 cm), W (of plug): 3 in. (7.6 cm)
Inscriptions: Robart Baird / his Horn made at leaK gorg Sep^t 1 ^{the} 1758 / I powder with my Brother ball a Hero like / do conquer all * If you should pros ed / Leak Gorg Steel not this Horn For Fear of / Shame For on nit Is the oners Name / JW his Pene
William H. Guthman

Robert Baird's name appears on the muster rolls as "Bard" in 1757 and "Baird" in 1758 and 1761. In all three campaigns he served under officers from Greenwich, Connecticut. During the 1757 campaign he served in Captain Stephen White's Company, Colonel Jonathan Hait's 9th Connecticut Regiment, which marched for the relief of Fort William Henry. During the 1759 campaign, Baird served in Captain Thomas Hobby's 4th Company, Colonel David Wooster's 3rd Connecticut Regiment, and during the 1761 campaign, he served in Captain Thomas Hobby's 6th Company, Colonel Nathan Whiting's (No. 12) 2nd Connecticut Regiment. Isaac Whelpley (No. 33) was a sergeant in that company, and his horn is also by the J. W. Carver. During the 1759 campaign, William Patterson (No. 61) served with Baird in the 4th Company, and his horn is similar to those carved by J. W., as is the horn carved for Christopher Palmer (No. 55). These last two horns, however, are not carved as well as horns by J. W., who in turn was not as capable an engraver as Richardson Minor (No. 52). There is, nevertheless, a definite relationship among the carvers of all these horns.

The planning of J. W.'s horns is as careful as that of Richardson Minor's examples. The name ROBART BAIRD is expertly executed with shaded copperplate letters, some of which have flowing wing-shaped devices as serifs. The wing motif is also used between words and at the end of phrases. The same style of lettering is also seen in the didactic verses.

J. W. apparently carefully copied the more elevated styles of calligraphy, rather than executing them freehand. His floral designs, consisting of naturalistic scrolling leaves, broken C-scrolls, delicately curling vines, and birds, evoke the baroque style. However, J. W. casually inserted recognizable non-artistic objects into this imagery, like crossed muskets, oriental-looking sailing ships, geometric designs, and birds and animals. The inspiration for these motifs was most likely illustrations in books and magazines and it is evident that he mixed motifs from many stylistic periods without any regard for consistency.

The lack of a border at the throat and butt ends is typical of J. W. horns, although most other Lake George School horns have them. This example is scalloped 5½ inches from the tip of the spout. Two raised rings ¼ inch apart and 2½ inches from the tip secured

32. *J. W.'s flowing floral-vine designs interwoven with wings and graceful birds and calligraphy. His signature is visible at upper left.*

the carrying strap. The 1½-inch spout is recessed ¾ of an inch to accommodate a metal ferrule which is now missing.

The pine plug is secured by six wooden pegs, three of which are reinforced by iron nails added later. Two holes at the plug end were intended for leather thongs on a carrying strap, and the plug is undercut next to the holes. The remainder of the plug is almost flush with the end of the horn and has a turned wooden knob at the center that extends ½ inch, presumably as a later attachment device.

BIBLIOGRAPHY: *Bulletin*, ASAC, No. 44 (March, 1981), 42–43; *Antiques*, 1978, 324–325; Milliman, *Bulletin*, Fort Ticonderoga Museum, Vol. XII (October, 1967), 183–185; Guthman, *Guns and Other Arms*, 146–147..

LAKE GEORGE SCHOOL
Signed by the J. W. Carver (w. 1758–1761)

33 *Sergeant Isaac Whelpley horn*

Fort No. 4 (Charlestown), New Hampshire, January 1, 1758
Horn, pine
OL: 14¾ in. (37.5 cm), w (of plug): 2¾ in. (6.9 cm)
Inscriptions: Sgt / I[sa]a[c] Whelpley his / horn [m]ade at Nº 4 ia, w1the AD 1758 / the roses of red, the grass is green, the day is / Past: which I have seen and a / fair Lades Buty decay est; Levi Burns
William H. Guthman

Isaac Whelpley served during the 1757 campaign as a sergeant in Captain Waterbury's 6th Company, Colonel Lyman's Connecticut Regiment, along with other horn owners who are mentioned above. During the 1761 campaign, Whelpley served as sergeant in Captain Thomas Hobby's 6th Company, Colonel Nathan Whiting's (No. 12) 2nd Connecticut Regiment. Robert Baird (No. 32) also served in this company.

The Whelpley horn is in poor condition but is included here because of its unusual philosophical verse, the excellence of its engraving, and the admirable state of preservation of the highly decorated lobe on its plug end. The decorative devices and calligraphy are virtually identical to those of the Baird horn but instead of muskets and ships, J. W. engraved a goblet and bowl. Severe scaling has obliterated some motifs on the spout and the body, especially the lettering, although the signature of J. W. has remained distinct. It is engraved just before the scalloping at the spout end and consists of two script W's and two J's.

Unlike the Baird horn, which has two holes at the butt end over a recessed portion of the plug, the Whelpley horn has a 1¼-inch lobe. It is beautifully scalloped and decorated with typical J. W. floral and geometric designs. There is a hole ⅜ inch in diameter in the center of the lobe for the carrying strap. The throat of the horn is scalloped 5 inches from the tip and has a raised ring 2¼ inches from the tip. The tip itself has splintered and is partially missing. The rounded pine plug extends ¼ inch beyond the horn and is secured with six wooden pegs.

BIBLIOGRAPHY: *Bulletin*, KRA, Vol. 12, No. 1 (Fall, 1985), 9; *Bulletin*, ASAC, No. 44 (March, 1981), 42; *Antiques*, 1978, 324–325; Guthman, *Guns and Other Arms*, 146–147.

LAKE GEORGE SCHOOL
Attributed to the J. W. Carver (w. 1758–1761)

34 *David Wheeler horn*

Lake George, New York, September, probably 1757
Horn, pine, iron, brass
OL: 10¼ in. (26 cm), w (of plug): 3½ in. (8.26 cm)
Inscriptions: David Whe[eler] / His Horn Made [at Lake . . .] / George Sept the [. . .]; 1776 *added later.*
William H. Guthman

David Wheeler (1726–1806) of Fairfield, Connecticut, served in Lieutenant Colonel Nathan Whiting's (No. 12) Company, Colonel Jonathan Bagley's Regiment, during the 1755 campaign. This regiment consisted of men and officers from New York, Connecticut (Colonel Whiting was from New Haven), Rhode Island, Massachusetts, and New Hampshire, and was ordered to garrison at Fort William Henry and Fort Edward. During the 1756 campaign, Wheeler served as clerk in Captain David Lacey's 7th Company (Fairfield), Colonel Andrew Ward's 4th Connecticut Regiment. In 1757, Wheeler enlisted in Captain Samuel Hubbell's 5th Company (Fairfield), Colonel Phineas Lyman's Connecticut Regiment. Also serving in that company was Thaddeus Bennett (No. 19); Sergeant Isaac Whelpley (No. 33) served in the 6th Company of that same regiment. During the 1761 campaign, Wheeler served in Captain Samuel Whiting's 4th Company (Stratford), Colonel Nathan Whiting's (No. 12) 2nd Connecticut Regiment; Robert Baird (No. 32) and Isaac Whelpley served in Captain Thomas Hobby's 6th Company (Greenwich) in this regiment. The date 1776 might have been added during Wheeler's service in the Connecticut regiments.

The Wheeler horn has been cut down by at least six inches, but is included because of the superior quality of its engraving and because two other men in the same regiments also owned J. W. horns. The

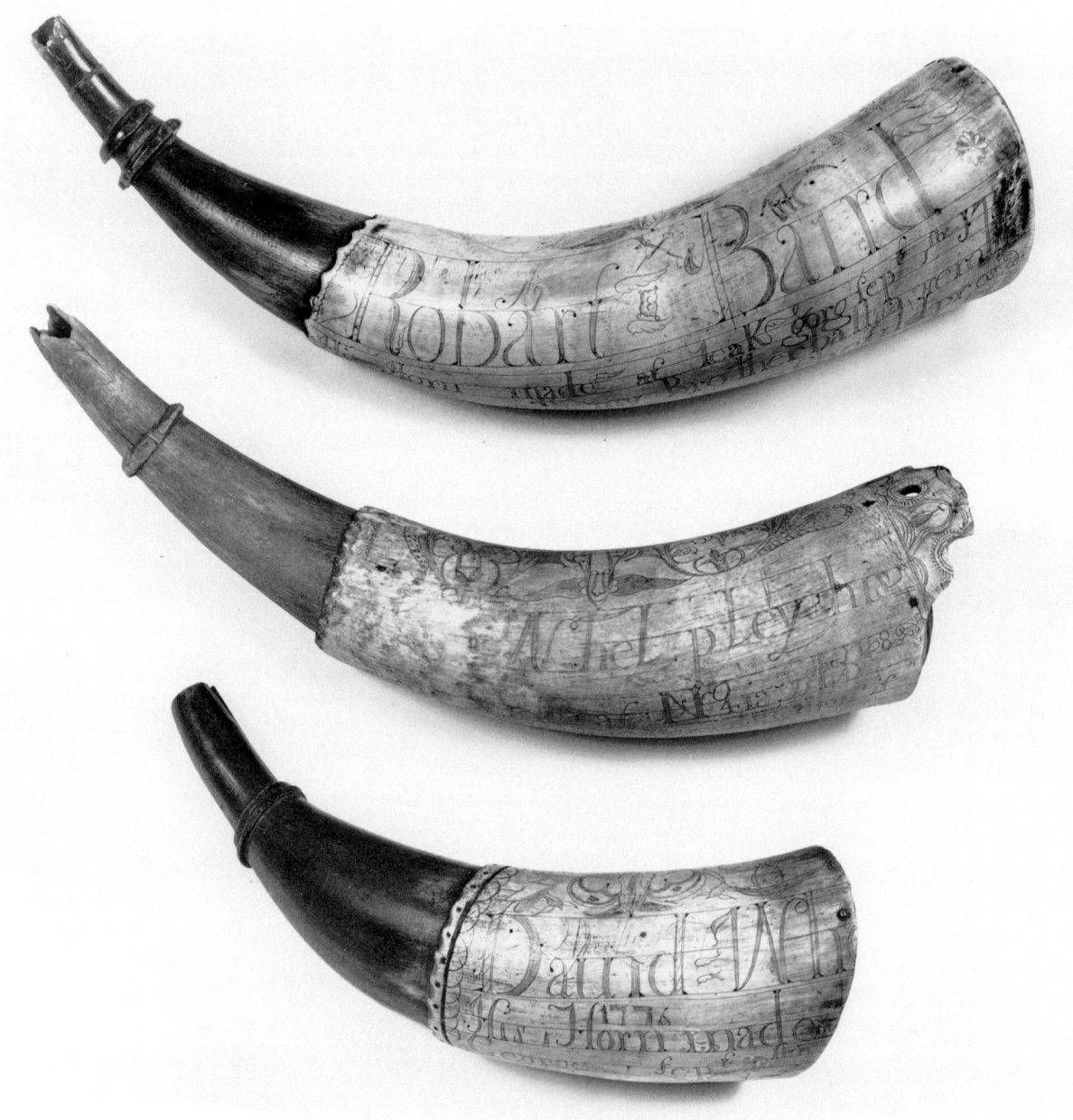

32 (Top, Robert Baird), 33 (middle, Isaac Whelpley), and 34 (bottom, David Wheeler): *three J. W. horns.*

calligraphy is typical of J. W.'s best engraving, utilizing his flowing style of lettering with winged serifs on some of the letters. The decoration includes his usual floral and geometric designs, as well as several graceful birds with elongated necks turned backward toward their tails, several birds in flight, and a wonderful animated lion above J. W.'s distinctive geometric device.

The throat is scalloped 5½ inches from the tip and is recessed from that point to the tip. A double carved ring is 2½ inches from the tip. The flat replacement pine plug, secured by six brass pins, contains an iron screw eye that held the carrying strap. The horn appears to have been cut down during the eighteenth century after it sustained damage to the plug end.

BIBLIOGRAPHY: *Antiques*, 1978, 324–325; Guthman, *Guns and Other Arms*, 146–147; *Bulletin*, ASAC, No. 44 (March, 1981), 43.

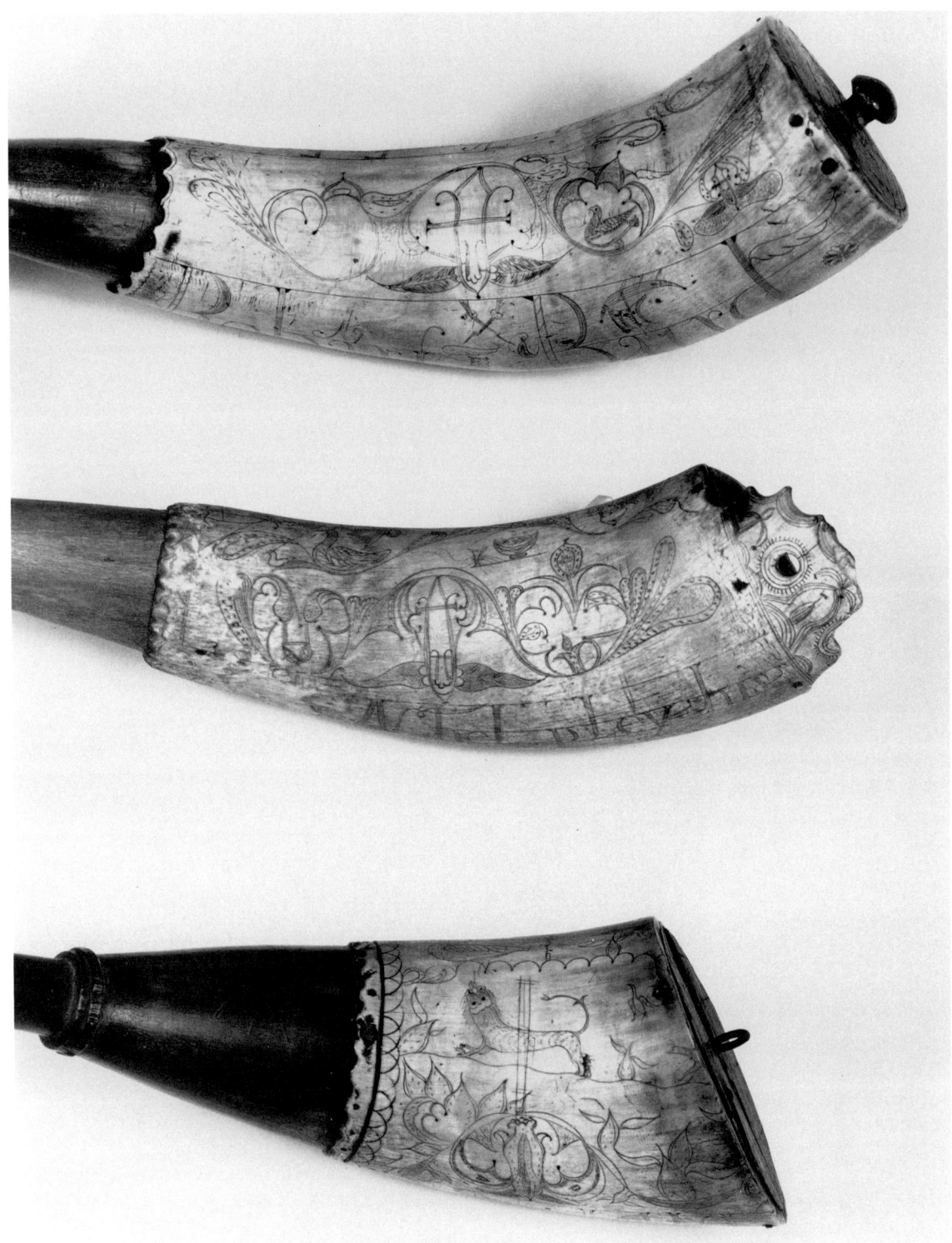

32 (Top, Robert Baird), 33 (middle, Isaac Whelpley), and 34 (bottom, David Wheeler): *reverse views showing J. W.'s fascinating stylized geometric designs.*

Catalogue

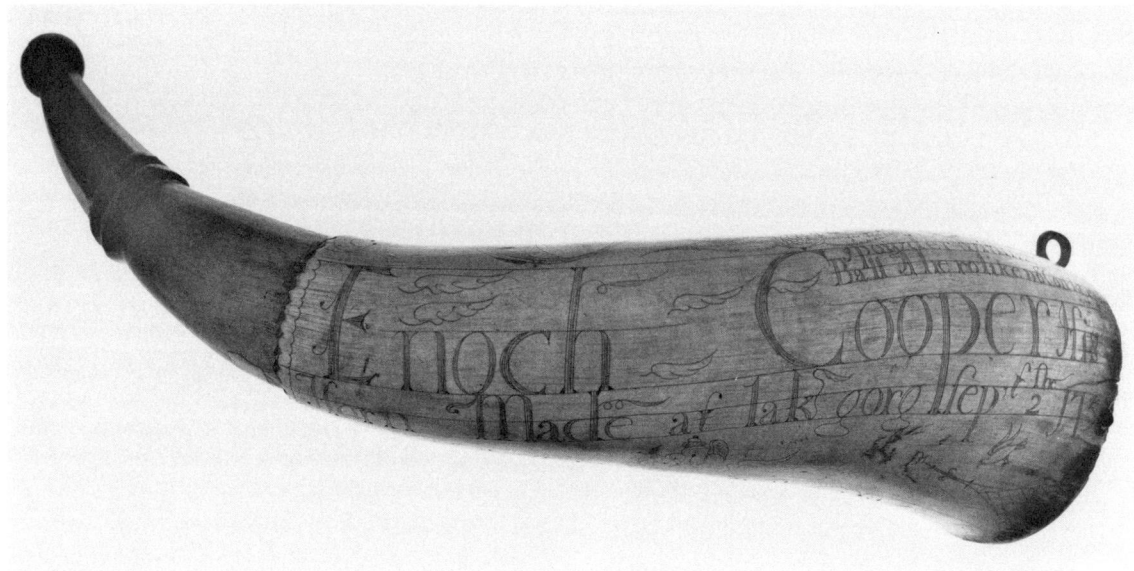

35. J. W.'s interwoven floral and bird compositions. The geometric design that serves as one of his signatures forms the central motif.

Catalogue only
LAKE GEORGE SCHOOL
    Signed by the J. W. Carver (w. 1758–1761)

35 *Enoch Cooper horn*

Lake George, New York, September 2, 1758
Horn, pine, iron, dark blue and reddish-brown pigments

OL: 15 in. (38 cm), W (of plug): 3 in. (7.6 cm)
Inscriptions: I Powder with my Brother / Ball A hero Like do Conker All / Enoch Cooper His / Horn Made at Lake gorg Sept the 2 1758 / J , W his Pen
Chicago Historical Society

Enoch Cooper is recorded as serving in only one campaign, that of 1758, in Colonel Phineas Lyman's 1st Company, Colonel Phineas Lyman's 1st Connecticut Regiment, from April 3 through November 15.

This is one of the most beautiful of the J. W. horns. It displays several of the geometric devices he utilized as signatures, the largest and most prominent of which is shaded with reddish-brown pigment and incorporated in a C-scroll border with a profile head, a running fox, and a goose. The carefully engraved name of the owner is neatly lined, and its contrasting shaded and light letters are ornamented with curling serifs and winged devices. J. W. engraved birds in different stages of flight and rest as well as profuse geometric and C-scroll devices. Besides his initials, J. W. used a signature device made up of a stylized "J" and "W" worked into a geometric motif and set off by a pair of spread wings.

No borders are present at either end of the horn. Crude scalloping demarcates the beginning of the recessed portion, 5½ inches from the tip. A raised double ring is 2⅞ inches from the tip, and a large raised lip forms the tip itself. The flat pine plug is secured by three wooden pegs and a wrought iron screw eye set at a right angle to the edge of the horn. The plug has traces of dark blue pigment.

BIBLIOGRAPHY: Grider, NYHS; Grancsay, No. 209, 48; *Antiques*, 1978, 320; Guthman, *Guns and Other Arms*, 142.

LAKE GEORGE SCHOOL
Attributed to the J. W. Carver (w. 1758–1761)

36 *Stephen Peck horn*

Lake George, New York, April 28, 1761
Horn, pine, iron, dark pigment
OL: 16¼ in. (41.3 cm), w (of plug): 3¼ in. (8.3 cm)
Inscriptions: Stephen Peck his / horn made in Amity April ye 28th AD 1761 / Nectorian cyder now with pork and beef gives many an aching stomach great relief / He that hasn't these or Money in his Purse / His case is bad and is / likely to be worse / I Powder With my Brother Ball / A Hero Like Do Conker All / The lion Ranges round The Wood / And Makes The Lesser Beasts his Food
Carol and the late Thomas J. Segal

Stephen Peck does not appear to have been a member of the Peck family of Fairfield County, Connecticut, although a Stephen Peck is listed in the Connecticut Revolutionary War rolls.

The phrase *made in Amity* in the inscription probably refers to the fact that the horn was produced in peacetime, because the French had surrendered at Montreal in 1760 and many of the Provincial forces raised in 1761 were occupation forces. The beautiful calligraphy is in the hand of J. W. and the three rhymes are expertly executed, decorated with winged serifs and dainty flowers. The lion's tail is similar to the graceful necks of J. W.'s birds.

A delicate floral border consisting of long-stemmed multipetaled flowers wound into a vine pattern precedes the scalloped edge 4¼ inches from the neatly turned tip. There is a 1⅞-inch-long recessed portion, after which the scalloped spout returns to its full diameter and becomes octagonal. The dark shading on the tips of the teeth of the fine incised sawtooth pattern that decorates the plug end and the V-shaped lines that outline the teeth are exceptional. The flat pine plug is secured by seven wooden pegs. An iron staple was driven through the horn into the plug to attach the carrying strap. A hole in the plug's center indicates the location of an earlier hanging device.

BIBLIOGRAPHY: Clements Library, 1984.

LAKE GEORGE SCHOOL
Attributed to the J. W. Carver (w. 1758–1761)

37 *Samuel Whitaker horn*

Crown Point, New York, November 4, 1761
Horn, pine
OL: 12½ in. (32 cm), w (of plug): 2¼ in. (5.7cm)
Inscriptions: Sam^{ll} Whitaker / His Horn Made at Crownpint Novr the 4 1761 / I powder with my Brother ball / A hero like do conquer all
New York State Education Department, New York State History Collection

A Samuel Whitaker is listed as having served for two weeks in Captain Uriah Stevens's Company in the Connecticut provincial militia during the relief of Fort William Henry in 1757. He is not recorded elsewhere.

The calligraphy is typical of J. W.'s work, with winged serifs and his geometric signature. It displays birds, curlicues, a running deer, two dogs, a winged demon with a human face, a sun, and a sloop with two jibs.

The replaced wooden plug is secured by seven wooden pegs. A geometric border decorates the plug end, with a simpler border at the beginning of the recessed portion. A raised carved ring rises in the middle of the recessed portion.

BIBLIOGRAPHY: Grancsay, No. 940, 73; Charles C. Adams, "Historical Collections and Allied Matters," *New York State Museum Bulletin*, No. 301 (March, 1934), 25.

Catalogue

36

37

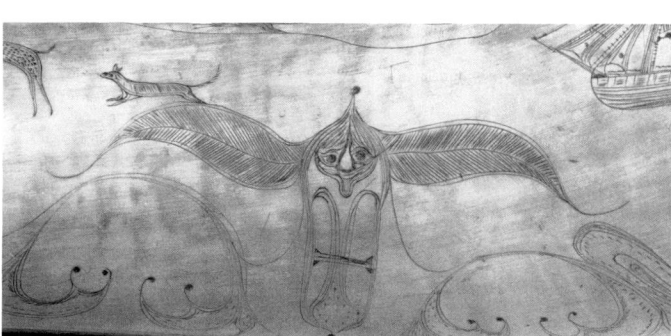

37. *A characteristic J. W. design incorporating geometric and naturalistic motifs.*

37. *J. W.'s confident mastery of the gravure produced graceful, flowing compositions such as this.*

121

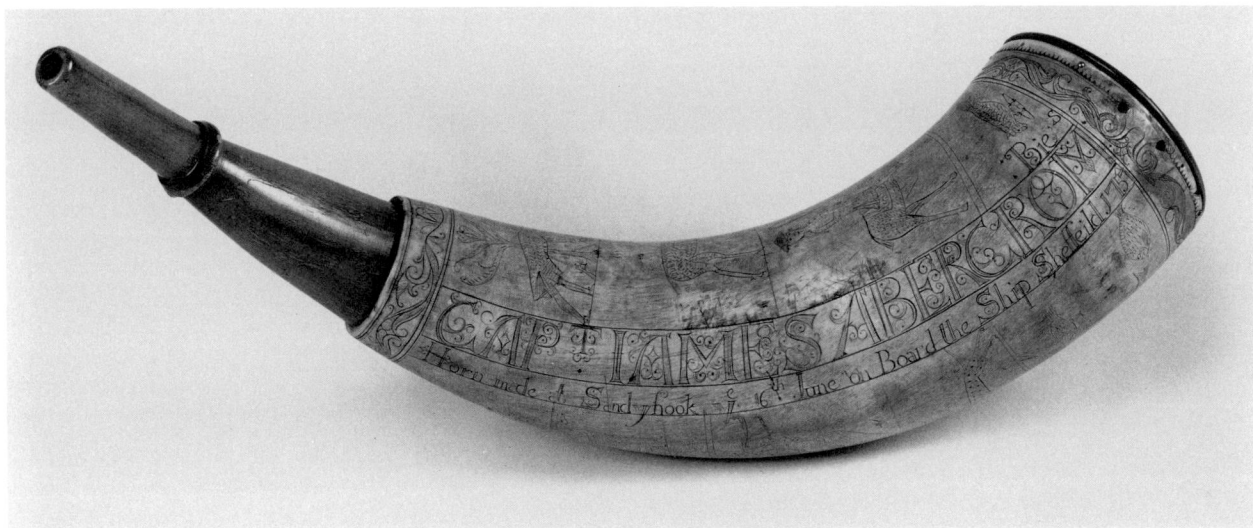

38

LAKE GEORGE SCHOOL

38 *Captain James Abercrombie horn*

Sandyhook, New York, June 16, 1757
Horn, cherry (replaced plug), brass
OL: 18 in. (45.7 cm), W (of plug): 3¼ in. (8.3 cm)
Inscriptions: CAPT. JAMES ABERCROMBIES / Horn made at Sandyhook Ye 16th June on Board the Ship Sheffield / 1757.
James E. Routh, Jr.

James Abercrombie was the son of General James Abercrombie, commander of British Forces in North America in 1758. British army records list Captain James Abercrombie, whose commission is dated February 16, 1756, as serving in North America in the 42nd Royal Highland Regiment. When his horn was made he was probably en route from New York City to Albany and Fort Edward, or possibly he was just arriving from or departing for England while his ship was anchored off Sandy Hook. Abercrombie died of a wound received at the Battle of Bunker Hill on June 24, 1775.

This is a fine horn, with decoration similar to the better work of Jacob Gay. The owner's name is executed in wonderful illuminated lettering, with C-scrolls curling from the uprights and crosspieces of the letters and with diamond-shaped crosspieces with C-scrolls. The secondary calligraphy is smaller and not so elaborate, as was common on Gay horns.

The horn displays a profusion of figural vignettes. Besides the usual deer, bear, birds, fish, and dog, there are also an elegantly dressed couple doing the minuet, a soldier scalping a naked Indian, a hunter firing at a bird in a tree, a mounted horseman, and a woodsman, probably a Ranger, with tomahawk and musket.

A neatly scrolled ½-inch-wide border precedes the recessed portion, which begins 5⅜ inches from the tip; there is a raised ring 1½ inches from the tip. A similar scroll border, ½ inch wide and with an amusing face, encircles the plug end, and the remaining ½ inch of the end of the horn is scalloped. The plug and the brass pins that secure it are modern replacements.

BIBLIOGRAPHY: Grancsay, No. 7, 41; duMont, pl. 13, 23; *Bulletin*, ASAC, No. 42 (Spring, 1980), 6, 12.

LAKE GEORGE SCHOOL
Signed by Jacob Gay (w. 1758–1787)

39 *John Pemberton horn*

Fort Edward, New York, March 3, 1759
Horn, cherry, iron
OL: 14 in. (35.6 cm), W (of plug): 3 in. (7.6 cm)
Inscriptions: JOHN PEMBERTON / his horn made at Fort Edward March Ye 3 1759 / JACOB / GAY ha / nd writ / Fort Edward / MOST TO BE MAD
William H. Guthman

Pemberton has not been identified.

Jacob Gay's illumination of letters is a distinctive feature of his work. A cartulary (book of charters) in Chertsey Abbey, England, executed in the fifteenth century, has illumination that Gay's closely resembles (see bibliography of calligraphy, Hector, pl. 13, p. 77). Gay hinged leaves, flowers, and geometric designs to

Catalogue

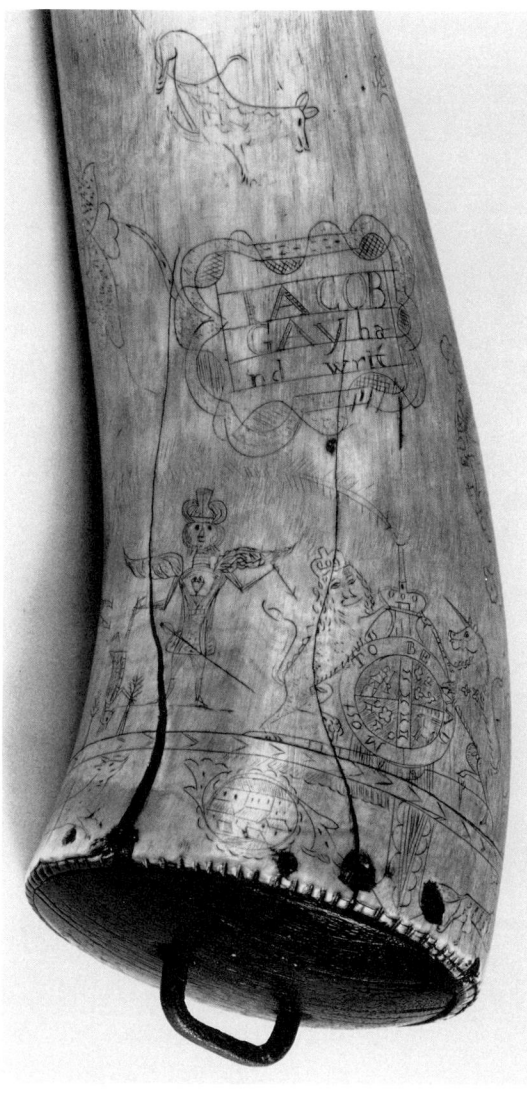

the uprights of most of his letters or their serifs. He filled the interiors of letters like "O," "P," and "Q" with flowers, leaves, or effigies, and sometimes placed strange profile faces on the uprights of letters or on the cross members of "H" or "M." Often the uprights of letters are deeply engraved with contrasting light and dark designs.

Unlike the lettering of most Gay horns, the Pemberton lettering (with the exception of the owner's name) is not of the best quality. The animals, in which Gay specialized, are expertly done but are smaller than ordinary and the soldiers are not so detailed nor so amusing as those on other Gay horns. Gay utilized cartouches quite often, and his name is here engraved within a scroll. In general, the lion and unicorn supports of his British coats-of-arms have animated expressions, but here they are subdued. The motto surrounding the heraldic devices of the coat-of-arms, *Most To Be Mad*, must be a satire on "Dieu Et Mon Droit," indicating that the horn's owner and many other soldiers were mad at being conscripted. Adjacent to the coat-of-arms is an Angel of Death wearing royal regalia, carrying a court sword, and holding a smaller person upside down in his right hand and an arrow in his left. Above Pemberton's name are several geometric designs and a vine border as well as a large geometric design attached to a flower.

The horn is scalloped where the recessed portion begins, 3¼ inches from the tip, but the scalloping is not up to Gay's usual standard. A raised ring is ¼ inch from the tip. A double line border, ⅛ inch wide and decorated with chevrons, encircles the butt ¾ inches from the plug. Animals and geometric designs fill this space, and the edge of the butt end is notched. The stained pine plug, which extends ⅛ inch beyond the

39. *Formations of soldiers and an abstract coat-of-arms characteristic of Jacob Gay.*

end of the horn, is secured by nine wooden pins. A 1-inch-long iron rivet is driven into the center of the plug to hold the carrying strap.

Extreme cold, arthritis, or some other affliction may have inhibited Gay's carving ability at the time he made this horn. The same lack of freedom is seen on the Goding horn (No. 40), which Gay carved seventeen days later.

LAKE GEORGE SCHOOL
Attributed to Jacob Gay (w. 1758–1787)

40 *William Goding horn*

Fort Edward, New York, March 20, 1759
Horn, pine, iron, brass
OL: 14½ in. (36.8 cm); w (of plug): 3 in. (7.6 cm)
Inscription: WILLIAM / GODING his horn made at Fort Edward / M[ar]ch 20 1759 / 1759; *a crude map of the forts above Albany includes several phonetically spelled locations.*
William H. Guthman

Goding has not been identified. His horn is almost identical to the Pemberton horn (No. 39), but has the addition of a small map. Most map horns depicting large areas in great detail were carved as souvenirs after the fact and many are relatively modern fakes. On the other hand, horns with simple maps like this that depict small areas are of the period and were made for practical use.

Like the Pemberton horn, the Goding horn is not the best work Gay was capable of producing, but is nevertheless covered with lively images. A formation of ten soldiers may be seen firing muskets at an unseen enemy, while above Goding's name is a cartouche showing two truncated men sitting at a table smoking long pipes; to their right is a white-tailed buck deer facing a doe. Under the map, which depicts the buildings at Fort Edward in some detail, is a compass which is not accurately placed in relation to the map. Above the compass are two dogs facing one another and below it is a hunter firing a musket at a deer.

The calligraphy of Goding's name is almost identical to that of Pemberton's. Profile faces sprout from the sides of the uprights and on the cross bars, and the geometric and floral appendages and shading of the letters are very close.

A floral device is placed just after the first name, WILLIAM, but no border demarcates the recessed portion, which extends 3¼ inches to the tip and ends in a raised lip. A ¼-inch double-line border with zigzag dotted lines decorates the plug end 1 inch from the end of the horn. A single feathered-line border finishes the edge. The 1-inch space between border and edging contains a geometric design, an elaborate scroll, birds, fish, and the date *1759*. Twelve brass pins secure the pine plug, which extends ¼ inch beyond the horn. An iron ring is attached to a hole at the edge of the horn for the carrying strap.

BIBLIOGRAPHY: Grancsay, No. 382, 54; Sotheby Parke-Bernet, Sale No. 4478Y, November, 1980, lot 1083.

40

Catalogue

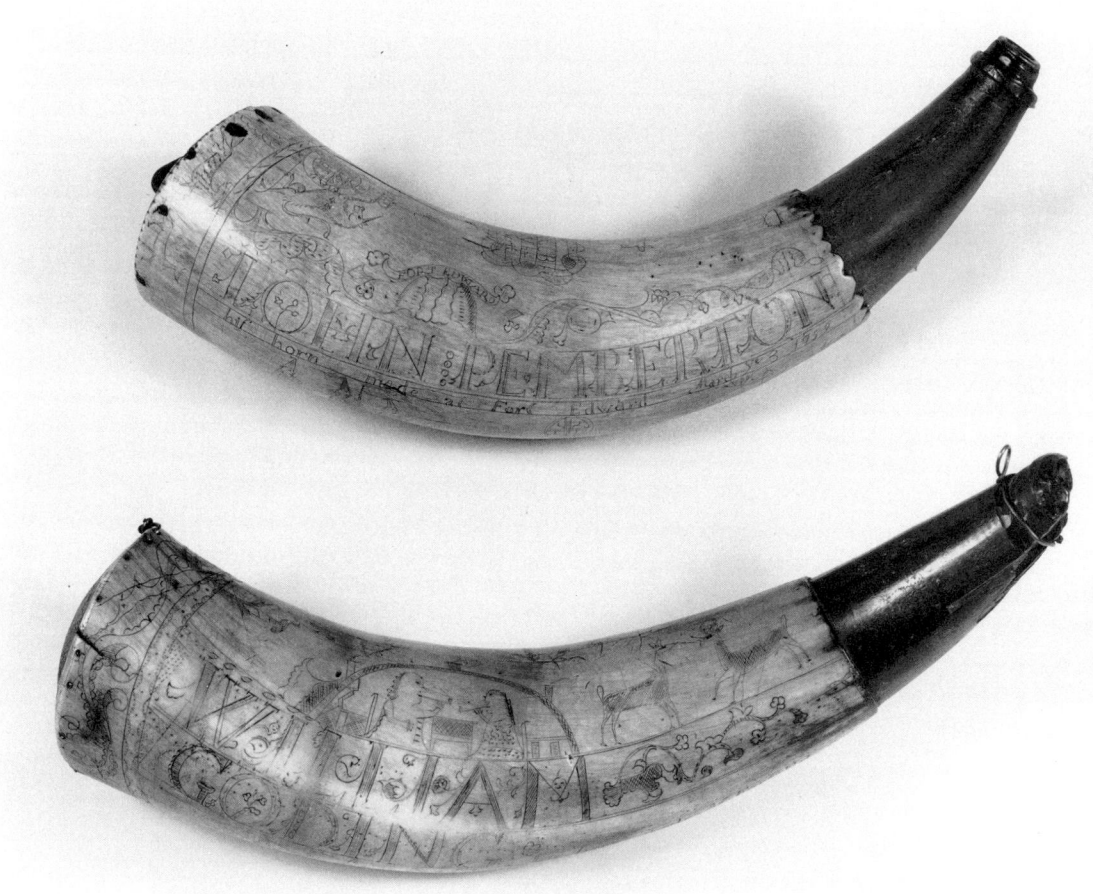

39 (Top, John Pemberton) and 40 (bottom, William Goding): *examples of Jacob Gay's interwoven illuminated calligraphy, floral designs, and vignettes.*

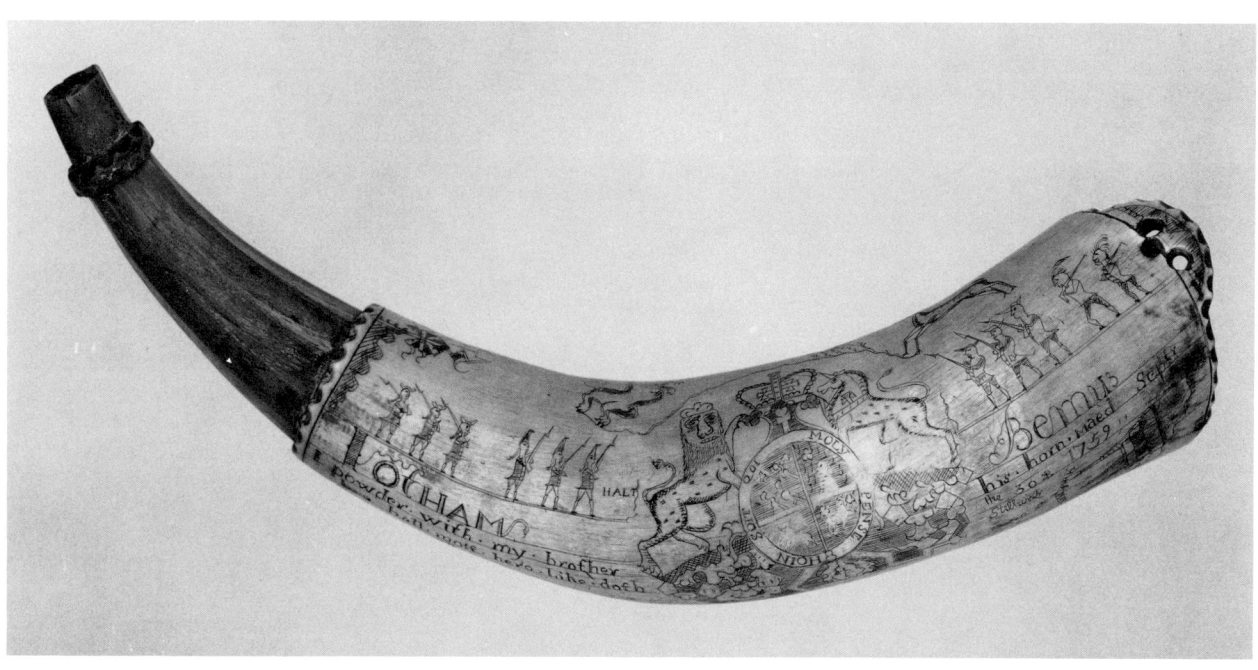

41

LAKE GEORGE SCHOOL
Signed by Jacob Gay (w. 1758–1787)

41 *Jotham Bemus horn*

Bemus Heights (between Stillwater and Saratoga), New York, September 30, 1759

Horn, pine, iron

OL: 15½ in. (39.4 cm), W (of plug): 3 in. (7.6 cm)

Inscriptions: IOTHAM BEMUS / I . powder . with . my . brother / ball . most . hero . like . doth / Conquer . all / JACOB GAY han / his . horn . maed Sep^{tr} / the 30 ∗ 1759 Stillwater / New York Albany Half Moon / Albany / ST LWTR HALT / SARETOG HOIN / FO MIL SOIT / AR QUI / FO EDWARD MOLY / H. BROOK PENSE / FO. W. H. / RO CARELONG / CRUPOT [Crown Point]

The Metropolitan Museum of Art, J. H. Grenville Gilbert Collection

Jotham Bemus (1738–1786) ran the only tavern on the road from Albany to Fort Edward; the location, on Bemus Heights, was four miles north of Stillwater, about halfway to Saratoga. It is very likely that Jacob Gay carved this horn in return for food, drink, and/or lodging. Gay may even have stayed at the tavern while carving horns for soldiers marching back and forth to the forts.

The Bemus horn is a fine example of Gay's work. The large British coat-of-arms, which is set within an elaborate geometric design that resembles a fort with two bastions, has lion and unicorn supports with animated expressions. The whole design is neatly outlined and shaded with cross-hatching.

To the left of the arms is a formation of six uniformed soldiers supplied with muskets and bayonets and marching in single file, captioned HALT. These soldiers appear to be challenging another formation to the right of the arms, which consists of three uniformed soldiers and two Indians decked out in feathered headdresses and equipped with muskets (the Indians' muskets do not have bayonets). None of the characters has an animated expression, simply incised eyes, noses, and mouths. Above the arms are three beautifully engraved animals—a prancing horse, a buck, and a small fox. There is also a plausible map showing landmarks for a short distance between Albany and Stillwater.

The owner's name is executed in shaded block lettering with vinelike serifs, while Gay's signature and the location *Stillwater* are in a vinelike script. The rest of the inscriptions are in block letters.

There is a scalloped border at the beginning of the recessed portion 4 inches from the tip. The recessed portion is worked into nine facets, which continue to a raised ring ¾ inch from the tip. The plug end is neatly notched and extends beyond the rounded plug ¹⁄₁₆ of an inch. An engraved and notched lobe extends ¼ inch beyond the edge of the horn, with two holes for the carrying strap. The plug is secured by four large rosehead nails.

BIBLIOGRAPHY: Grancsay, No. 68, 43; *Antiques*, 1978, 319; Guthman, *Guns and Other Arms*, 141.

LAKE GEORGE SCHOOL
Attributed to Jacob Gay (w. 1758–1787)

42 *Philip Bunker horn*

Lake George, New York, November 11, 1759

Horn, iron, pine, sealing wax, deer hide, dark pigment

OL: 15½ in. (39.4 cm), W (of plug): 3⅛ in. (7.9 cm)

Inscriptions: PHILIP BUNKER / HONI SOI MALI PENS / DIEU ET MON DROIT / his horn Mead / Novm 11 1759 / LAKE GORG / FT. CARELONG [Carillon] / FT CROUN POINT

Carol and the late Thomas J. Segal

41. *A plausible map shows landmarks for the short distance between Albany and Stillwater; Jacob Gay's signature may be seen to the right.*

Catalogue

42

According to family tradition, Bunker served in Captain Andrew Fuller's Company, Colonel Jonathan Bagley's Massachusetts Regiment, during the campaigns of 1758 and 1759. He is listed as a minuteman drummer in 1775.

This is a good example of Jacob Gay's French and Indian War style, with the carving emphasized by dark pigment. The three soldiers are amusing, although not so highly developed as those on other Gay horns. The animals are animated and abundant, including the lion and unicorn on the coat-of-arms. The lettering is of medium quality: the uprights of the name PHILIP are embellished with faces and flowers, but the crosspieces are plain. The only embellishment on the name BUNKER is a face on the letter "K," and the geometric device above it is similar to J. W.'s signature device. The crude map, which is like others on some Gay horns, may, like them, have been added later. The date, also roughly inscribed, was probably added by that later hand.

A rough scalloped edge precedes the recessed portion 5¼ inches from the tip, and a scalloped ring is 2¼ inches from the tip. There is no border at the butt end. Three large wooden pegs secure the flat pine plug, which is still covered by the original beige sealing wax. Two rosehead nails secure a piece of the original hide thong that held the carrying strap.

BIBLIOGRAPHY: Grancsay, No. 129, 45.

LAKE GEORGE SCHOOL
   Signed by Jacob Gay (w. 1758–1787)

43 *John Mills horn*

Crown Point, New York, November 4, 1760
Horn, pine
OL: 14 in. (36 cm), W (of plug): 3¼ in. (8.3 cm)
Inscriptions: John Mills / his horn made at Crown Point / November the 4 1760 / J G
William H. Guthman

This horn was purchased from a direct descendant of the original owner. Since the descendant lived in Chelsea, Massachusetts, the likelihood is that Mills served in a Massachusetts provincial regiment. However, a John Mills also served in Captain Alexander Todd's Company of the New Hampshire provincial regiment stationed at Crown Point in 1760.

This is one of Jacob Gay's finest horns, with every detail of consistently high quality. The lettering of the name incorporates all of Gay's decorative techniques: beautifully shaded letters with strong contrast; delicate scrolling; and amusing faces on the uprights and in the "O," as well as on the crossbar of the "M." The guidelines for the lettering are decorated, and the lowercase lettering of the location and date is artfully embellished. Above the name and facing in the opposite direction is a hunter shooting at two beautifully engraved deer; what appears to be a dog is running

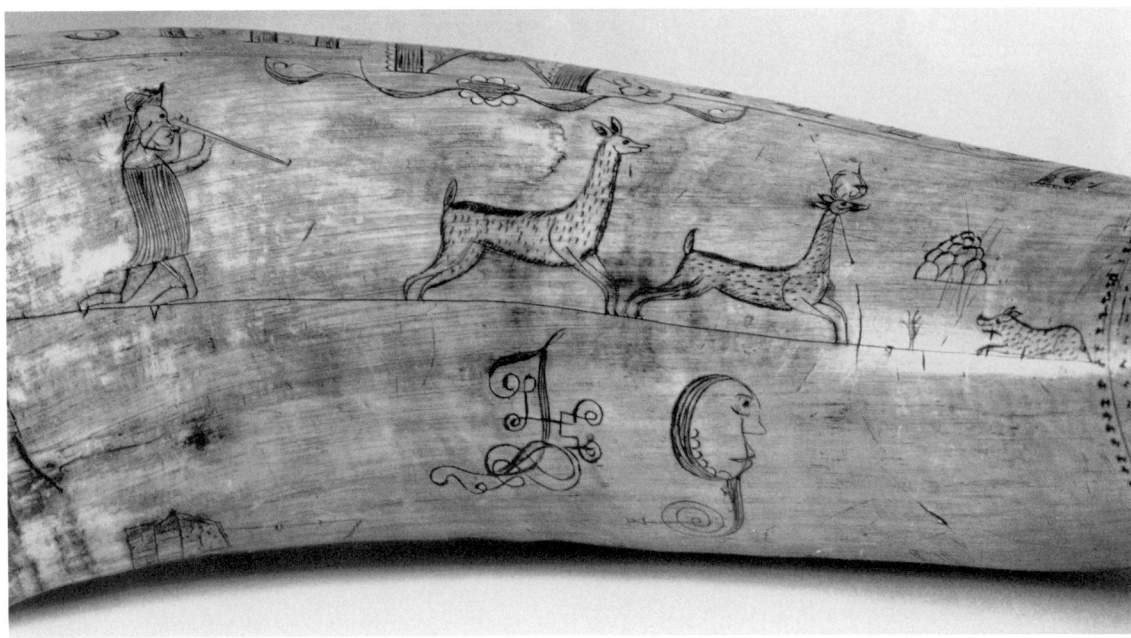

43

toward the deer. The hunter's face beams with Gay's typically delightful monkeylike expression, also seen on the Pemberton and Goding horns (Nos. 39 and 40). Below the deer are Gay's initials, large and ornamented. Inside the "G" is the severe profile of a man, who appears to be an Indian.

The scalloping at the border of the recessed portion is deeply and expertly done. Just before the scalloping is a deeply carved raised ring, and before this is a ¼-inch-wide series of lines and dots that encircle the throat. The recessed portion begins 3¼ inches from the tip. The only border at the butt end consists of four incised lines. The rounded pine plug, which extends ½ inch beyond the horn, is secured by four wooden pegs. The carrying strap was secured through two holes at the butt end that formerly extended through the plug. There may originally have been an extension lobe.

BIBLIOGRAPHY: *Antiques*, 1978, 323; *Bulletin*, ASAC, No. 44, 41, 48, 49; *Bulletin*, KRA, Fall, 1985, 9; Guthman, *Guns and Other Arms*, 145.

LAKE GEORGE SCHOOL
Attributed to the Miller-Tribble, or Lyme, Carver (w. 1758)

44 *Thomas Miller horn*

Fort Edward, New York, September 20, 1758
Horn, pine
OL: 15½ in. (39.4 cm), w (of plug): 3 in. (7.6 cm)

Inscriptions: Thomas Miller / of Lyme: His Horn Made : at / Fort: Edward: September: Y^e 20 1758 / I powder: With: my: Brother / Ball: a. Hero: Like: Do: Conquer all; N *in several places and* N S *added in a different hand at a later date.*
William H. Guthman

Thomas Miller of Lyme, Connecticut, served in Captain James Harris Jr.'s 5th Company, Colonel David Wooster's 2nd Connecticut Regiment, during the 1756 campaign. During the 1758 campaign, Miller served as drummer in Captain Timothy Mather's 11th Company (Lyme), Colonel Eleazor Fitch's 3rd Connecticut Regiment. A company muster roll dated October 19, 1758, is endorsed at Fort Edward. Miller again served as drummer in Captain Zebulon Butler's 9th Company (Lyme), Colonel Fitch's 4th Connecticut Regiment, during the 1759 campaign, and again in Captain Butler's 8th Company, Major General Phineas Lyman's 1st Connecticut Regiment, during the 1761 campaign.

The carver of the Miller and John Tribble (No. 45) horns was exceptionally talented. His style combines elements of the styles of J. W., Jacob Gay, and the Selkrig-Page Carver. His calligraphy combines characteristics of both J. W. and Gay: stylized thick-and-thin copperplate lettering that utilizes shading for emphasis; C-scroll and wing serifs and C-scrolls on bars and crossbars of letters; and careful lining. His animals and birds are similar to Jacob Gay's, and his incised decoration, although larger and more deeply cut, resembles J. W.'s. The compound style adopted by the Miller-Tribble Carver is strong evidence that carvers were aware of one another's styles and readily copied them. However, this carver's winged angels and his use of heavy dots to pick out designs and borders are unusual features. Although other horns have drums as decorative devices, this horn has no other accoutrements illustrated, and the inference is strong that the drum referred to Miller's function as a drummer.

A heavily dotted border precedes the nicely scalloped edge at the beginning of the recessed portion 5¾ inches from the tip of the spout. A ⅛-inch raised ring is 2⅝ inches from the faceted raised circular tip, and the throat is faceted octagonally from the raised ring to the spout. The slightly raised pine plug is secured by three wooden pegs. A finely engraved ¾-inch floral and vine border at the plug end is exceptionally large for Lake George School horns. Two holes pierce the butt and wooden plug for the carrying strap.

LAKE GEORGE SCHOOL
Attributed to the Miller-Tribble, or Lyme, Carver (w. 1758)

45 *John Tribble horn*

Fort Edward, New York, September 5, 1758
Horn, oak, iron
OL: 14½ in. (36.8 cm), w (of plug): 3 in. (7.6 cm)
Inscriptions: John Tribble: of / Lyme: His: Horn Made: At: / Fort. Edward: September / Y^e 5th A.::. ¹⁷⁵⁸ / Don't. Take. This: Horn. For. Fear. of / Shame. for. on. it. Stands. Y^e. owners name
William H. Guthman

John Tribble of Lyme, Connecticut, served in Captain Joshua Abel's 6th Company (Norwich), Colonel Eliphalet Dyer's 3rd Connecticut Regiment, during the 1755 campaign. His enlistment ran from September 11 to November 25, and when it expired Trib-

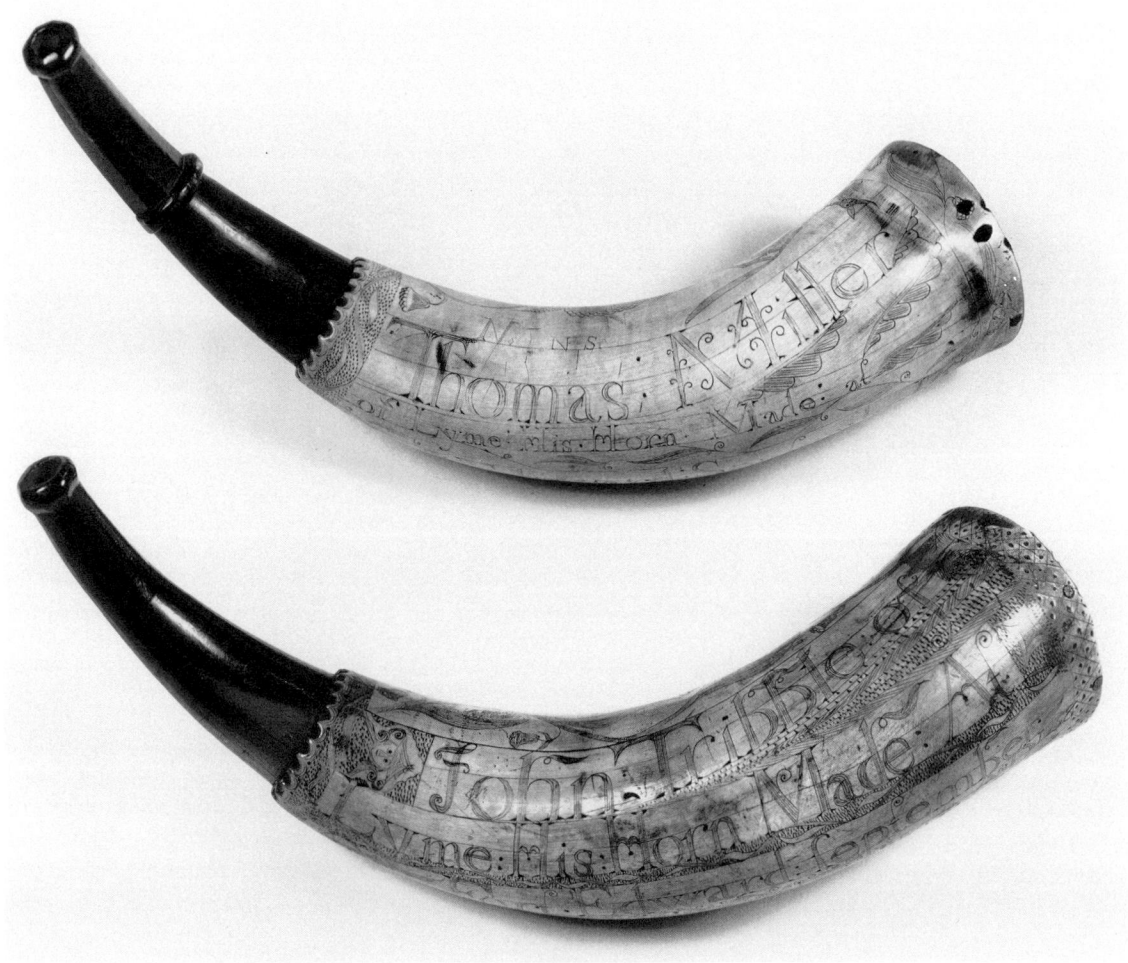

44 (Top, Thomas Miller) and 45 (bottom, John Tribble): *the Lyme, or Miller-Tribble, Carver combined strong, intricate calligraphy with baroque motifs.*

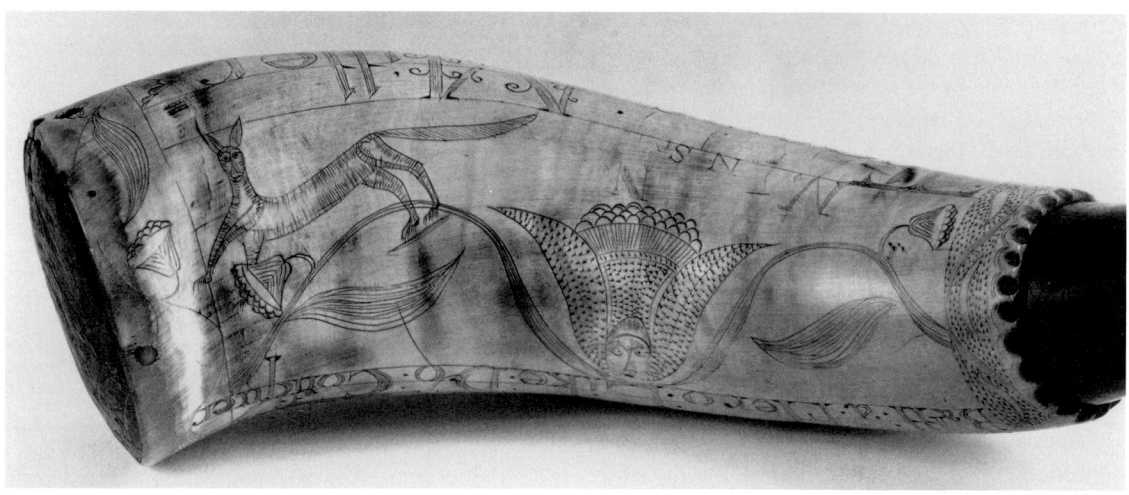

44

44

45

ble enlisted as a corporal in Major Nathan Payson's (No. 29) Company, Colonel Bagley's Regiment, consisting of men from New York, Connecticut, Rhode Island, Massachusetts, and New Hampshire. This enlistment began November 25, 1755, and ended May 21, 1756.

Captain Noah Grant assumed command of the company on December 20, 1755, and after Tribble's enlistment expired in 1756, he reenlisted as a private in Grant's company for the period June 9 to December 3. Zebulon Waterman (No. 27) was a private in this same company. However, a muster roll of the company, dated at Fort William Henry on October 13, 1756, lists Tribble as a corporal.

Tribble does not appear on the rolls again until 1759, but his horn is inscribed at Fort Edward in September of 1758. During the 1759 campaign he is listed as serving as a private in Lieutenant Colonel Joseph Spencer's Company (East Haddam), Colonel Nathan Whiting's (No. 12) 2nd Connecticut Regiment. He again served under Colonel Spencer during the campaign of 1760, in Colonel Whiting's 2nd Connecticut Regiment. Hobart Spencer (No. 47) served in Tribble's company in 1759 and 1760. During the 1762 campaign, Tribble served in Captain Zebulon Butler's 8th Company (Lyme), Colonel Phineas Lyman's 1st Connecticut Regiment. The 1st Regiment participated in the disastrous Havana Expedition, during which most soldiers were killed or died of disease. Tribble is listed as either dead or having deserted.

The carver of this horn also carved the Thomas Miller horn (No. 44) at Fort Edward during the same month. The two horns are quite similar in design, although Tribble's has neither animals nor fish, and instead of winged angels it has two separate vignettes showing busts of Elizabethan characters; heavy dots take the place of incising in carrying out patterns. A dotted vinelike border precedes the neatly scalloped edge where the recessed portion of the throat begins 4½ inches from the slightly faceted raised spout. No raised ring was provided between the border and the spout. The slightly rounded oak plug, secured by eight wooden pegs, is probably a replacement. A large screw eye is in the center of the plug and a ½-inch border around the plug end of the horn consists of cross-hatching filled in with heavy dots.

OPPOSITE

44 (Top): *elements of the styles of the Selkrig-Page Carver, J. W., and Jacob Gay combine here*; 44 (middle): *the Miller-Tribble, or Lyme, Carver employed prominent dark dots to emphasize his designs*; 45 (bottom): *the Miller-Tribble Carver relied on heavy dots and curlicues instead of incising.*

LAKE GEORGE SCHOOL
Attributed to the Spencer-Hitchcock Carver
(w. 1757–1762)

46 *Josiah Walker horn*

One of the forts, Albany to Lake Champlain, 1757–1759
Horn, pine, brass, black pigment
OL: 13½ in. (34.3 cm), W (of plug): 2¾ in. (7 cm)
Inscriptions: Lieut, Josiah Walker; I B Smith *in another hand at a later date.*
William H. Guthman

Josiah Walker is listed as serving as first lieutenant in companies raised in Fairfield and Greenwich, Connecticut, in the following campaigns: in 1757, in Captain Samuel Hubbel's (Fairfield) Company, Colonel Phineas Lyman's Connecticut Regiment; in 1758, in Captain Thomas Hobby's (Greenwich) 5th Company, Colonel David Wooster's 4th Connecticut Regiment; in 1759, in Lieutenant Colonel James Smedley's (Fairfield) 2nd Company, Colonel David Wooster's 3rd Connecticut Regiment.

Like the calligraphy of the Hobart Spencer and Ebenezer Hitchcock horns (Nos. 47 and 48), this resembles the work of J. W. except that it lacks winged serifs. The three flamingolike birds with their necks gracefully turned toward their tail feathers are also like J. W.'s motifs. These are used in combination with floral vines to make a border underneath the formation of soldiers whose buttons and large eyes are accented with dark pigment. The two groups of ten soldiers each—which are almost identical to those on the Spencer and Hitchcock horns except that there are no mounted troops here—are facing each other in firing position. The thick-and-thin calligraphy is almost identical to the copperplate engraving on the other two horns, as well. There is a neatly executed abstract British coat-of-arms (which is not present on the other two examples) surrounded by a delicately scalloped border surrounded by floral decoration similar to that of J. W.

There is a neat scalloped edge decorated with dots and triangular elongated devices of a familiar Lake George School type preceding the recessed portion at the throat 3 inches from the tip. There is a raised ring ¼ inch from the tip. The spout may have been shortened slightly at some point. The slightly rounded pine plug, which extends beyond the horn only 1⁄16 of an inch, is secured by eight tiny wooden pegs. There is a border of semicircles and elongated dotted triangles at the plug end also. A small round brass staple secures an oval brass ring that held the carrying strap.

BIBLIOGRAPHY: Richard Oliver auction catalogue, Kennebunk, Maine, June 23, 1988; brochures for above auction.

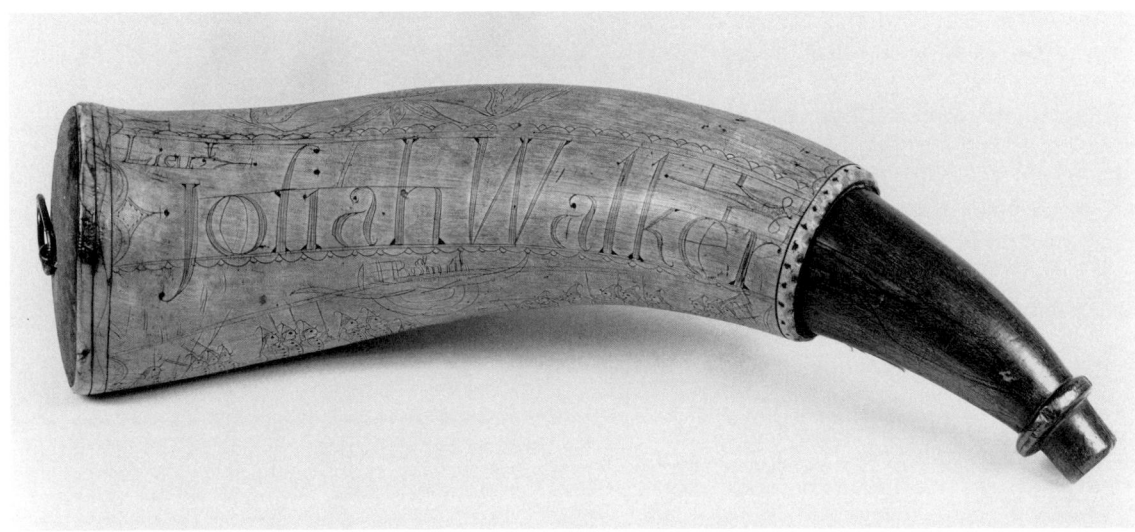

46

46. *Soldiers' buttons and eyes match here; the border below the soldiers resembles those of J. W.*

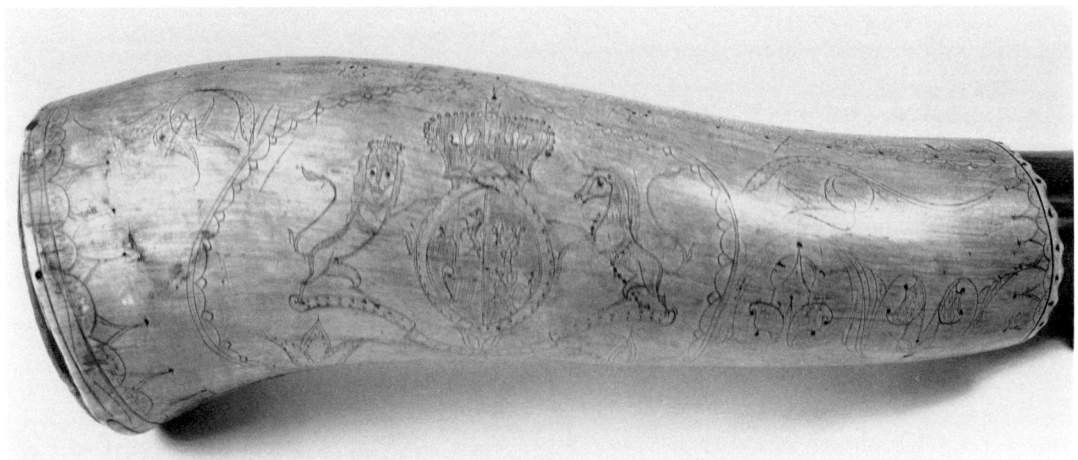

46. *The British coat-of-arms, a ubiquitous motif in pre-Revolutionary America, is here delicately rendered and surrounded by floral decoration in the style of J. W.*

Catalogue

LAKE GEORGE SCHOOL
Attributed to the Spencer-Hitchcock Carver
(w. 1757–1762)

47 *Hobart Spencer horn*

Crown Point, New York, November 1, 1759
Horn, pine, iron
OL: 16 in. (40.6 cm), W (of plug): 3 in. (7.6 cm)
Inscriptions: Men of might they take Delight / In gun and sword that they may fight / Hobart Spencer's / Horn Dat'd Crown point Nov,^{r} 1st = 1759

William H. Guthman

Hobart Spencer (1742–1806) of East Haddam served as a private under his father, Lieutenant Colonel Joseph Spencer, in the latter's 2nd Company (East Haddam), Colonel Nathan Whiting's (No. 12) 2nd Connecticut Regiment, during the 1759 and 1760 campaigns. John Tribble (No. 45) served in this same company during both campaigns, but the Spencer and Tribble horns are completely different and are attributed to two different hands.

The Spencer horn has typical Lake George School calligraphy, with thick-and-thin copperplate letters

47

47. *Formations of soldiers and catchy rhymes are typical of the Lake George School.*

47. A rare vignette of mounted troops with drawn pistols firing at marching soldiers.

and shaded accents. The calligraphy is equal in quality to that of the Miller-Tribble Carver (Nos. 44 and 45) or of J.W., although the letters do not have winged serifs as J.W.'s do.

The formations of soldiers are not equal in detail to those of the Selkrig-Page Carver (Nos. 23–27), but are about on a par with formations carved by Jacob Gay in this period. Although these soldiers are not drawn in detail, they are accurately drawn in the positions of "ready," "present," and "fire." One soldier at the end of the line looms twice as large as the others. The engraver may have been indicating a particular person or emphasizing the practice of placing the tallest soldier at the rear. Another exceedingly rare vignette illustrates a group of mounted troopers with their holster pistols drawn and held in firing position. As mounted troops were rarely used in frontier fighting, the inference is that the carver inserted motifs he knew from militia drilling at home or perhaps from pictures in books and magazines. The balance of the decoration consists of a competently drawn ship and a fish.

Borders at throat and butt are sparse and linear. The merely adequate scalloping that precedes the recessed portion 5½ inches from the spout is followed by a ¼-inch raised ring 2½ inches from the tip. The rounded pine plug, extending ⅜ inch beyond the horn, is secured by a single iron nail and by the two prongs of an iron staple driven through the horn into the plug to secure the carrying strap.

BIBLIOGRAPHY: Grider, NYHS; Grancsay, 68.

Catalogue only
LAKE GEORGE SCHOOL
    Attributed to the Spencer-Hitchcock Carver
    (w. 1757–1762)

48 *Ebenezer Hitchcock horn*

Crown Point, New York, October 17, 1762
Horn, pine
OL: 14½ in. (36.8 cm), w (of plug): 2½ in. (6.4 cm)
Inscriptions: Ebenezer Hitchcock / Horn made at Crown Point Octr 17th 1762: in the 12th year of his age / Be: These Thy: art to bid Contention Cease / Chain up Starn Way and give the Nations Peace / O'er subject lands Extend thy gentle sway / and teach with iron Rod the French Dogs to obey
James E. Dresslar

Although the age of the owner is given as twelve years, an Ebenezer Hitchcock is listed as serving from March 15 to December 3, 1762, in Captain Amos Hitchcock's 6th Company (New Haven), Colonel Nathan Whiting's (No. 12) 2nd Connecticut Regiment. The soldier serving in that campaign was probably Ebenezer Hitchcock (1719–1764) of New Haven, but the owner of this horn was probably the Ebenezer Hitchcock born March 4, 1750/1, to Captain Amos Hitchcock (1724–1791). In other words, Captain Hitchcock, as an officer of some consequence, had the horn made for his son as a souvenir. The moralistic doggerel verse tends to reinforce this surmise.

This horn was carved by the same hand as the

Catalogue

48. *Another rare view of mounted soldiers enlivens this horn; the horses appear to be posing for the artist while their riders focus on the enemy.*

Hobart Spencer horn (No. 47), hence the name assigned the carver. Spencer served in Colonel Whiting's regiment at Crown Point during the 1759 and 1760 campaigns. The calligraphy and the formations of soldiers are nearly identical and the soldiers, both mounted and on foot, are amusing for their highly stylized uniforms, flying flags, and prancing horses. The expressions on the faces of the men are identical to those of the horses, an additional comic feature.

Although badly worn, the floral design is extremely fine and recalls horns by J. W. These designs, carved as borders under the soldiers, extend the length of the horn and end at the recessed portion 3¼ inches from the tip. An incised double line encircles the edge of the recessed portion and the tip has a raised ring. A winglike vine border surrounds the plug end. The pine plug extends ⅜ inch beyond the horn and is secured by four wooden pegs. A large threaded hole in the center of the plug suggests a missing wooden screw like those used on artillery priming horns; the horn could have been filled from this end, and undoubtedly the wooden screw had an eye for securing the carrying strap. Two holes above the large screw hole indicate the previous location of an iron staple.

BIBLIOGRAPHY: Grancsay, No. 441, 56; Grider, NYHS.

49

LAKE GEORGE SCHOOL
Attributed to the Cleaveland-Carril Carver
(w. 1758–1761)

49 *Aaron Cleaveland horn*

Fort Edward, New York, September 29, 1758
Horn, brass, iron, black pigment
OL: 15¼ in. (38.7 cm), w (of plug): 3 in. (7.6 cm)
Inscriptions: LIEU^T A:C: WAS BOR^N the yer 1727 /ARON. CLEAVELAND^S. HORN Md / FORT. EDWARD Y^e 29 1758 SEP^T / HONI MAL PENSE / ET MON DRT
The Connecticut Historical Society, Gift of Hiram Holt

Aaron Cleaveland (1727–1785) of Canterbury, Connecticut, served as lieutenant in Captain Coit's Company, Colonel John Dyar's Regiment, in August of 1757, and marched for the relief of Fort William Henry. For the entire campaign of 1758, he served as first lieutenant in Major Israel Putnam's 3rd Company, Colonel Eleazor Fitch's 3rd Connecticut Regiment. He served in militia officerships in the 1760s and 1770s and was captain of a company that answered the Lexington Alarm in 1775. His personal friendship with Israel Putnam made him something of a local hero and political figure.

This horn was probably carved by the same man who made the John Carril and William Cotton horns (Nos. 50 and 51). All three examples have deeply incised carving filled in with heavy black pigment for contrast. Two borders of linked ovals spaced two inches apart extend the length of this horn; the ovals are rather small at the throat end and gradually expand in size as they approach the butt end. The block lettering is large and neatly shaded. All three horns have comparable misspellings of the mottoes on the coats-of-arms and the lion support always faces the viewer with a comical expression on his face. In May of 1848 the secretary of the Connecticut Historical Society accessioned this example, inscribing the accession information in brown ink across the body of the horn.

All three horns have a British coat-of-arms with a fiery griffin attacking it. The Cleaveland horn has a hunter taking aim and firing at the griffin, as well as a soldier presenting arms and two uniformed riders on a saddled horse approaching the soldier. These scenes take place amidst several trees where a deer may also be seen.

A ring of intertwined demilunes is engraved around the throat at the point where a scalloped edge demarcates the recessed portion 3½ inches from the tip. A small raised ring is ¼ inch from the tip. A ring of linked ovals encircles the plug end and the plug is covered by a brass cap secured to the horn by four brass nails. A 1-inch iron staple is driven into the brass cap at the top portion of the butt.

BIBLIOGRAPHY: Grider, NYHS; Grancsay, No. 190, 47.

Catalogue

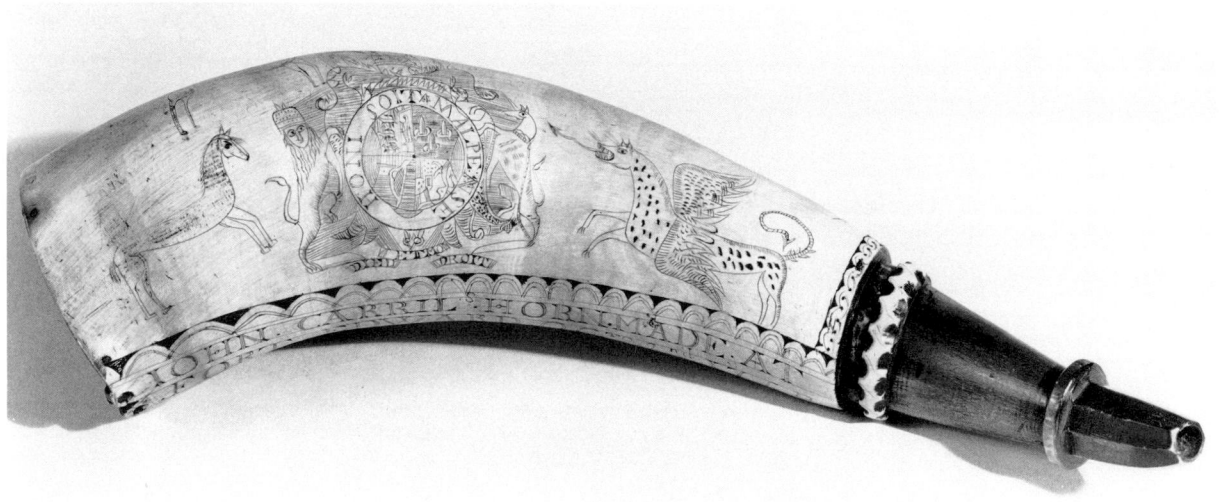

50

LAKE GEORGE SCHOOL
Attributed to the Cleaveland-Carril Carver
(w. 1758–1761)

50 *John Carril horn*

Fort Edward, New York, October 2, 1758
Horn, walnut, black pigment
OL: 12¼ in. (31.1 cm), W (of plug): 3 in. (7.6 cm)
Inscriptions: JOHN.CARRIL^S. HORN. MADE. AT / .FORT. EDWARD. OC^{to}. 2nd 1758 / HONI SOIT MALPENSE / DIEU ET MON DROIT
Private Collection

A John Carril is listed in the New York muster rolls as serving in Captain Wright's company during the campaign of 1760.

This example and the Aaron Cleaveland horn (No. 49) were carved three days apart at Fort Edward. They have many of the same motifs, including that of the British coat-of-arms being attacked by a griffin; to one side of the arms is a horse with a deer above it. The Carril horn is devoid of other decoration and is slightly inferior in quality to the Cleaveland example. The William Cotton horn (No. 51) was probably carved one year later by the same engraver.

The neatly scrolled, deeply incised border blackened with pigment is carved at the point where the recessed portion begins, 4½ inches from the tip. A scalloped raised ring, also blackened, is carved ¼ inch from this first border, and a plain ring is 1¼ inches from the tip, after which the horn is faceted hexagonally to the tip. There is no border at the plug end, and a minute double lobe with a hole in each lobe extends ¼ inch from the end of the horn. The walnut plug, which is probably a replacement, is slightly rounded and is notched under the lobes to permit the thongs from the carrying strap to pass through. The plug is secured by four square-head nails.

BIBLIOGRAPHY: Grancsay, No. 157, 46.

LAKE GEORGE SCHOOL
Attributed to the Cleaveland-Carril Carver
(w. 1758–1761)

51 *William Cotton horn*

Lake George, New York, September 9, 1759
Horn, pine, brass, iron, bone
OL: 13¾ in. (35 cm), W (of plug): 2½ in. (6.4 cm)
Inscriptions: WILLIAM : COTTON . his . horn . made / Sep$^{t\,the}$ 9 1759 / HONI SOIT QUI MAL PENSE / AW
Carol and the late Thomas J. Segal

William Cotton is listed as serving in Captain John Pickering's 1st Company, Colonel John Hart's New Hampshire Regiment, raised for the 1758 campaign against Crown Point. He does not appear in the records again, but obviously served in the 1759 campaign.

Except for the unusual format of the carving of the owner's name, in which block capital letters appear on a shaded background, this horn conforms to the pattern of the Cleaveland-Carril Carver (Nos. 49 and 50). The remainder of the calligraphy and the lion, unicorn, and fiery griffin are identical to those on the related horns, as is the British coat-of-arms in a quatrefoil. Other decoration includes a beautiful C-scroll vine with comical birds within and on top of the vine,

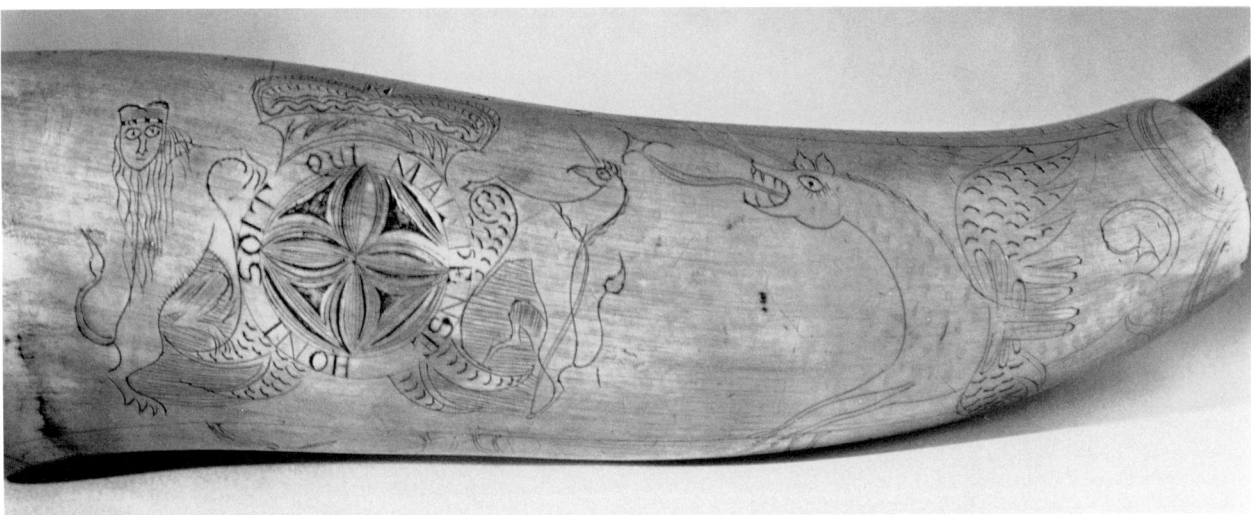

51

a device seen on other Lake George School horns. A prancing deer is passing between two gracefully rearing horses.

A ring of multiple lunettes precedes the recessed portion 4 inches from the tip. A raised ring is 1¾ inches from the tip, and another raised ring is at the tip. A border of simple intertwined lunettes encircles the plug end. The flat pine plug is secured by seven brass pins placed methodically within the center of each lunette in the plug-end border. A wrought-iron nail driven into the plug secured the carrying strap. A later bone stopper is in the spout.

BIBLIOGRAPHY: Clements Library.

LAKE GEORGE SCHOOL
Attributed to Richardson Minor (1736–1797)

52 *Richardson Minor horn*

La Gallette, Quebec, August 29, 1760
Horn, pine, iron
OL: 16 in. (40.6 cm), W (of plug): 3 in. (7.6 cm)
Inscriptions: Richardson Miner's Horn / Dat'd at Le Galatte Aug. 29 1760 / Sir i hope you hant forgo / ᵗ Alway to strike when Th'Iron hot
William H. Guthman

Richardson Minor was a Stratford, Connecticut, silversmith and clockmaker who served during the 1755 campaign in the 7th Company, 2nd Connecticut Regiment, and during the 1760 campaign as regimental armorer of that same regiment. La Gallette was located on the St. Lawrence River near Montreal.

This is a superb example of the Lake George School carving style. Because Minor was a trained metalworker, this horn is attributed to his own hand, as is the Tambling horn (No. 53). This example's deeply incised floral design of stylized tulips is meticulously laid out and executed with the skill of a confident professional. The blend of fine and deep accents is impressive; the lettering is precise, with fine copperplate letters; the serifs are deeply cut, as are the punctuation points; and the proportions and shading exceed those of all other known horn carvers. The phonetic spelling of Minor's last name is not unusual for this period.

The horn is scalloped 5½ inches from the tip. The throat is faceted octagonally, and has three rings spaced along it. The 3-inch-long wooden stopper has a notched finial holed for a guard thong. The plug is neatly rounded, almost flush with the end of the horn, and is secured with four iron tacks. An iron rivet is driven through the side of the butt into the upper end of the plug ⅛ inch behind the edge and is secured with a square iron nut. The rivet is bent at a right angle as it enters the horn so that it can penetrate the plug. The upright portion of the rivet is 1⅜ inches high, with an ⅛-inch slot for securing the carrying strap. Undoubtedly only a metalworking artisan would have contrived such an elaborate arrangement.

BIBLIOGRAPHY: Theodore Offerman sale, American Art Association/Anderson Galleries, New York, November 11–13, 1937; James W. Flanagan sale, Parke-Bernet Galleries, New York, April 21–22, 1944; Grancsay, No. 588, 61; Crosby Milliman, "An Exhibition of American Engraved Powder Horns of the Colonial and Revolutionary Periods," *Bulletin*, Fort Ticonderoga Museum, Vol. 12, No. 3 (October, 1967), 185; *Antiques*, 1978; Guthman, *Guns and Other Arms*, 148; *Bulletin*, ASAC, March, 1981, No. 44, 47.

Catalogue

52

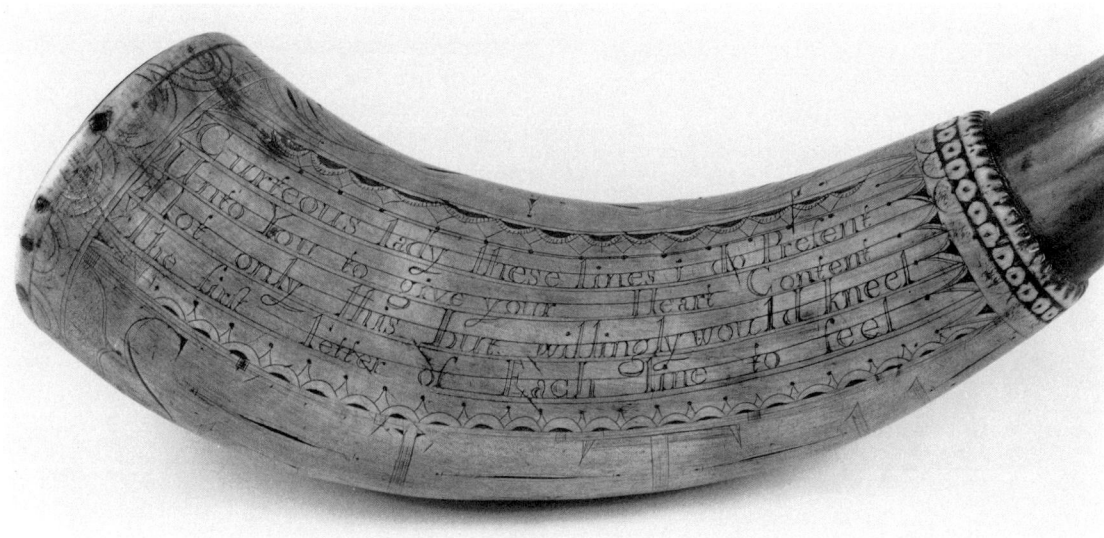

53

LAKE GEORGE SCHOOL
Attributed to Richardson Minor (1736–1797)

53 *Stephen Tambling horn*

Crown Point, New York, August 4, 1761
Horn, pine, iron, black pigment
OL: 15½ in. (39.4 cm), w (of plug): 3 in. (7.6 cm)

Inscriptions: Stephen:Tambling / His Horn Dat'd Crown Point Aug.ᵗ 4ᵗʰ. 1761 / Curteous Lady these lines I do present / Unto You to give your Heart Content / Not only this but willingly would kneel / The first letter of Each line to feel

James E. Routh, Jr.

Stephen Tambling was born in Windham, Connecticut, on April 1, 1738. He is listed during the 1758 campaign as a private in Captain Joseph Canfield's 11th Company, Colonel David Wooster's 4th Connecticut Regiment. No other record of his service survives.

The attribution to Minor (No. 52) arises from the almost identical style and quality of the calligraphy of this horn and Minor's own, as well as the very similar incised floral design of stylized tulips and geometric designs—resemblances first called to my attention by

Jim Routh. As with Minor's own horn, the work here is like that of J. W., but the carving and surface engraving are on a more accomplished level. The design motifs of the Minor horns resemble the geometric device J. W. used as a signature, but their enhancement with dark pigment makes them more vivid.

Above the verse is a deeply incised lunette border, with a smaller version below. A coggled border, also deeply incised, precedes the scalloping at the start of the recessed portion 4¼ inches from the tip. There is a double raised ring 1 inch from the tip. Lightly engraved lunettes encircle the plug end and the slightly rounded plug, secured by seven round wooden pegs, extends ⅛ inch beyond the horn. A 1¼-inch iron staple is driven into the center of the plug to secure the carrying strap.

BIBLIOGRAPHY: Grancsay, No. 832, 69.

ATTRIBUTED TO THE LAKE GEORGE SCHOOL
Possibly by J. W. (w. 1756–1761)

54 *George Batterson horn*

Lake George, New York, October 17, 1758
Horn, pine, red sealing wax, cork, iron, wool yarn
OL: 16½ in. (41.9 cm), W (of plug): 3¼ in. (8.3 cm)
Inscriptions: George Batterson / His horn Made at lake gorge Octr Ye 17 AD 1758
The Connecticut Historical Society, Gift of Charles S. Bissell

No record survives of this man's service in the French and Indian War, although a George Batterson of Fairfield, Connecticut, is listed in Revolutionary War rolls.

The calligraphy is so similar to that of J. W. that an attribution to him is justified. While birds and an abstract lake are not found on any other J. W. horn, these are palpably in his manner. The flowing floral designs are heavily dotted in the manner of the Lyme, or Miller-Tribble, Carver (Nos. 44 and 45).

The marvelous soldiers and Indians fighting among trees along the shore of Lake George impart a graphic idea of the tactics used in frontier warfare. The tiny combatants, wielding pistols, muskets, swords, and tomahawks, resemble scenes in Samuel Blodget's perspective plan of the Battle of Lake George, published in 1755. Although the people are not so expertly executed as those on horns by Jacob Gay or John Bush, they are clearly differentiated into Indians, Rangers, and provincial light infantry. The soldier in a mitre cap wielding a spike tomahawk and a pistol is quite unusual because light infantry almost never used pistols.

A neatly carved scalloped and dotted edge decorates the beginning of the recessed portion 5⅞ inches from the tip of the horn, a style of scalloping associated with J. W.'s horns. There is a raised ring 1⅝ inches from the tip of the spout, which is octagonal. The cork stopper may be original. There is no border at the plug end; the plug is rounded and has a petallike

Catalogue

54. *A vividly descriptive scene of Woodland warfare: the combatants, peering from behind trees and bushes, brandish swords, tomahawks, muskets, and pistols.*

54. *Two soldiers confront one another with swords and pistols amidst heavily dotted floral and plant motifs.*

raised knob at the center which was at one time covered with red sealing wax. The plug, secured by thirteen round wooden pegs, has a wrought-iron nail driven into it to hold the carrying strap. Traces of red yarn adhere to the nail.

55

LAKE GEORGE SCHOOL

55 *Lieutenant Christopher Palmer horn*

Fort Edward, New York, October 24, 1758
Horn, pine, brass, iron
OL: 12¾ in. (32.4 cm), W (of plug): 3 in. (7.6 cm)
Inscriptions: Lieu^t Christopher Palmer / HIS HORN MAD^e at fort edward october ye 24th 1758
William H. Guthman

Christopher Palmer of Stonington, Connecticut, was a second lieutenant during the 1758 campaign in Captain John Denison's 12th Company, Colonel Eleazor Fitch's 3rd Connecticut Regiment. He had also served during the 1755 campaign as an ensign in Colonel Eliphalet Dyer's 1st Company, Colonel Eliphalet Dyer's 3rd Connecticut Regiment, and in the 1756 campaign as adjutant in Colonel David Wooster's 2nd Connecticut Regiment.

This horn was carved by a skillful engraver who used bold, sophisticated calligraphy for the owner's name. It is not professional quality engraving, but the bold flourishes suggest a keen desire to employ a stylish and up-to-date manner. The lettering of the name is similar to that on the William Patterson horn (No. 61). The balance of the lettering includes many Lake George School effects, including the winged crossbar in the two "H's" and the "A," the copperplate lettering of the second line, and the block lettering of the third line. The decorative floral design of stylized tulips is of high quality and the "big fish, little fish" motif is extremely humorous. A smaller calligraphic device is the copperplate scroll connecting the stems of the stylized tulips.

The horn is recessed 4½ inches from the tip of the spout, but the actual tip is hidden under a tapered brass ferrule 1¾ inches long, which appears to be an early addition made from the ramrod pipe of a musket. The usual scalloping preceding the recessed area is absent and in its place is a wriggly line incised between two straight lines. The recessed area ends with a turned ring that provides an unusually wide area for the carrying strap. The plug end is decorated with a simple border consisting of two straight lines ¼ inch apart between which are groups of five closely spaced hash marks. The flat plug is secured by two wooden pins. In view of its weakness, the tack with a brass head and an iron shank is probably not the first attachment device on the plug.

BIBLIOGRAPHY: *Antiques*, 1978, 328; Swayze, 65–66; Guthman, *Guns and Other Arms*, 150; *Bulletin*, KRA, Fall, 1985, 8; Guthman, *U. S. Army Weapons, 1784–1791*, 91.

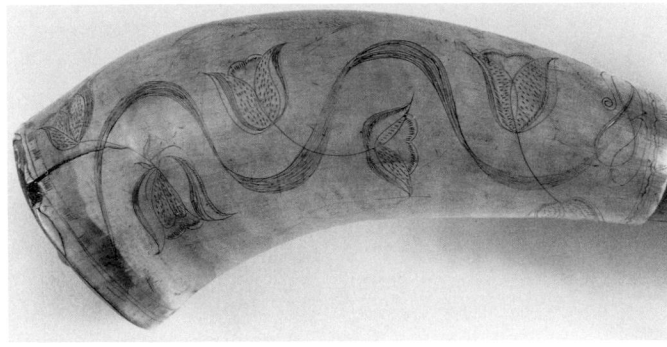

55. *Boldly delineated tulips are linked by a shaded scrolling line in the free-flowing manner of the Lake George School.*

56

56. *The word* WAR, *so often found on Lake George School horns, in a simplified version.*

LAKE GEORGE SCHOOL

56 *Lieutenant Joseph Smith horn*

Lake George, New York, September 25, 1758
Horn, wood, black pigment
OL: 15¼ in. (38.7 cm), W (of plug): 3 in. (7.6 cm)
Inscriptions: a Cage for a pare^t / Lieu^t × Joseph × Smiths Horn / me^{a}de at lake gorge Sep^{te} ^{Y} 25 A^D 1758 W A R
William H. Guthman

Joseph Smith is listed as a second lieutenant in Captain Sommersbee Gilman's 4th Company, Colonel John Hart's New Hampshire Regiment. A portion of the regiment under Hart's command was detached for the second Louisburg Expedition, during which Hart died. The balance of the regiment, under the command of Lieutenant Colonel John Goffe, joined the expedition against Crown Point. There is no other record of Smith's service.

The Smith horn combines features of several Lake George School carvers. The calligraphy of the name, place, and date is similar to that of J. W., with winglike serifs on some of the letters. The motto WAR, found on horns engraved by the Selkrig-Page Carver, is not so stylized as that master's mottoes. And instead of an incised design above and below the motto, there is a simple cartouche composed of a tightly drawn, wavy line. The letters of the word WAR are separated by swords, and below is a floral design reminiscent of that on the Aaron Page horn (No. 25). Two winglike devices and two small round prickly objects appear

143

above the swords between the letters of WAR. The spelling of the word *mede* by the addition of a superscript "a" is distinctive.

The symbolism of the drawings above Smith's name is not clear. A pine tree and a caged parrot with a separate caption in a cartouche are followed by a floral design. This same parrot design appears on other Lake George School horns. Another, later, hand crudely incised a sailing ship and the letter *M* on the reverse side of the horn.

The edge of the butt end of the horn is neatly and closely notched, creating the only border. The black-painted wooden plug held by four wooden pegs is slightly rounded and protrudes about $1/16$ inch. A hole in the center of the plug contains the stump of a wooden knob. The throat is recessed $4\frac{1}{2}$ inches from the tip and is notched like the butt end. Three neatly carved raised rings $1\frac{1}{2}$ inches from the tip of the spout secured the carrying strap.

BIBLIOGRAPHY: *Antiques*, 1978, 330; Guthman, *Guns and Other Arms*, 152; *Bulletin*, KRA, Fall, 1985, 10; *Bulletin*, ASAC, March, 1981, 49.

LAKE GEORGE SCHOOL

57 *John Stiles horn*

Lake George, New York, October 17, 1758
Horn, pine
OL: 15 in. (38 cm), W (of plug): $2\frac{1}{2}$ in. (6.4 cm)
Inscriptions: john Stiles 1758 / John Stile^s : Ho^{r}n A 1758 / Made At Lake / Geo^{r}ge Oc^t : 17.
William H. Guthman

John Stiles is recorded as serving from September 11 to December 8 of 1755 in Colonel Eliphalet Dyer's 1st Company (Windham), Colonel Eliphalet Dyer's 3rd Connecticut Regiment. Christopher Palmer (No. 55) served as ensign in the same company. From March 29 to November 17 of 1758, Stiles served in Colonel Nathan Whiting's (No. 12) 1st Company (New Haven), Colonel Nathan Whiting's 2nd Connecticut Regiment. Nathaniel Selkrig (No. 24) and Aaron Page (No. 25) also served in this company.

The work on this horn resembles that of several known carvers, but cannot be attributed to any one of them. The calligraphy is good but not of the highest quality: some of the letters were left out and had to be added in superscript and the words are staggered to permit pleasant floral and geometric designs to become part of the inscription. The stylized flowery vines including tulips and winglike devices are similar to those of J. W. An abstract British coat-of-arms has a moon-shaped face peering out of the cartouche, and the arms are supported by an animated lion and unicorn. A profile head and shoulder with horns represents either an Indian or a cuckold, in the manner of the Memento Mori Carver. The idea is that while the soldier is on duty at a frontier fort, his wife is being unfaithful at home. The view of an unidentified fort may be Fort Edward or Fort Ticonderoga. Several houses on the bank of what looks like a river running past the fort suggest Fort Edward.

A nicely scalloped edge precedes the recessed portion $4\frac{1}{4}$ inches from the tip. There is a well executed octagonal double-raised ring $2\frac{1}{4}$ inches from the tip, and the remainder of the spout is octagonal. A simple border precedes the end of the butt, where the flat

57

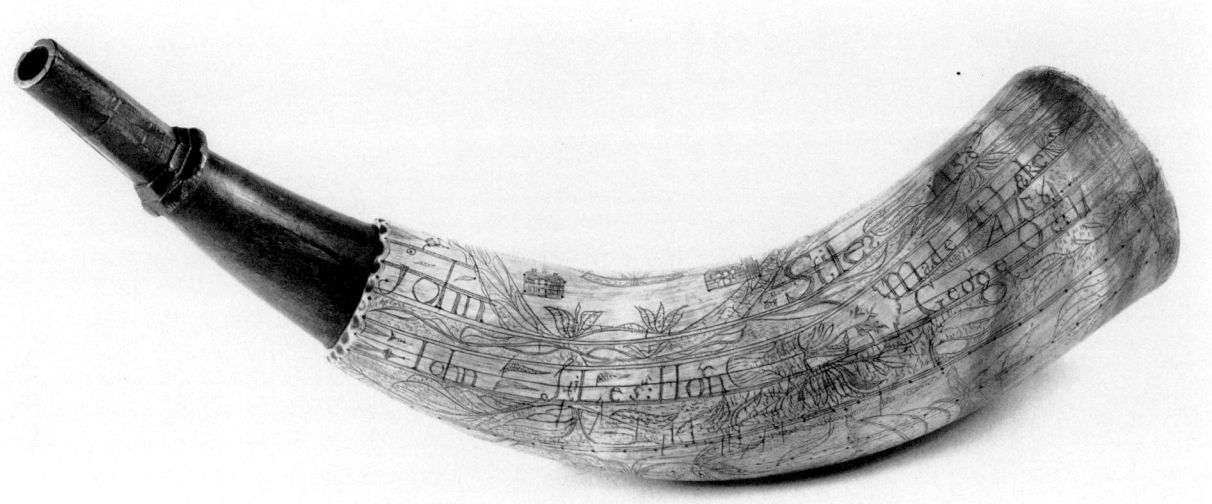

Catalogue

57. *Three vignettes seen on many Lake George horns: a cuckold, a stylized British coat-of-arms, and a view of a fort.*

pine plug is held by friction alone. It appears to have been flat when it was made, because it bears an incised sunburst. A hole in the center of the plug held a nail for securing the carrying strap.

This horn was painted many years after it was made. A later owner then steamed it to remove the paint (not a recommended treatment) and rubbed shoe polish on the surface (also not recommended).

LAKE GEORGE SCHOOL

58 *John Miles horn*

No location cited, May 5, 1759
Horn, pine, iron, brown paint
OL: 15½ in. (39.4 cm), w (of plug): 3 in. (7.6 cm)
Inscriptions: John Miles / his Horn: May 5th, / 1759
James E. Dresslar

John Miles (1738–1815), a cabinetmaker of Milford, Connecticut, is listed during the 1759 campaign as a private in Captain Joel Clark's 4th Company (Southington), Colonel Nathan Whiting's (No. 12) 2nd Connecticut Regiment. During the 1762 campaign he served as sergeant in Captain Eldad Lewis's 7th Company (Southington), Colonel Nathan Whiting's 2nd Connecticut Regiment.

The handsome calligraphy on this horn reflects the better style of the Lake George School with thick-and-thin copperplate letters, shading, and deeply cut serifs. The calligraphy is similar to that of the Spencer-Hitchcock Carver (Nos. 46–48) and to that of J. W. The beauty of this horn resides in its simplicity: the carving is fine but not ornate; there is an interesting geometric device at the end of the inscription; and a scalloped edge at the border of the recessed portion is modestly emphasized with fine incised lines. The border is 6 inches from the tip of the spout,

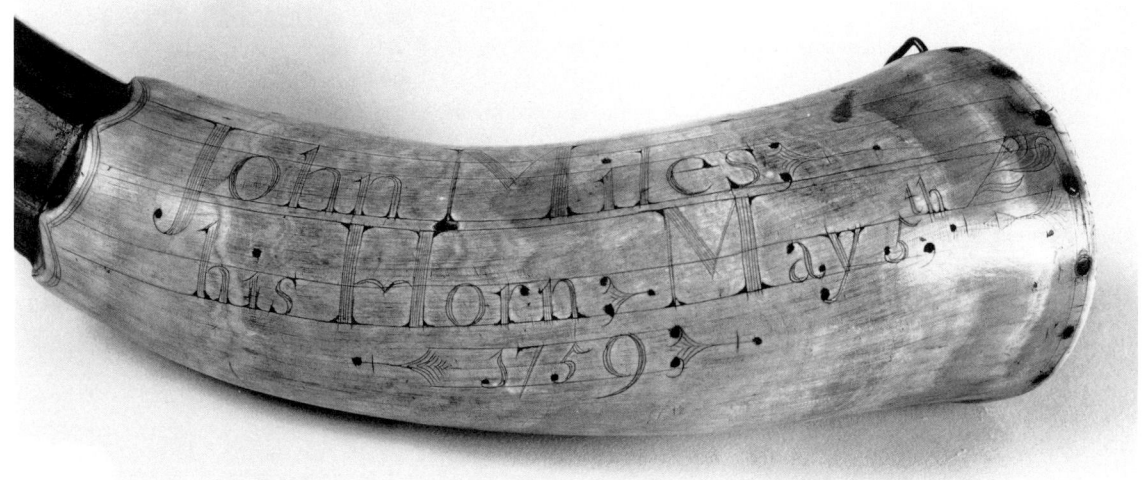

58

which is septagonal in section and has a septagonal raised ring 2 inches from the tip. A neat double-line border surrounds the butt end. The slightly rounded plug retains traces of brown paint, extends ⅜ inch beyond the horn, and is secured by ten neatly cut wooden pegs. An iron staple is driven into the plug through the top edge of the horn.

LAKE GEORGE SCHOOL

59 *Captain Nathaniel Porter horn*

Crown Point, New York, September 17, 1759
Horn, pine, varnish, hemp
OL: 16 in. (40.6 cm), W (of plug): 2¾ in. (7 cm)
Inscriptions: Captn Nathaniel Porter His Horn / Made at Crownpoint September ^{ye} 17 1759
William H. Guthman

During the 1755 campaign Nathaniel Porter (born 1727) of Lebanon, Connecticut, served as lieutenant in Lieutenant Colonel Nathan Whiting's (No. 12) 2nd Company, Colonel Elizur Goodrich's 2nd Connecticut Regiment. It was Colonel Whiting's company that turned the tide of the Battle of Lake George during the French ambush at Rocky Gulch. When his enlistment expired, Porter reenlisted for the period November 25, 1755, to April 26, 1756, as lieutenant in Captain Israel Putnam's (No. 15) company, which was garrisoned at Fort Edward. Putnam, along with the famous Robert Rogers (No. B), was performing the duties of a scout, a category later named "Ranger."

During the 1756 campaign, Porter became captain of the 4th Company, Colonel David Wooster's 2nd Connecticut Regiment. Christopher Palmer (No. 55) was adjutant of this regiment. Porter is not listed in the official rolls for later campaigns, but his own account books show that he was active as a captain during the 1758 campaign, and his horn attests to his duty at the taking of Crown Point during the 1759 campaign.

This horn has an interesting history: it was listed in the accessions records of the Essex Institute of Salem, Massachusetts, as having the inscription *Capt. Nathaniel de Venter His Horn*, and Rufus Grider drew a sketch of the horn in 1890 (New-York Historical Society) using that name. Stephen Grancsay later listed the horn in his book in the same manner. When the Essex Institute deaccessioned the horn in the 1950s, historian Harold Peterson acquired it and illustrated it in his *Arms and Armor in Colonial America*, using the same inscription. Grider and Grancsay also cited the date as 1758, but Peterson corrected it to 1759. Under adequate light and with the aid of a strong magnifying glass, the name Captain Nathaniel Porter is clearly visible. Why the wrong name was

recorded in the first place and repeated in the literature is a mystery.

The carver of the Porter horn incorporated the styles of several of the better Lake George School artists. The heavy dotting accenting geometric and foliate designs recalls the work of the Miller-Tribble Carver (Nos. 44 and 45). The neatly incised zigzag border above the owner's name is found on horns by John Bush, and the calligraphy is similar to that of Bush, J. W., and the Selkrig-Page Carver.

The abstract British coat-of-arms, the marching soldiers, and the mounted officer are unlike any of the formations depicted by other carvers. Indeed, mounted soldiers, like those on the Hobart Spencer horn (No. 47), are rare. The Porter horn also has three different sailing ships, a fort (probably Fort Edward), and an Indian peering out from behind a flower.

The horn is nicely scalloped where the recessed portion begins 5 inches from the tip, and has a simple triangular-line border at the edge of the scalloping. There are two raised rings ½ inch apart 3 inches from the tip, and the spout is faceted from the second ring to the tip. The flat pine plug, which may have been trimmed at some point, is secured by seven small wooden pegs. The butt is also decorated with a triangular border. Two holes ¼ inch apart and ⅛ inch from the edge of the butt end extend through the plug for securing the carrying strap. An old hemp-rope fragment remains between the double rings on the spout. A small horn measure that was attached to the other end of the rope when the horn was being offered for sale has disappeared.

BIBLIOGRAPHY: Grider, NYHS; Grancsay, 50; Peterson, *Arms and Armor*, 238; *Bulletin*, ASAC, March, 1981, 37–38.

LAKE GEORGE SCHOOL

60 *James Meldrum horn*

Crown Point, New York, November 17, 1759
Horn
OL: 17 in. (43 cm), W (of plug): 3¼ in. (8.3 cm)
Inscriptions: JAMES MELDRUM IN THE 42ᵈ OR ROY–¹ HIGHLAND REG–ᵀ HIS POWDER HORN / MADE AT CROWN POINT NOVEMBER YE 17 AND 1759 / A TROOP OF HORSES / ROYAL HELENDERS / NEW YORK, SCNECCIDY, HALF MOOD, STILL/WATERS, SARATOGA, FORT EDWARD / SOUTHBAY, TICONDEROGA, CROWN POINT AND LAKE CHAMPLAIN / NEW HEAVENE TOWN / ADAM/EVE / HOI SOIT QUI MAL YIPENSE / DIET ET MON DROIT
Yale University Art Gallery, on permanent loan from University Archives

Meldrum was an enlisted man in the British 42nd or Royal Highland Regiment, which served in America from 1756 to 1767 and again during the Revolution. This is the only known horn by one of the better Lake George School carvers that was made for a British regular. The hand is quite close to that of the maker

60

60. *An extremely rare view of New Haven, Connecticut.*

of the Hobart Spencer horn (No. 47), which was also carved at Crown Point, sixteen days before this one.

The work on this horn is good but not of the best quality. The ornate and unusual map, decorated with compasses at both top and bottom, is difficult to follow because it turns back upon itself. All the soldiers and the figures of Adam and Eve have amusing expressions, but are not so animated as Jacob Gay's. The vignette of Adam picking an apple from the Tree of Knowledge as Eve sits with another apple and the Serpent slithers around the tree trunk has a good deal of visual appeal, as do the soldiers. The addition of a sketch of New Haven Green, and of floral designs, vines, and patterned trees makes for a densely decorated horn.

The lion and the unicorn on the elaborately carved coat-of-arms are cartoonlike, but the arms are well done. Save for Crown Point and Fort Edward, forts are not depicted in detail. There is a vignette of a portion of a palisade with five defenders peering over the top with their muskets pointed at the enemy.

A simple looped border precedes the recessed portion 4½ inches from the tip. There is a raised ring 2½ inches from the tip, which has a raised lip. The plug is missing, the end of the horn is badly chipped, and the entire horn has suffered seriously from insect infestation.

BIBLIOGRAPHY: Grancsay, No. 578, 61; George Dudley Seymour, "Henry Caner, 1680–1731, Builder of the First Yale Building," *Old-Time New England*, 15 (1924–25), 99–124.

60. *A unique and detailed view of the 42nd Royal Highlands Regiment's uniforms.*

Catalogue

61

LAKE GEORGE SCHOOL

61 *William Patterson horn*

Probably Lake George, New York, 1759
Horn, wood, black paint, iron
OL: 15½ in. (39.4 cm), w (of plug): 3¼ in. (8.3 cm)
Inscriptions: William / Paterson / 1759
William H. Guthman

William Patterson, Jr. (1729/30–1761), of Stratford, Connecticut, served in Captain Thomas Hobby's 4th Company, Colonel David Wooster's 3rd Connecticut Regiment, during the 1759 campaign. Robert Baird, whose horn is signed by J. W. (No. 32), also served in this company, and the two horns share many stylistic features, although the Patterson horn is even more closely related to the Christopher Palmer example (No. 55). During the 1758 campaign, Patterson served in Captain Nathaniel Evert's 12th Company, Colonel Daniel Wooster's 4th Connecticut Regiment, and during the 1761 campaign he served as a sergeant in Colonel Nathan Whiting's (No. 12) 2nd Connecticut Regiment. Three men who owned horns made by J. W. also served in that regiment.

This horn's lettering displays the free-flowing copperplate script and bold calligraphic ornament seen on the Christopher Palmer horn. The Patterson horn also has a letter "S" ornamented with the profile head of an Indian attached to a finial and an Indian head as part of an illuminated letter "W," motifs that both Jacob Gay and the Memento Mori Carver incorporated into their lettering. Gay often attached profile heads to crossbars and serifs, while the Memento Mori Carver used them for borders.

A motif that appears on both the Patterson horn and examples by J. W. is the stylized calligraphic device found near the spout, adjacent to the border. This device is closely related to that J. W. used almost as a signature. The carver of the Patterson horn embellished it with other geometric designs like those incorporated in the sailing ship and the cartouche bordering it. The fish swimming beneath the ship are quite like those on the Palmer horn.

A crudely engraved border consisting of a series of zigzag lines interspersed with dots precedes the sharply scalloped border at the start of the recessed portion, which begins 6 inches from the tip and is broken only by a raised ring 1½ inches from the tip. The wooden plug is neatly rounded, painted black, and secured to the horn by eleven wooden pegs. The plug extends ⅜ inch beyond the horn. A crude border of two double lines ¼ inch apart with wavelike half circles joined into a continuous line decorates the butt end. A small iron staple driven into the center of the plug secured the carrying strap.

BIBLIOGRAPHY: *Antiques*, 1978, 329; Guthman, *Guns and Other Arms*, 151.

62

LAKE GEORGE SCHOOL

62 *Alexander Oviatt horn*

Crown Point, New York, September 25, 1759
Horn, pine, iron
OL: 15½ in. (39.4 cm), w (of plug): 3¼ in. (8.3 cm)
Inscriptions: Alexander * Oviatt * His * Ho^{r}n: / Made+at Crownpoint * Sep$^{t the}$ 25 1759 / the horn is don So Make no More Fon
Private Collection

Alexander Oviatt, Jr. (1737/8–1810), of Milford, Connecticut, served during the 1759 campaign in Major David Baldwin's 3rd Company (Milford), Colonel Nathan Whiting's (No. 12) 2nd Connecticut Regiment. In 1756 Baldwin, then a captain, had acquired a horn carved by John Bush (No. 14). Also serving in Baldwin's company during the 1759 campaign was Corporal Nathaniel Selkrig, whose horn had been carved in the John Bush style in 1758 by the Selkrig-Page Carver (No. 24). Oviatt is also listed as serving in the 1761 campaign in Major David Baldwin's 3rd Company, Colonel Nathan Whiting's 2nd Connecticut Regiment.

Oviatt's horn is similar in style to horns by J. W., but the lettering is definitely by a different hand. Although the loose vinelike decoration is close to J. W.'s, the neat floral borders at the throat and plug ends are not found on J. W. horns. The combination geometric and floral device prominently displayed on the Oviatt horn, however, strongly resembles J. W.'s signature device, as does a smaller floral design.

The lettering is carefully planned but not as expertly executed as J. W.'s: in two instances the carver did not allot sufficient space for his inscription, as may be seen above, and his letters are shaded, but not expertly. The rhyme *the horn is don So Make no More Fon* is crudely incised and unshaded, but it is plotted between three layout lines and is not, therefore, a contemporary or later addition. A crude engraving of a man firing a musket at a large letter "T" is a later addition, however.

The horn is neatly scalloped 5½ inches from the tip. An engraved floral border circles the throat just before the scalloping, and the recessed portion extends from just beyond the scalloping 4¾ inches to the tip. There are raised rings 1⅞ inches from the tip and at the tip. The flat pine plug was secured by nine wooden pins, three of which are missing; an iron nail has supplemented the pins. The plug end has a ¼-inch floral border identical to that at the throat and the wooden pins are driven through this border. Several holes in the plug indicate the previous locations of carrying-strap attachments.

BIBLIOGRAPHY: *Bulletin*, Fort Ticonderoga Museum, 1967, 186; Swayze, 74–76.

63

LAKE GEORGE SCHOOL

63 *John Vaughan horn*

Probably Lake George or the St. Lawrence River Valley, September 20, 1764

Horn, pine, iron

OL: 16¾ in. (42.5 cm), w (of plug): 2⅞ in. (7.3 cm)

Inscriptions: I Powder:With:my Brother:Ball / A Heroe: Like do:Conquor: all / John Vaughan:His:Horn:Made: Sep^r :20th 1764 / Steal:not:this:Horn:by:Day:nor Night / For:the:owners:name:Stands:fare:in:Sight / The Devil / ile have / one of them / Gard your head / I do my lad / J G Hyde

Private Collection

Rufus Grider sketched this horn in 1887, when it was in the possession of the Oneida Historical Society in Utica, New York. Grider made a notation on the sketch saying that during the Revolution John Vaughan was a sergeant in Colonel Van Ness's New York Regiment. Colonel Peter Van Ness did command the 9th Regiment of the Albany County Militia, but although there are an Edward Vawn and a Richard Vawn, there is no John Vaughan listed on the rolls.

From 1763 to 1767, a British lieutenant colonel named John Vaughan was stationed in North America as part of the British 46th Regiment. Prior to that, from 1759 to 1761, Vaughn had served in North America as commander of the Royal Welsh Volunteers, a light infantry regiment. Conceivably this horn may have been made for him, as his regiment was serving at Fort Ticonderoga and the Lake George area in 1763.

With his drawing, Grider included a poem written by General C. W. Darling, Corresponding Secretary of the Oneida Historical Society. Dated December 26, 1887, the poem was presumably composed for the drawing:

Behold this ancient powder horn
The owners name was John Vaughan
Who carved upon its oval face
The record of a by gone race
In Seventeen Sixty five A.D.
This horn was found within a tree
Placed there by one who lost his life
By Mohawk gleaming scalping knife

Among the decorative devices on the horn are a spike tomahawk, a ball-headed war club with spike, and a bow and arrow. There are also two swords, a bayonet, a shovel, a felling ax, a drum and drumsticks, and a Union Jack. The weapons and accoutrements are all appropriate to the frontier wars.

The calligraphy is adequately executed, although much of the spelling is phonetic, and the decoration is imaginative, if difficult to interpret. The curious central motif shows a dueling scene: two swordsmen face each other, while the Devil stands to one side saying "ile have one of them." To the right of the duelers is the British flag. There are also a large batlike creature and a man, who appear to be engaged in a sexual act.

A lightly incised border precedes the scalloping at the beginning of the recessed portion 4 inches from the tip. There is a raised carved ring 2¼ inches from

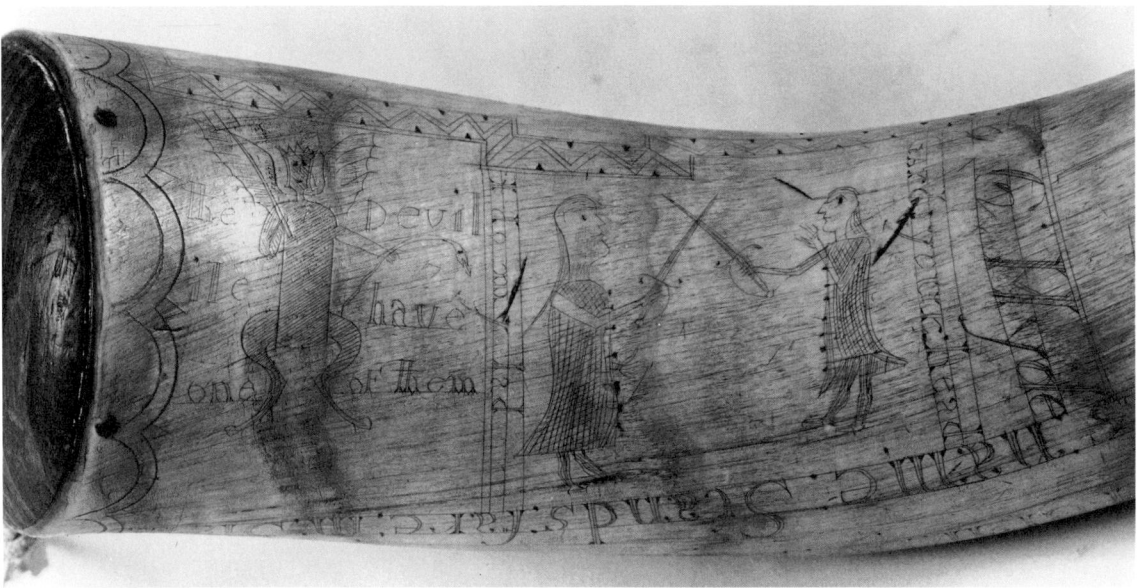

63. A most unusual scene showing a duel, with the Devil waiting to claim the loser.

the tip, which has a raised lip. A double-lunette border surrounds the plug end and the rounded plug, extending ¼ inch beyond the horn, is secured by five wooden pegs. A 1-inch iron staple is driven into the center of the plug, and initials that are too worn to read are carved into it.

BIBLIOGRAPHY: Grider, NYHS; W. M. Beauchamp, "Rhymes from Old Powder-Horns," *Journal of American Folk-Lore*, 2 (1889), 117–122, and 5 (1892), 284–290 (the publisher of the JAF made fifty copies of the Vaughan horn illustration, as well as of two other horn drawings that Grider had made, and gave them to Grider to distribute to friends. One of these prints, with pencilled notes by Grider, is in the writer's possession); Stewart Culin, "Powder Horns of Men Who Made Our History," *The Philadelphia Press*, Sunday, February 20, 1898, 33; Gilbert Thompson, "Historical Military Powder Horns," *The Society of Colonial Wars of the District of Columbia, Historical Papers*, No. 3 (1901), reprinted in *The American Monthly Magazine*, 20 (1902), 1005–1028, and also in *The Journal of the Military Service Institution of the United States*, 33 (1903), 248–263; Swayze, 89–90.

LAKE GEORGE SCHOOL

64 Giles Barns horn

New Haven, Connecticut, January 27, 1766
Horn, pine, brass, iron, varnish
OL: 15½ in. (39.4 cm), W (of plug): 3¼ in. (8.3 cm)
Inscriptions: Giles Barns / his Horn Made In / New Haven ★This / 27th Anno ue Domini:1766 / Day of January And In The fifth year of His / Majestys Reign / R H / The Brigg Lively / Of Sixteen Guns / Commanded by / Capt Anby / The Kings Coat of Arms / THE CONNECTICUT COAAND / DIEU ET MON [DROIT]
William H. Guthman

Giles Barns is unidentified, but may have been a sailor on the brig *Lively*. The style of the carving is related to that of Lake George horns of a few years earlier; it resembles the style of J. W., as well as of the Lieutenant Joseph Smith horn (No. 56).

The horn has many unique features. The raised plaque at the spout end with the slogan *The Kings Coat of Arms* depicts the lion and unicorn supporting the Connecticut arms. In this somewhat abstract depiction, the "Droit" from "Dieu et mon Droit" is missing and the three vines of Connecticut are incongruously placed within the shield. The calligraphy of the name, place, and date are in a competent copperplate style with featherlike serifs on some of the letters. The words *And In The fifth year of His Majestys Reign* are engraved in exceedingly fine script. The scene of New Haven Green is quite competent and extremely rare (the only other view of this subject known to me appears on the Meldrum horn, No. 60). The engrav-

Catalogue

64

ing of the ship, although worn, is well executed and shows eight cannon on the port side.

There is a professionally carved floral rococo border above the name, as well as next to the scalloped border at the beginning of the recessed portion 4½ inches from the tip. The raised plaque, which apparently served as a stop for the carrying strap, is spaced 1 inch from the scalloping, extends toward the spout 1⅞ inches, and is 3½ inches across at its widest point. A brass charger with an iron spring and an iron screw fits over the spout, covering ½ inch of it and extending beyond it 1¼ inch; this is undoubtedly a nineteenth-century addition. There is no border at the plug end. Two wooden pegs securing the plug are visible, with the rest hidden by nine brass nails that circle the butt. The brass nails were probably added along with the nineteenth-century charger. The flat plug has a modern screw eye in the center and an older deep hole ½ inch away from the screw eye.

BIBLIOGRAPHY: Lindsay, *The New England Gun*, 132–134; *Antiques*, 1978, 331; Guthman, *Guns and Other Arms*, 153; Swayze, 97–98.

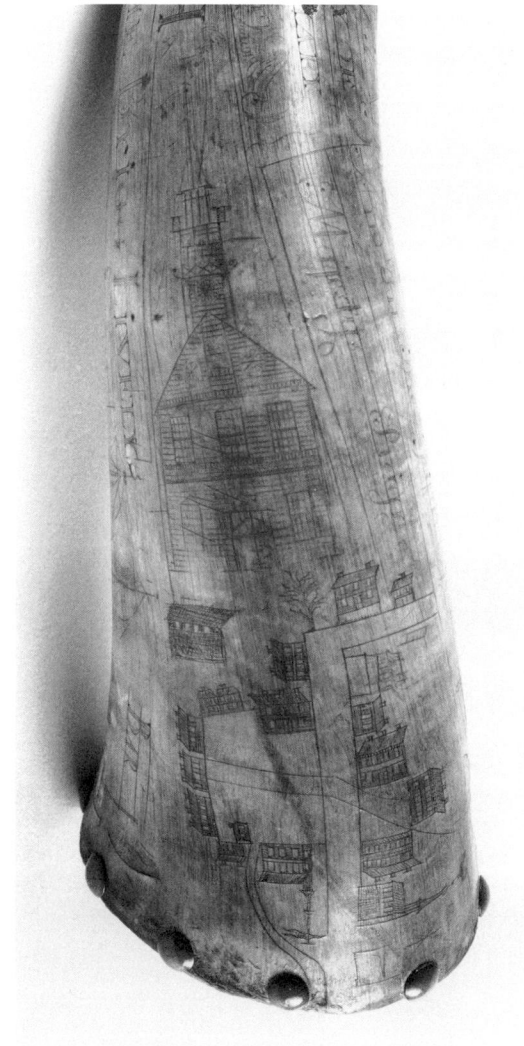

64. Another rare view of New Haven Green.

PRE-SIEGE OF BOSTON SCHOOL
Possibly by Nicholas Edgecomb Pickett
(1737–1809)

65 *Nicholas Edgecomb Pickett horn*

Probably Marblehead, Massachusetts, September 12, 1766
Horn, cherry, iron, pewter, black pigment
OL: 18½ in. (47 cm), w (of plug): 3 in. (7.6 cm)
Inscriptions: Nicholas:Edgcomb:picket: Septmbar: Ye 12: 1766 / Mary:picket: Nicholas:picket:John:picket
Private Collection

Nicholas Edgecomb Pickett was a merchant of Marblehead, Massachusetts, who served in Colonel John Glover's Massachusetts Regiment at Cambridge in 1775 and 1776. Of Pickett's five children, three were born before this horn was made: Mary (May 31, 1761), Nicholas (September 5, 1762), and John (September 7, 1766), and undoubtedly these are the names and faces shown on the horn. Each face is perfectly round, as if drawn with a compass. The two with borders at their outer edges are wearing crowns and are presumably the two sons, Nicholas and John. The third, uncrowned, face representing Mary has long hair within the outer circle instead of an abstract border. Mary is wearing beads, while the two sons wear gorgets. The artist has shaded the shoulders and chests of the figures to suggest garments and has emphasized each figure with dark pigment.

Pickett's name and the date are contained within a simple cartouche, above which are two vignettes. The first depicts the full-length figure of a girl holding onto a half-length boy who is double her size. The second shows a full-length woman holding a man or boy by the hand, with her other hand on her hip. The first vignette probably represents Mary holding her brother Nicholas, and the second very likely symbolizes their mother, Mary Green Pickett, restraining the youngest child, John. At the plug end are a crudely engraved hunter with a musket and dog representing Pickett himself and suggesting that this is a hunting horn.

The children's names are contained in a line cartouche with a wriggle-work border, below which is an impressive view showing a house, a church, and a fort, all probably derived from actual views of Marblehead and the harbor; this is supplemented by a profile view of the fort. Five different sailing vessels with accurate rigging, some vines, and a number of animals are also shown, some of which may have been inspired by magazine illustrations. The animals, which have animated expressions, include lions, camels, elk, a boar, and a zebra carrying a mounted figure.

This horn, by a capable but by no means profes-

65

sional carver, is delightful because of its varied motifs. Its calligraphy is adequate—the script lettering is evenly spaced and well planned—and it is difficult to resist the conclusion that Pickett engraved the horn himself.

There is no border or recessed portion, but there is a double raised ring 2¼ inches from the tip. An abstract vine design and a single tree are placed in front of a geometric border at the butt end. Three of five original wooden pegs securing the plug are missing and two iron nails have been added. A ¼-inch

Catalogue

pewter band binds the butt end where the horn has split and a 6-inch crack runs from the end of the horn to the word *Septmbar*. The rounded cherry plug extends ¾ inch beyond the horn and has a 1-inch iron staple in its center.

BIBLIOGRAPHY: Swayze, 99–101.

PRE-SIEGE OF BOSTON SCHOOL
Attributed to Hugh Tolford (w. 1770–1772)

66 *Tilton Bennet horn*

Chester, New Hampshire, December 31, 1770
Horn, pine, iron
OL: 12½ in. (31.8 cm), W (of plug): 2⅝ in. (6.8 cm)
Inscriptions: TILTON:BENNET:his:horn: / MADE: IN : CHESTER . DECEMBER. / BY, ME . H. T. LET . NONE . ROB . IT. 31 : 1770 / T. B. DEc^R. 1770
William H. Guthman

No published military record for Tilton Bennet exists but he was probably a member of a local militia company, which would have been the reason for the carving of this horn.

The collection of Carol and the late Thomas Segal contains an almost identical horn engraved TITUS CAVE 1772 (No. 67); Thomas Holt's horn (No. 68) is also by the same hand, that of Hugh Tolford. The workmanship of these horns is quite close to that of Jacob Gay, and they might be mistaken for Gay's work if they were not signed by H. T. Tolford's own horn is illustrated in John duMont's book (p. 70, pl. 96). Robert F. Trent discovered that Tolford's horn, dated

66. *This carver copied Jacob Gay's engraving style but supplied his own imaginative characters, such as the violinist and the birds atop various perches.*

66. *A horse-drawn chariot—another uncommon subject.*

at Chester on November 14, 1770, is in a hand identical to this one.

The records and published histories of Chester, New Hampshire, list only one resident who might plausibly be identified as the carver. Hugh Tolford

(born 1747–alive 1794) was of Scotch-Irish extraction and, according to Trent, may have served in the militia during this period.

Hugh Tolford was not so accomplished an engraver as Jacob Gay but, if Gay is to be identified as the man of that name residing in Allenstown, New Hampshire, Tolford may have trained with him. H. T.'s calligraphy is almost as good as Gay's—delicate scrollwork outlines each letter of his inscriptions and amusing faces peer out from the enclosures of "O's" and from the upper extremities of the uprights of other letters. Many of his serifs trail off into scrolls, and some of his letters are deeply shaded for emphasis. However, his calligraphy is not so deeply engraved as Gay's nor so skillfully disposed.

The Bennet horn is profusely decorated with wonderfully imaginative animals whose execution is on a par with Gay's although they lack Gay's whimsical expressions. The detailed and realistic depictions form many odd vignettes: one character plays a fiddle for a donkey, another with whip in hand is seated in a chariot drawn by two mules (Britannia was sometimes depicted allegorically, seated in a chariot). There are also birds perched in trees with tentaclelike branches; a turtle with the tail of a long snake in its jaws (which also appears on the Titus Cave horn); a deer with nursing fawn; animals standing on pedestals as if they were statues or perhaps stuffed, including an unusual geometric horse; and an abstract British coat-of-arms with lion and unicorn supports and various scrolled and shaded borders.

Neat scalloping preceded by an intricate ½-inch-wide geometric border decorates the area 3 inches from the tip where the recessed portion begins. There is a raised ring at the tip, which is plugged with an old and crude wooden stopper. Another border, ¼-inch wide and consisting of shaded chevrons and dotted lunettes, decorates the plug end. Trees, animals, birds, and single curved waves are incorporated into this border. Five small wooden pegs originally secured the flat pine plug, which is now recessed ¼ inch inside the horn and held by both friction and a circle of iron wire secured through two holes at the end of the horn. Several crudely incised letters on the plug are now illegible.

PRE-SIEGE OF BOSTON SCHOOL
Attributed to Hugh Tolford (w. 1770–1772)

67 *Titus Cave horn*

Chester, New Hampshire, January 11, 1772
Horn, brass
OL: 13¾ in. (34.9 cm), W (of plug): 3¼ in. (8.3 cm)
Inscriptions: Titus Cave / His.Horn.Made / In.Chester.Jan^r. 11th.1772.
Carol and the late Thomas J. Segal

There is no listing for Titus Cave in military records either of this period or of the Revolution. The name CAVE may be phonetic for a different spelling of the name.

This is another horn by Hugh Tolford, one of four known. All are similar in style to the work of Jacob Gay, as is discussed in the entry for Tilton Bennet (No. 66). This example illustrates Tolford's fine engraving: the calligraphy is neatly executed and the letters are intricately embellished and shaded so that the lettering stands out in a very visually appealing

Catalogue

way. The inscription is framed in interesting, varied, and deeply carved borders of scrollwork. This horn is decorated with the same assortment of animals, birds, and reptiles as the other Tolford horns, including the turtle with a snake in its jaws seen on the Bennet horn. Like the Holt horn (No. 68) this also has a Lorelei with a heart as a navel and an abstract British coat-of-arms with a lion and a unicorn supporting a flaglike heraldic device.

There is neat scalloping preceded by the geometric border seen on the other Tolford horns. The scalloping begins ¾ of an inch from the brass charger that was secured over the spout with brass tacks sometime in the nineteenth century. At that same time a brass cover was placed over the wooden plug and fastened to it with six brass screws. Although the wooden plug is not visible, it is probably the same shape as the brass cover, which would make it slightly rounded, extending out from the end of the horn about ½ inch. There are brass rings attached both to the center of the brass plug cover and to the charger.

BIBLIOGRAPHY: Grancsay, No. 162, 47.

PRE-SIEGE OF BOSTON SCHOOL
Attributed to Hugh Tolford (w. 1770–1772)

68 *Thomas Holt horn*

Chester, New Hampshire, March 1, 1772
Horn, pine, brass, pewter, paper
OL: 15 in. (38.1 cm), W (of plug): 2⅜ in. (6 cm)
Inscriptions: Thomas.Holt / His.Horn.Made / In.Chester.
  March / 1772.
William H. Guthman

This is the fourth Hugh Tolford horn known. There is no military listing for Thomas Holt and because the horn is so greatly worn, the "l" in Holt could be an "i," making the name Hoit. This horn is quite similar to the Tilton Bennet horn (No. 66), with the almost identical coat-of-arms consisting of the lion and the unicorn supporting a cartouche with the British HONI SOIT QUOI MAL Y PENS in a circular border surrounding a flaglike device and the DIEU ET MON DROIT in a ribbonlike device supporting the lion and the unicorn. The ribbon comes together at the base of the coat-of-arms in a reverse crown design below which is a floral and scroll swag.

Unlike the other Tolford horns, this displays a ship with a whimsical crew and the British Union Jack. There are many animals and birds, as well as the turtle with a snake in its jaws that is a Tolford trademark. The Cave horn (No. 67) also exhibits the Lorelei with the heart-shaped navel, as well as the coat-of-arms.

68. *The Lorelei with heart-shaped navel is one of this carver's signatures.*

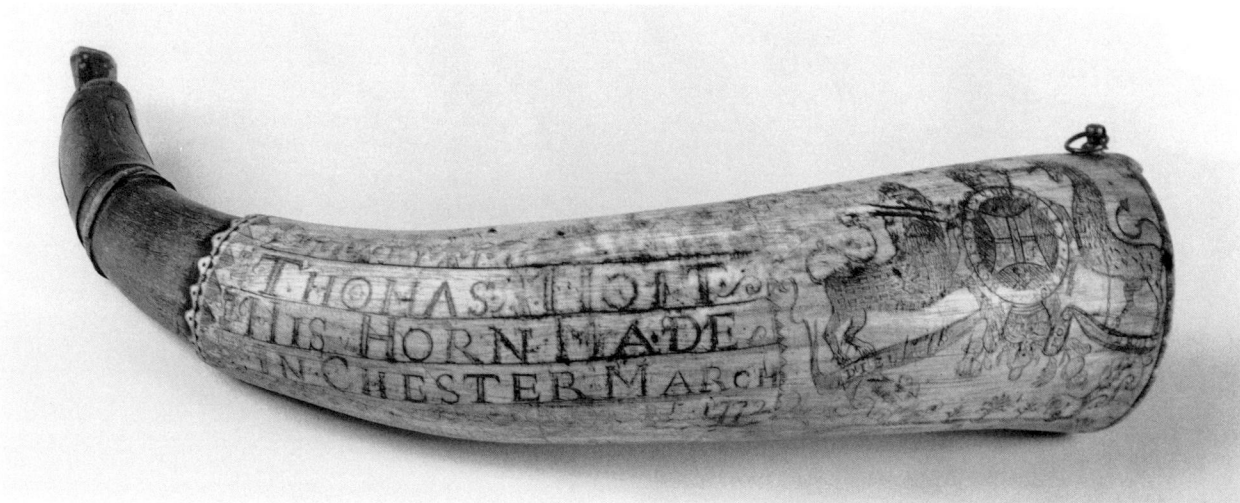

A neatly carved scalloped edge decorates the beginning of the recessed portion 4¼ inches from the tip of the spout; this is preceded by Tolford's typical geometric border. There is a brass ring at the beginning of the recessed portion and 1¾ inches from that is a raised ring. Just ¼ inch from that is a pewter inlay protecting the spout, consisting of a ring of pewter ¼ inch from the raised ring, with four branches, at different widths apart, ranging from ½ inch down to ¼ inch, extending to another pewter ring at the tip of the spout. An apparently original 3-inch tapered pine stopper is inserted into the spout.

A neat demilune border encircles the base of the horn and there is an extension lobe ¾ of an inch long and ¼ inch deep with two holes for the carrying strap; there is also a brass ring between the two holes. An old piece of paper pasted onto the flat wooden plug states in brown ink, "This powder horn was picked up at Bunker Hill by John Hardy, a drummer, and given to me by his nephew, Captain J. Hyatt, aged 80, July, 1844." It is signed "Daniel Brooks." The paper and calligraphy appear to be of the 1844 period, but the story might have been a romantic tale made up to accompany the gift.

PRE-SIEGE OF BOSTON SCHOOL
   Signed by Jacob Gay (w. 1758–1787)

69 *Hamilton Davidson horn*

Probably New Hampshire, 1772
Horn, brass, iron
OL: 15 in. (38 cm), w (of plug): 2½ in. (6.4 cm)
Inscriptions: HAMELTON DAVIDSON / His Horn Made By J. G. 1772 / Jacob Gauy / HONI SUIT QUI MALY PENSE / DIEU ET MON DROIT
William H. Guthman

The name of the owner is of Scotch-Irish derivation and, according to local genealogical history, suggests someone who lived near Jacob Gay in Allenstown, New Hampshire. However, the only person of that name in the vicinity was the owner of an iron mill in Windham and Charlestown who was born in 1787 and died in 1847. Perhaps an unrecorded member of the Davidson family was the owner, for it seems unlikely that the horn would have been made for a member of a British regiment stationed in America.

This horn, which represents the very best of Gay's work, is one of the few genuine American powder horns that depicts a historical scene. Its inscriptions support the theory that Gay was functionally illiterate and either spelled words phonetically or copied inscriptions that were written out for him. The names of several animals are spelled phonetically, as is his own name, in one of several phonetic variants he used. However, the inscription on the British coat-of-arms is almost entirely correct.

The Davidson horn is one of the unquestionable masterpieces of American horn engraving. The owner's name is expertly carved in large, beautifully shaded letters with graceful geometric and floral devices on the uprights and crossbars, as well as faces within the "O's." The name is contained within an elaborate rococo cartouche. To the left of the cartouche is a vignette of two men fencing, and to the right is the British coat-of-arms. Above are fourteen fish and animals, two of which are labeled. A hunter is depicted shooting a bird perched atop a tree. The lowercase lettering of the date and Gay's name is neatly done. The major vignette, however, is a scene Gay adapted from either the Henry Pelham or the Paul Revere print of the Boston Massacre. Gay has reversed the scene and given all the characters the

69

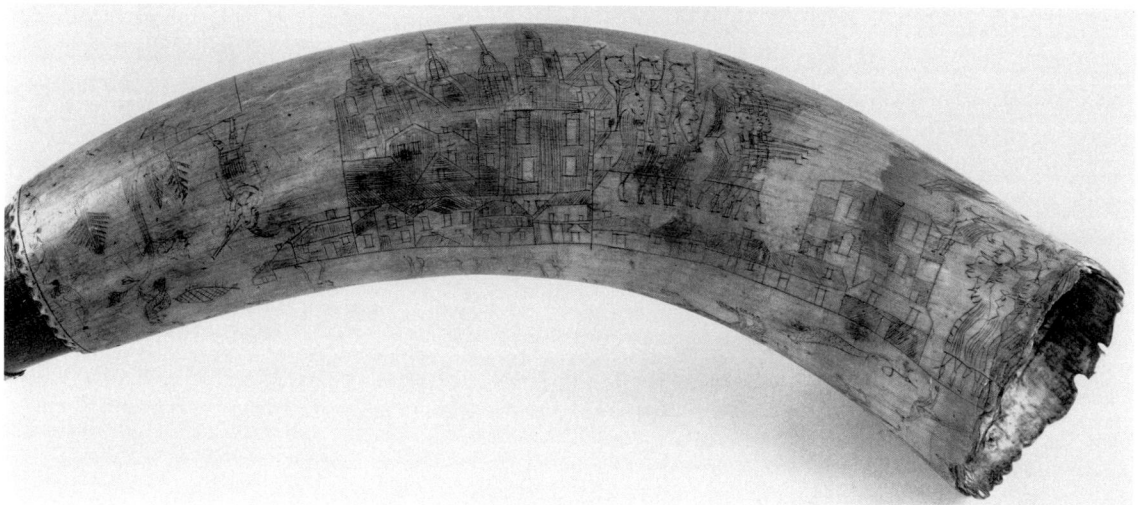

69. *Jacob Gay copied Paul Revere's* Boston Massacre *on horn.*

simian physiognomies he employed for the creatures on most of his horns.

The plug end of the horn has suffered damage, probably as a result of the loss of the plug, and a small portion of the Boston Massacre scene has been lost as a consequence. There is neat scalloping at the beginning of the recessed portion 1½ inches from the tip. A nineteenth-century brass-and-iron powder charger or nozzle is affixed to the spout end, and it is impossible to tell how much of the spout was trimmed to install it. The charger is secured by iron screws and has an iron spring and lever. There is no border at the plug end, although there are traces of scalloping, an unusual feature in this location which is also seen on the Connor horn (No. 94).

BIBLIOGRAPHY: duMont, 74; Lindsay, *The New England Gun*, 132; *Bulletin*, KRA, Fall, 1985, 7.

PRE-SIEGE OF BOSTON SCHOOL
Signed by Jacob Gay (w. 1758–1787)

70 *Jonathan Clark Lewis horn*

Probably Groton, Massachusetts, February 27, 1773
Horn, pine, iron
OL: 15¾ in. (40 cm), W (of plug): 3¼ in. (8.3 cm)
Inscriptions: I Powder With My Brot[her] / Ball Most Hero Like Doth / Conquer ALL / JONATHAN CLARK / LEWIS / his horn Made By JG / Febr the 27:1773
William H. Guthman

A story in *The Boston Chronicle* for November 23, 1763, relates that "Mr. Jonathan Clark Lewis and wife belonging to Boston were picked up in a long boat with 11 other passengers of the ship Providence which sprang a leak. The ship was bound from Colerain to New York. The 13 were drifting in the long boat for 8 days."

Jonathan Clark Lewis (1744–1781), an English-born merchant of Groton, Massachusetts, served during the Revolution as aide-de-camp to Brigadier General Oliver Prescott of the Massachusetts militia. His importance can be inferred from the facts that records consistently referred to him as "Mr. Lewis" and that his gravestone is embellished with armorial designs. Gay's engraving a horn for a resident of northern Essex County, Massachusetts, suggests that in the years before the Revolution he may have traveled to militia training sites to engrave and sell horns.

This is another example of Gay's best work. He had tremendous engraving ability, and his finest achievements seem to have involved close study of an engraved design source. The magnificent, beautifully illuminated lettering seen here, which must have been copied from a cipher book or a printed document, is placed on a large, high quality horn. The cartouche enclosing the inscriptions is simpler than that of the Davidson horn (No. 69), but has four extremely elaborate scrolls at either end of the surround. Gay covered the balance of the body with exceptionally large, graceful animals suited to the horn's size. These include a turtle and a bird atop a tree with a fox attempting to climb the trunk. The animals have the curled snouts that are peculiar to Gay's work and a number have the first letter of their name engraved near them.

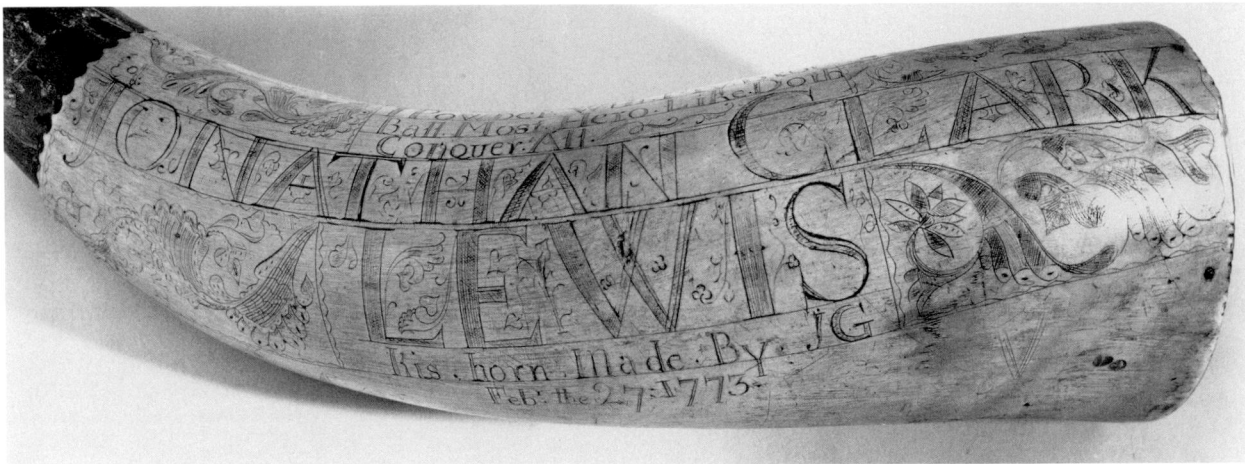

70

There is neat scalloping at the beginning of the recessed portion of this horn 4½ inches from its tip. A deeper recessed channel ½-inch wide is 2½ inches from the tip, followed by a raised ring. There is no border. A slightly rounded pine plug of indeterminate age is jammed into the spout and secured by six wooden pegs. A large iron screw in the center of the plug appears to be a barrel-tang screw for a musket. There is no border at the plug end, although there is informal notching which probably postdates the horn's engraving. An incised circle 1¼ inches in diameter surrounds the screw in the plug. Additional decoration within the circle has been severely abraded.

BIBLIOGRAPHY: Grider, NYHS; Grancsay, No. 531, 59; *Antiques*, 1978, 320; Guthman, *Guns and Other Arms*, 142; Green, Samuel A., "Inscribed Powder Horns," *New England Historical and Genealogical Register*, 1896, vol. L, 43–45.

LEXINGTON ALARM SCHOOL

71 *John Parker horn*

Probably eastern Massachusetts, 1775
Horn, pine, varnish, pewter
OL: 18 in. (45.7 cm), W (of plug): 2⅝ in. (6.8 cm)
Inscriptions: JOHN PARKER HIS HORN 1775
William H. Guthman

The rush to provide equipment at the outbreak of the Revolution undoubtedly produced many horns and it seems likely that this one was made for a minuteman, although it is far more elaborate than most Lexington Alarm examples.

Many men with the name John Parker are listed as serving in New England and New York in 1775. A John Parker was captain of a company of Rangers in Colonel Timothy Bedel's Regiment of Rangers raised by the Colony of New Hampshire in 1775. A John Parker is listed in the 1st New York Regiment for 1775, and another was a minuteman from the town of Coventry, Connecticut, in 1775. Nine men of that name are listed in the Massachusetts rolls for 1775.

The most famous was the John Parker (1729–1775) who commanded the detachment of the Lexington militia that engaged the British at Lexington on April 19, 1775. Parker was a joiner, having served in the Louisburg campaign of 1758 and the Quebec campaign of 1759. From the reminiscences of the Reverend Jonas Clark (1730–1805) of Lexington (quoted in Frothingham, *History of the Siege of Boston*), it is known that Parker regularly borrowed books from the minister's library. He also served the town of Lexington as assessor, constable, and tax collector.

No proof that this John Parker was the owner of this horn exists, but he was exactly the kind of ambi-

Catalogue

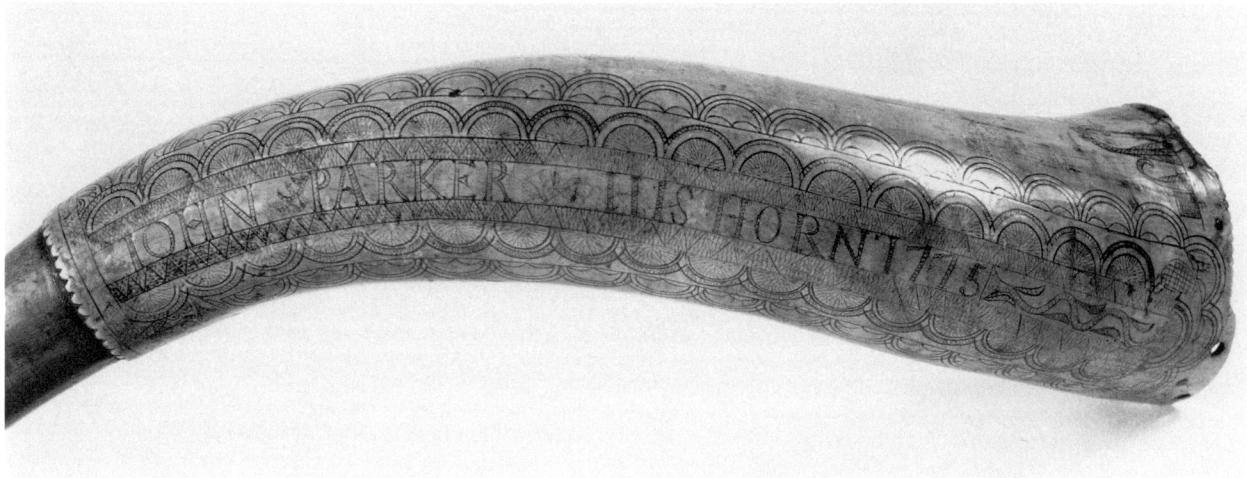

tious artisan with a strong interest in reading and self-improvement who might have owned it. The inspiration for the decoration is a drawing made during one of Captain James Cook's voyages to the Pacific Ocean in the early 1770s. It depicts a New Zealand war canoe with eighteen natives paddling—one is in the stern holding a spear, two are standing amidships, and one is standing in the bow brandishing a war club. The drawing was made by William Hodges, an artist-officer who accompanied Cook, and an engraving was made from the drawing by Sydney Parkinson, a draughtsman, engraver, and artist who also took part in the Cook expedition. The engravings appeared in different editions of the published accounts of Cook's voyages. The 1777 edition advertises drawings by Hodges and engravings by "masters." In addition, the September, 1773, issue of the London publication *Gentleman's Magazine* included a woodcut of the war canoe (p. 56 and pl. 10). An explanatory text states that the scene was delineated by Parkinson, who explained that the canoes were propelled by eighteen paddles, carried up to twenty-two men, were between fifty and sixty feet long, and were adorned with fine spiral filigree of curious workmanship. At the top of the bow was a head with heart-shaped tongue and mother-of-pearl eyes. Both bow and stern flew streamers of feathers that almost touched the water.

The carver of the Parker horn transformed the print source in surprising ways. The New Zealanders are translated into Woodland Indians dressed in a mixture of Indian and white-man's clothing. English halberds are substituted for the spears. The native standing in the bow is peering into the distance through a telescope. The number of paddlers is reduced to five and the commanders to three.

Perhaps the most extraordinary transformation is in the decoration of the canoe. The hull is now covered with the double-scroll motif favored by Woodland tribes and the feathers are from an eagle or a turkey. The ornamental head on the bow is that of a bird with a long neck. The stern is decorated with a peculiar mixture of Polynesian, American Indian, and rococo motifs. This carving is of the utmost importance because it demonstrates that horn engravers did not necessarily use print sources in a slavish manner. The adaptations reflect the experiences of New Englanders, who were exposed to colonial, English, and Native American design traditions and synthesized them in a self-conscious manner.

The design at the stern of the canoe is duplicated in an amoebic-looking pattern that extends toward the border of the spout. The engraving of the name and date is executed in neat block lettering within a rectangular cartouche decorated with various lunettes. A series of intertwined circles forms the beginning of the border of the spout, ending at the top and bottom of the cartouche. A double line encircles the border at the beginning of the finely scalloped recessed portion 5½ inches from the tip. A raised ring is 2¾ inches from the tip, a ⅜-inch pewter collar (an early repair) is ⅛ inch from the tip, and the tip itself has a raised lip. The plug end displays a deeply carved, three-masted ship, an abstract tree, and curlicue vines that are cane-shaped. A simple lunette border and fine notching surround the butt end and are carried around an extension lobe for the carrying strap, which is 1½ inches wide and ¼ inch long. The entire horn is covered with old varnish. The flat pine plug is secured by five wooden pegs.

BIBLIOGRAPHY: *Bulletin*, ASAC, 1981, 40; *Man At Arms*, Vol. 2, No. 4 (July/August, 1980), 43–45.

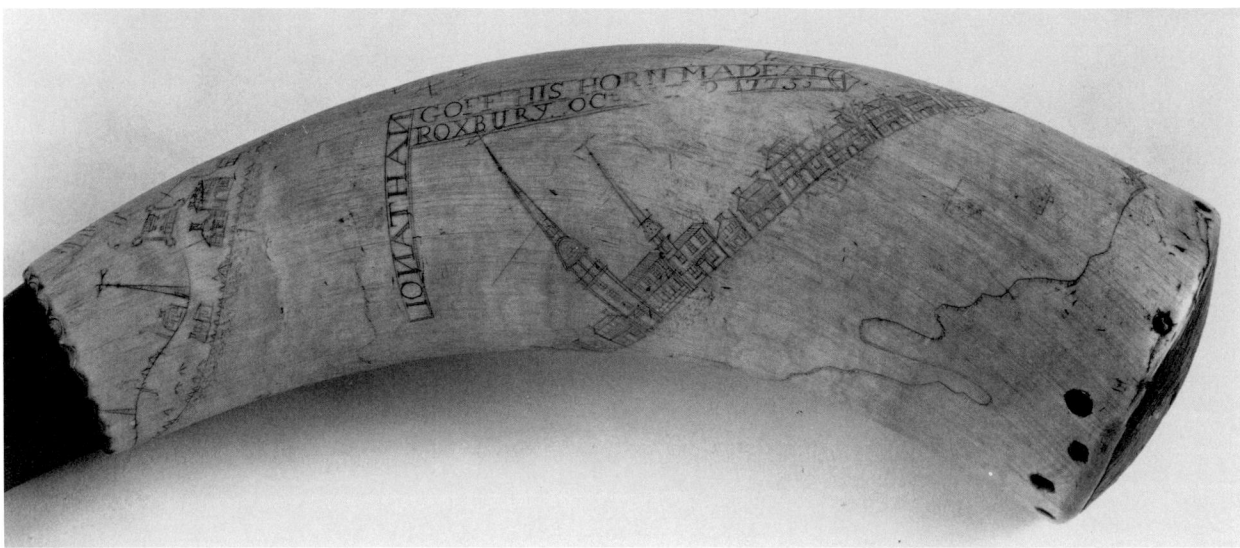

72. British ships in Boston Harbor—a typical Siege of Boston scene.

SIEGE OF BOSTON SCHOOL

72 *Jonathan Goff horn*

Roxbury, Massachusetts, 1775
Horn, pine
OL: 15 in. (38 cm), W (of plug): 2¾ in. (7 cm)
Inscriptions: JONATHAN / GOFF HIS HORN MADE AT / ROXBURY OCt [illegible] AD 1775
Private Collection

Jonathan Goff served as a private in Captain Return Meigs's 4th Company (Middletown), General Joseph Spencer's 2nd Connecticut Regiment, from May 8 to December 19, 1775. The men marched by company to Boston and took post at Roxbury until their enlistments expired in December. Goff then served as a private in Captain Robert Warner's Company (Middletown), Colonel Samuel Wyllys's 3rd Connecticut Regiment, from January 24, 1776, to December 31, 1780. He was promoted to corporal on November 1, 1780. The regiment saw action at various locations throughout the war, including Danbury, Stony Point, and along the Hudson. The Continental line was reorganized on January 1, 1781, and the number of regiments was reduced although the number of men remained the same. The eight Connecticut regiments were reduced to five, and the 3rd Regiment became part of the 1st, Colonel John Durkee commanding. Goff served as a corporal in Captain Robert Warner's Company from January 1 to December 31, 1781.

This is a fine horn with detailed views of Boston and the surrounding fortifications done in neat car-

72. Depictions of fortifications around Boston were a popular Siege of Boston theme.

tographic vignettes and outlines with ships, forts, and a vignette illustrating Charlestown, Boston, and Roxbury. It is, however, sparsely engraved in comparison with many others.

There is neat scalloping but no border at the beginning of the recessed portion 3⅝ inches from the tip, and there is a raised ring 1½ inches from the tip. The butt end has no border. The slightly rounded plug extends ¼ inch beyond the horn and is secured by eight round wooden pins. Two ¼-inch holes, ¾ inch apart and ¼ inch from the rim of the horn, extend through the wooden plug to hold the carrying strap.

SIEGE OF BOSTON SCHOOL

73 *Captain John Pennoyer horn*

Charlestown, Massachusetts, October 10, 1775
Horn, pine
OL: 15 in. (38 cm), W (of plug): 2⅞ in. (7.3 cm)
Inscriptions: CAP:ᵗ JOHN: PENNOYER / SETLER:FOR:THE 26 REDGᵗ / IN N:º 3 HIS HORN MADE OCTOᵇʳ Yᵉ 10ᵗʰ / In The year of Our Lord 1775: XIII DAY / new & old boston / CHARLSTON
Private Collection

Captain John Pennoyer of Sharon, Massachusetts, served as a second lieutenant in Colonel John Patterson's 26th Massachusetts Regiment. Patterson assumed command of the regiment in April, 1775, and his troops were stationed at various posts around Boston, including No. 1, the redoubt between that post and No. 2, and No. 3. The troops were shifted around so often during the summer and fall of 1775 that it is difficult to pinpoint the location of a specific regiment unless a surviving manuscript or horn supplies a definite location and time for a particular unit.

Pennoyer is not listed with the rank of captain, but of second lieutenant. He served in Patterson's regiment from April to November of 1775 and then served as a second lieutenant in the 15th Continental Infantry from January 1 to December 31, 1776. He is not listed after that date.

The word SETLER, a phonetic spelling of sutler, adds another mysterious aspect to Pennoyer's identity. The general definition of a sutler is a person who earns his living selling food and liquor to the troops, but Pennoyer probably simply distributed provisions obtained from a commissary.

The Roman numeral XIII, with a hand pointing to a map of the area, is a third inexplicable aspect of the horn. The vignettes of "old" and "new" Boston and Charlestown, where No. 3 was situated, are enlivened by sketches of horses' heads and necks to illustrate Charlestown Neck. The map of the vicinity of the camps is filled with animated birds and animals that represent rivers, roads, and other features of the landscape. This is the only known horn with this conceit—vignettes of Boston, Charlestown, fortifications, and ships are far more typical of Siege of Boston horns. The quality of the engraving here is good but not exceptional. The graphics are superb, however, and represent the unusual imaginative efforts that went into horns of this period.

A simple border precedes a nicely scalloped edge at the beginning of the recessed portion 3¼ inches from the tip. The plug end has no border, but there is a

73

Drums A'beating, Trumpets Sounding

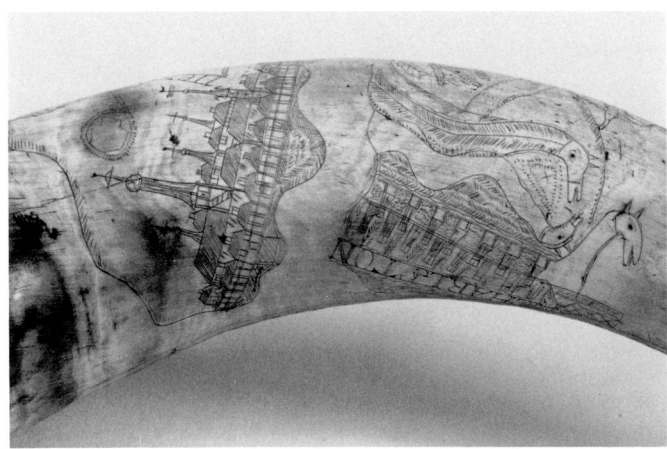

73. This carver emphasized the geographic area of Charlestown Neck by adding horses' necks to his view.

simple decorative device that appears as a finial at the beginning of the line cartouche enclosing the inscription. The slightly rounded plug was originally held by friction only and appears to have been glued in place at a later time.

BIBLIOGRAPHY: Grider, NYHS; Grancsay, No. 656, 63; Swayze, 138–140; duMont, 81.

SIEGE OF BOSTON SCHOOL

74 *Jabez Arnold horn*

Roxbury, Massachusetts, October, 1775
Horn, pine, iron, brown stain
OL: 13 in. (33 cm), W (of plug): 3 in. (7.6 cm)

Inscriptions: JABEZ / ARNOLD / oct. ye . 1775 / MADE [at] ROCKS[bury] / LIBERTY. / HIS.HORNE . / LIBERTY / Boston 1775 / Roxbury
Private Collection

Jabez Arnold of Haddam, Connecticut, served as a private in Captain Samuel Wyllys's 2nd Company (Hartford), Colonel Joseph Spencer's 2nd Connecticut Regiment, from May 5 to December 18, 1775. This company, like many others recruited after the Lexington Alarm, marched to Boston and took up quarters in Roxbury.

This horn is similar in format to a number of other Siege of Boston examples. The name, date, and other inscriptions are in rectangular cartouches set next to one another at forty-five-degree angles, a feature also seen on the Goff horn (No. 72) and others not included in the exhibition. Although the calligraphy here is mediocre, it is legible. However, the view of Boston and environs (similar to views on other horns) is graphically portrayed and the buildings are engraved with great attention to detail. The trees, vegetation, and fort scenes are nearly identical to those on the Pennoyer horn (No. 73), although the two horns were carved by different hands.

A simple border precedes the scalloping at the beginning of the recessed portion 3¾ inches from the tip and there is a neat raised ring 1⅝ inches from the tip. The inscriptions *Boston 1775* and *Roxbury* within simple rectangular cartouches serve as borders at the plug end. The slightly rounded plug, secured by one large iron nail and two wooden pins, has an iron screw eye in its center.

BIBLIOGRAPHY: Grancsay, No. 30, 42; duMont, 81; William D. Coakley, "A Horn from the Siege of Boston," *The Canadian Journal of Arms Collecting*, Vol. 13, No. 3, 92–97;

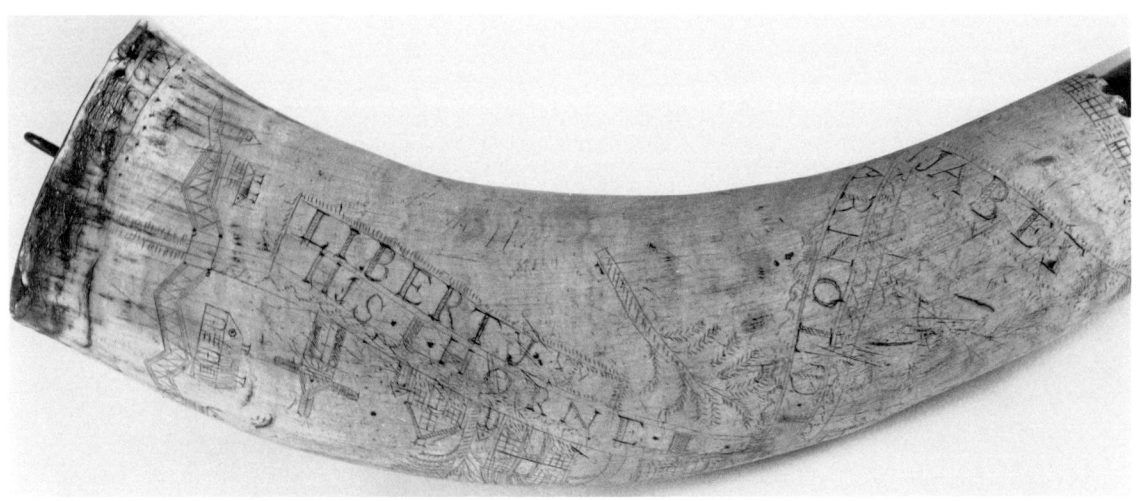

74

Catalogue

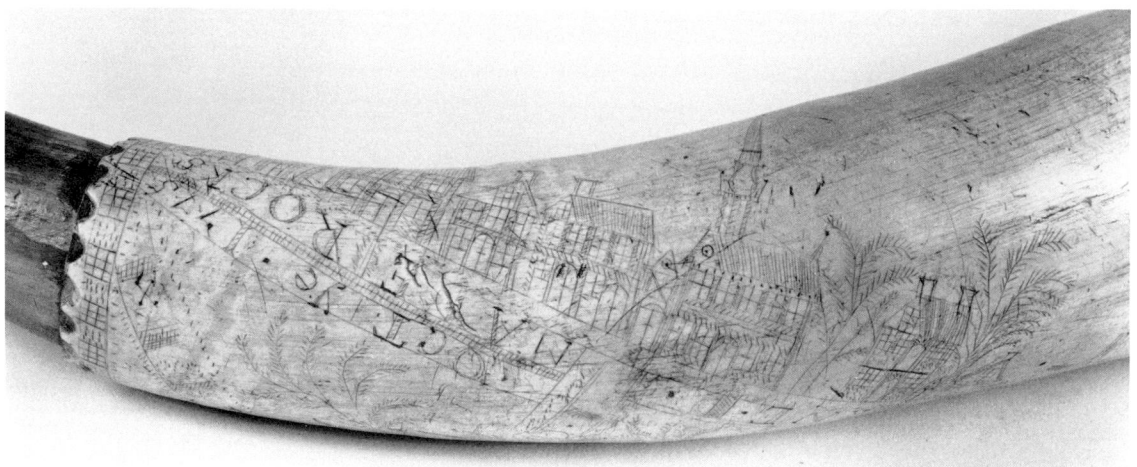

74. *Town views were popular powder-horn decorations during the Siege of Boston.*

James Dean Auction, Keeler Art Galleries, New York, March 1918; Bale-Poillon Sale, Elder Coin and Curio Company, New York, 1918; Walpole Galleries, New York, April 12, 1922.

SIEGE OF BOSTON SCHOOL

75 *Obadiah Johnson horn*

Cambridge, Massachusetts, 1775
Horn, pine, iron, leather

OL: 17¾ in. (45 cm), W (of plug): 3 in. (7.6 cm)

Inscriptions: OBADIAH : JOHNSONS : HORN : / MADE: AT : CAMBRIDGE : 1775 / F.S.ᵘ POINT [probably Lechmere Point redoubt] / F[ort]. No. I / F[ort]. No II / F[ort]. N. III / F[ort]. P[owde]r HILL / F[ort]. Pl[oughed] HILL / F[ort]. Wi[nter] HILL / F, B[unker], Hill Boston, Roxbury / WILLIAM / HOVEY of / MANSFIELD:

Yale University Art Gallery, Bequest of Janet Smith Johnson in memory of her husband Frederick Morgan Johnson

Obadiah Johnson (1735/6–1801) of Canterbury, Connecticut, was lieutenant colonel of the minute-

75. *Forts around Boston and ships in Boston Harbor were popular Siege of Boston themes; here* F. W. HILL *indicates Fort Winter Hill,* F. B. HILL *stands for Fort Bunker Hill, and so on.*

men from that town when they were called for seven days during the Lexington Alarm of April, 1775. He served as major in General Israel Putnam's 3rd Connecticut Regiment, was captain of the 4th Company, and was stationed at Cambridge during the Siege. His commission was dated May 1, 1775, and he was discharged the following December 16th. He then served as lieutenant colonel in Colonel Andrew Ward's Connecticut Regiment throughout 1776, participating in the New York, White Plains, and Trenton campaigns. He was a colonel of militia regiments in 1777 and 1778.

This is a well-carved Siege of Boston horn, with typical maps showing the fortifications, rivers, and harbors in and around Boston. There are vignettes of Boston and Roxbury, three ships, and one extremely large man-of-war guarded at the stern by a soldier standing at attention with the butt of his musket on the ground and the muzzle in the air. There are also two smaller ships, another man-of-war, a single-masted vessel, vine and floral borders, forts and windmills, some trees, and a silly looking bird. Two flintlock muskets are set butt-to-muzzle at the bow of one of the ships.

The inscription with William Hovey's name is contained within a deeply carved double-line square border filled with dots and surrounded by lunettes and dots. Hovey's identity is not known, nor is it known whether he was the carver or a later owner of the horn.

A neatly carved border of intertwined lunettes precedes the recessed portion, which is octagonal, 4¾ inches from the tip; the last 3 inches of the spout, also octagonal, is raised and the edge that faces the plug end is scalloped so that it matches the opposite edge. A leather band encircles the spout end. The raised plug, extending ¼ inch beyond the horn and held by five wooden pegs, has an iron staple driven into its center to secure the carrying strap.

SIEGE OF BOSTON SCHOOL

76 *Christopher Andrus horn*

Roxbury, Massachusetts, November 14, 1775
Horn, pine, iron, rope, string, buff leather
OL: 14¾ in. (37.5 cm), W (of plug): 2⅞ in. (7.3 cm)
Inscriptions: CHRISTOPHER / ANDRUS / HIS / HORN / ROXBURY / CAMPS / NOVEMBER / 14 1775 / Tom Gage
Private Collection

The only listing for a Christopher Andrus serving in the Revolution is one for a man serving in Captain Ebenezer Lathrop's Company, Colonel Jonathan Latimer's Connecticut Militia Regiment, raised to reinforce General Gates at the Battle of Saratoga in 1777. It is important to note, however, that records for participants in the Siege of Boston are incomplete. Two soldiers with the last name Andruss served in the 2nd Company of General Spencer's 2nd Connecticut Regiment camped at Roxbury in 1775. Jabez Arnold (No. 74) also served in that company.

The carving of the Andrus horn is not of the highest quality, but the carver conveyed his imaginative ideas

Catalogue

76

with clarity and humor. The calligraphy consists of large multilined block letters, some of which have lined highlights. The words are separated by pillars that are crosshatched, and the entire legend is enclosed in a rectangular three-lined cartouche. Ships, forts, trees, animals, and birds are all executed with accuracy, and there is also a Lorelei. A vignette of the Devil holding a serpent bears the legend TOM GAGE and is related to the portrait on the Howland (or Rowland) horn (No. 105). No vignettes of the Boston area are shown with the outline views of forts.

A simple border precedes the scalloping at the beginning of the recessed area 3¾ inches from the tip; a scalloped transition introduces deeper recessing 2¾ inches from the tip. There is a raised ring ½-inch wide and scalloped on both sides 1½ inches from the tip, and the spout is octagonal from that ring to the tip. The tip itself is tightly bound with string, probably to close up a split, and has a carved wooden stopper that may be original. There is no border at the plug end. The slightly rounded plug, which extends ⅜ inch beyond the horn, is secured by seven tiny round wooden pegs. A buff loop with an old rope attached is fastened to an iron screw in the center of the plug.

SIEGE OF BOSTON SCHOOL
    Attributed to the Crain-Arnold Carver (w. 1775)

77 *Elisha Crain horn*

Cambridge, Massachusetts, November, 1775
Horn, pine, brass, iron, black pigment
OL: 17 in. (43 cm), W (of plug): 3⅜ in. (8.6 cm)
Inscriptions: ELISHA:CRAIN, HIS, HORN, MADE / IN, THE, CAMPS, IN, CAMBRIDGE / NOVEMBER AD 1775 / EUNICORN-LION / MONDROIT NOVEMBER / CAMBRIDG / F[ort]. W[inter]. HILL / F[ort]. P.L.[oughed] Hill / F[ort]. PR[ospect] Hill F[ort]. B[unker]. Hill / Charlston / Boston / Roxbury / F[ort]. N[o]. 1 F[ort]. N[o]. 11 L. T. F. / F[ort]. N[ook]. Hill
Private Collection

Elisha Crain or Crane served as a private in Captain Experience Storrs's 2nd Company (Mansfield), Colonel Israel Putnam's 3rd Connecticut Regiment, from May 17 to December 16, 1775. John Arnold (No. 78) also served in this company. Some of the men took part in the Battle of Bunker Hill, although the lists are so incomplete that it is difficult to identify the participants.

It seems likely that either John Arnold or Elisha Crain carved these two horns, for their lettering, phonetic spelling, poor planning, and quality of engraving are identical. Both examples have dotted borders around the name, place, and date, and both have a sketch of a tall-case clock—the time on Arnold's is 12:45, while that on Crain's, which is more elaborate, is 6:15. Above Crain's clock is a coat-of-arms with a lion and unicorn supporting a circular design surmounted by a half-face instead of a crown. Inside the circular design are the words EUNICORN-LION and there is compasswork in the center. A banner below contains the abbreviated motto MONDROIT and the month NOVEMBER.

Outlines of Boston, Boston Harbor, and environs are neatly drawn in great detail and include individual features of forts and vignettes of Boston, Charlestown, and Roxbury. There are five ships in the harbor. Two flintlock muskets and several plants are used as fillers.

A finely carved border precedes scalloping at the beginning of the recessed portion 5 inches from the

77 (Bottom, Elisha Crain) and 78 (top, John Arnold): *the Crain-Arnold Carver employed a tall-case clock as a central motif.*

tip, a scalloped double raised ring rises 3¼ inches from the tip, and another is at the tip itself. The rounded plug extends ¼ inch beyond the horn, is painted black, and is secured by five brass pins and four tiny iron pins. A large half-round iron rivet is driven into the center of the plug to secure the carrying strap.

BIBLIOGRAPHY: *Bulletin*, Fort Ticonderoga Museum, 12, No. 3 (October, 1967), 189–190; duMont, 79; Swayze, "Elisha Crain's Siege of Boston Map Horn," *Man At Arms*, 1, No. 3 (May/June, 1979), 40–43.

SIEGE OF BOSTON SCHOOL
    Attributed to the Crain-Arnold Carver (w. 1775)

78 *John Arnold horn*

Cambridge, Massachusetts, 1775
Horn, pine, iron, black pigment
OL: 15½ in. (39.4 cm), W (of plug): 3½ in. (8.9 cm)
Inscriptions: JOHN: ARNOLD:HIS:HORN:MADE:AD / 1775: MADE:IN:CAMPS:IN:CAMBRIDG / CLOCK / F[ort]C[obble] hill / L[echmere]. P[oint]. / Boston / Ch[arlestown] / F[ort]. B[unker]. Hill / F[ort]. Pr[ospect]. Hill / F[ort]. Wi[nter]. Hill / F[ort]. No. I / F[ort]. No. II / F[ort]. No. III
William H. Guthman

Catalogue

John Arnold served as a private in Captain Experience Storrs's 2nd Company (Mansfield), Colonel Israel Putnam's 3rd Connecticut Regiment, from May 8 to December 10, 1775. Elisha Crain (No. 77) also served in this company, and the almost identical Crain horn is by the same hand as this example. Arnold is listed in the Connecticut records as having lost his gun at the Battle of Bunker Hill. His record between 1775 and 1780 is incomplete, but he is listed as an ensign of the militia in 1780.

Both the Arnold and Crain horns illustrate a tall-case clock, although each shows a different time (this one shows 12:45). Both also have similar calligraphy and reveal the carver's limited ability to plan the available space. On the Arnold horn the "c" is squeezed between the "o" and "k" in CLOCK, the "H" is squeezed between the "O" and "N" in the name JOHN, and the letter "M" is squeezed in between the "A" and "B" of CAMBRIDGE. Three of the ships are identical to three out of five on the Crain horn. In place of the Crain horn's coat-of-arms this one shows two uniformed soldiers standing at parade rest with muskets at hand. Several other muskets and vines fill out blank areas.

A neat border precedes the scalloped edge at the beginning of the recessed portion 3½ inches from the tip, a single scalloped ring rises 1¾ inches from the tip, and the remainder of the spout is worked into seven facets that correspond to the scallops on the ring. There is no border at the plug end. The slightly rounded plug extends 3/16 of an inch beyond the horn and is secured by five small wooden pins and four later large round tacks. There are traces of black pigment on the plug and there is an iron screw eye inserted at the top edge of the horn ½ inch from the rim.

SIEGE OF BOSTON SCHOOL
Attributed to the Simsbury Carver (w. 1775)

79 *Elijah Case horn*

Roxbury, Massachusetts, November 7, 1775
Horn, pine, iron, pigment
OL: 15½ in. (39.4 cm), w (of plug): 2⅞ in. (7.3 cm)
Inscriptions: Elijah . Case . of . Simsbury . his . horn / Made . at . Roxbury . November . 7th 1775 / Great Britton Please . to . Condesend / Our Liberty . Once . More . Defend / LIBERTY:AND:PROPERTY
William H. Guthman

Elijah Case, Jr., is listed as serving in Captain Elihu Humphrey's 4th Company (Simsbury), Colonel Jedidiah Huntington's 8th Connecticut Regiment, from July 11 to December 18, 1775. Originally stationed on Long Island Sound, the regiment was ordered to the Boston camps on September 14 and took post in Spencer's Brigade at Roxbury.

This horn is one of three known examples made by the Simsbury Carver, whose identity is as yet unknown. (The Asa Willcocks horn, No. 80, and an example owned by the Simsbury Historical Society, not in the exhibition, are the other two horns recog-

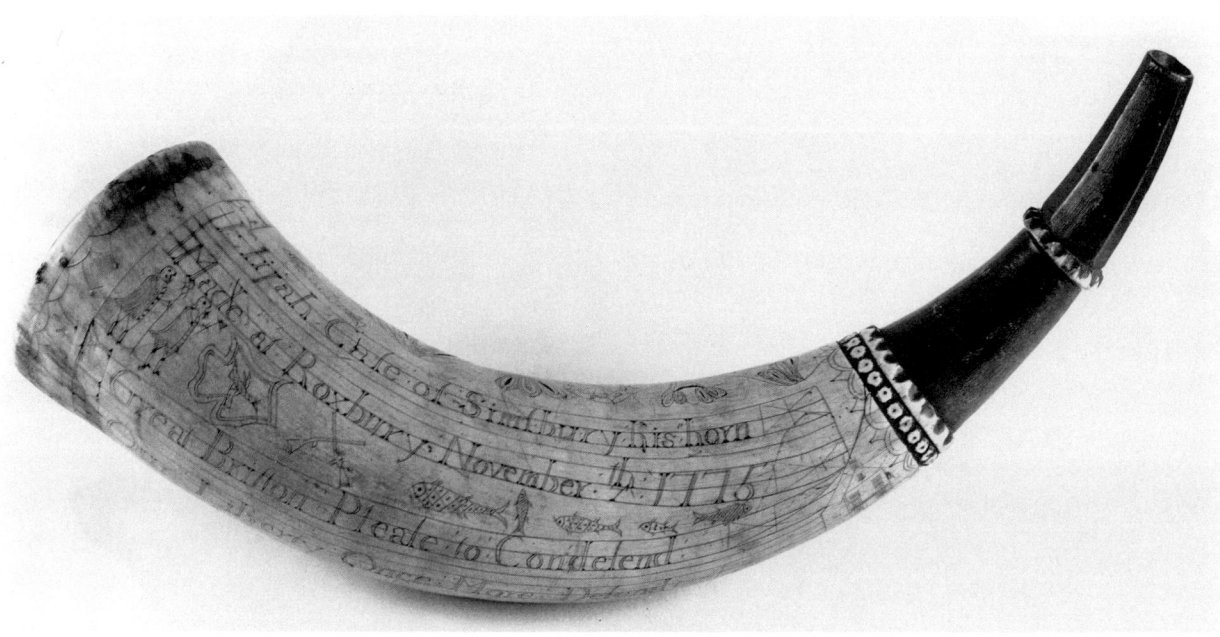

79. Distinctive geometric designs and simply drawn but properly attired formations of soldiers in the work of the Simsbury Carver.

nized as being by this maker.) The Case horn is above average in all aspects of its carving: the beautifully executed incised geometric designs in the detached border above the neatly spaced and consistently rendered name and place are similar to those of Richardson Minor (No. 52). There are four separate vignettes of formations of soldiers engaged in different types of military activity. Each soldier holds a symbolic object, including muskets, swords, halberds, and a flag. The soldiers' faces are expressionless and although their uniforms are not detailed, the designs of their coats and headgear show extremely interesting variations. Instead of repeating his lovely geometric designs, the carver filled most of the space with fish, crossed muskets, a house, and a tree. The fish are of various sizes, but all have the same shape and pleasant geometric patterns that suggest scales. This same carver made two other Simsbury horns with identical design motifs (see No. 80).

There is a border of dots next to the scalloping at the beginning of the hexagonally faceted recessed portion 4½ inches from the spout tip, and there is a scalloped raised ring 2 inches from the tip. A simple incised border decorates the plug end. The plug itself, slightly rounded and extending ¼ inch beyond the horn, has a rosehead nail with a ¼-inch head driven into its center to fasten the carrying strap. Secured by six round wooden pins, the plug retains traces of dark reddish-brown pigment.

BIBLIOGRAPHY: Sotheby's sale no. 4590Y, April 29–May 1, 1981, lot no. 542.

SIEGE OF BOSTON SCHOOL
Attributed to the Simsbury Carver (w. 1775)

80 *Asa Willcocks horn*

Roxbury, Massachusetts, November 11, 1775
Horn, pine, iron, brass, pigment
OL: 12½ in. (31.8 cm), W (of plug): 2⅝ in. (6.8 cm)
Inscriptions: ASA WILLCOCKS of NH—His horn / Made AT Roxbury Novr ye 11 1775 LIBERTY AND PROPERTY
William H. Guthman

Asa Wilcox is listed as serving in Captain Seth Smith's New Hartford, Connecticut (New Hartford is a short distance from Simsbury), company of minutemen that answered the Alarm of April, 1775. After that three-day period of service, Wilcox served from May 9 until December 18, 1775, when the enlistments again expired, in Captain Abel Pettibone's (Simsbury) 7th Company, Colonel Joseph Spencer's 2nd Connecticut Regiment, which was stationed at Roxbury. He served as drummer from September 13, 1776, until December 25, 1776, in Captain Philip Burr Bradley's (Ridgefield) Battalion, Brigadier General James Wadsworth's (Durham) Connecticut State Brigade.

The State of Connecticut was required to provide seven battalions to reinforce Washington's army; they participated in the Battle of Long Island, the retreat across New York, the Battle of Harlem Heights, the fall of Fort Washington, and the Battle of White Plains. Wilcox again served in Lieutenant Colonel Samuel Canfield's (New Milford) Connecticut Militia Regiment at West Point in September, 1781. At this time he was listed as residing at Middletown, Connecticut.

Catalogue

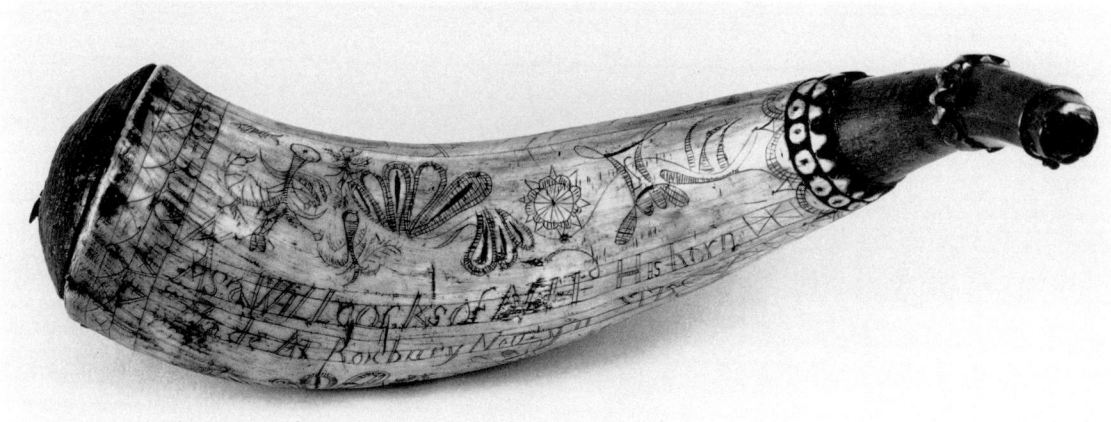

80. Unusual sawtooth and dot bands encircle the Simsbury Carver's horns just before the recessed portion.

This horn is identical stylistically to the Elijah Case horn (No. 79), with the same high quality incised geometric designs. The style of the calligraphy is also the same but it appears that the Wilcox carver was rushed to complete this horn, because the lettering is careless, almost sloppy. The formations of soldiers are the same, too, and appear in almost the same positions, but they are not so skillfully carved. It is again as if the engraver was rushed.

The Wilcox horn lacks the patriotic rhyme of the Case example but does have the same kind of fish and musket decorations. It also has the phrase LIBERTY AND PROPERTY engraved as a border encircling the plug end of the horn. There is a third Simsbury horn with identical geometric designs, but it is not so profusely decorated and seems even more rushed than the Wilcox horn.

Although their dimensions are different, the Wilcox and Case horns are treated identically at the spout end. The recessed portion of this example begins 3½ inches from the tip of the spout, with the scalloped raised ring 1½ inches from the tip. The rounded pine plug, which was probably originally secured by four round wooden pins and is now held by two iron nails, extends almost 1 inch from the base of the horn. Three square-cut nails are driven into the center of the plug, which bears traces of the same reddish-brown paint that appears on the Case horn. A piece of brass wire is inset into a channel at the tip of the spout and wound tightly to hold an old split together.

SIEGE OF BOSTON SCHOOL
Attributed to the Morley-Smith Carver (w. 1775)

81 *George Morley horn*

Charlestown, Massachusetts, December 17, 1775
Horn, iron, reddish-brown paint
OL: 16½ in. (41.9 cm), w (of plug): 2¾ in. (7 cm)
Inscriptions: George Morley / his horn Charlestun / Camp No 3 Decr Ye 17th AD 1775 / LIBERTY
William H. Guthman

George Morley enlisted on May 18, 1775, in Captain Oliver Hanchett's 10th Company (Suffield), General Joseph Spencer's 2nd Connecticut Regiment, and was discharged December 17, 1775, the date that is seen on the horn. This regiment marched to Boston by companies and served during the Siege. Morley is not listed in the records again.

This and the very similar Smith example (No. 82) are the most beautiful of the Siege of Boston horns, and are almost certainly the work of a professional engraver. Because this horn has little wear, and because it is dated on the day of Morley's discharge, it seems reasonable to assume that it was a souvenir rather than a functional accoutrement.

The lettering is done in the highest quality copperplate engraving embellished with flourishes and a rococo border. The date forms the upper portion of a border below the inscription, which is placed within an arched cartouche; the horn must be turned in the opposite direction to read the word LIBERTY, which is divided in order to fit into the cartouche properly. The engraver, who was obviously literate, took artistic license with his word construction in order to create a more pleasing design. The arched cartouche, which recalls a gravestone, may have been a symbol for "Liberty or Death." The engraver continually switched the orientation of the horn as he added a church and two men-of-war, one of which has a crowned Neptune holding a trident leaning out from the stern. Both ships are flying the British colors. The engraver was careful to leave a generous amount of space around each motif, a practice not followed by most makers.

A handsome scroll border encircles the horn just before the scalloping at the beginning of the recessed portion 5¼ inches from the tip, and there is an equally expert carved ring 1⅞ inches from the tip. Like some other Siege of Boston horns, the Morley example has a recessed portion at the butt end as well as at the tip. Preceded by a zigzag engraved border and scalloping and beginning 2½ inches from the plug, it ends within ⅝ inches of the plug. The flat plug, painted a reddish-brown, has an iron staple driven into its center; it is held by friction only.

BIBLIOGRAPHY: Swayze, 127–129; *Antiques*, 1978, 331; Guthman, *Guns and Other Arms*, 153.

SIEGE OF BOSTON SCHOOL
Attributed to the Morley-Smith Carver (w. 1775)

82 *Lieutenant Simeon Smith horn*

Charlestown, Massachusetts, December 12, 1775
Horn, hardwood, pine, iron
OL: 18 in. (45.7 cm), w (of plug): 3 in. (7.6 cm)
Inscriptions: Lieu:t Simeon Smith / his horn Charles Town / Camp No 3 Dec:r ye 12th AD 1775 / Grait Liberty inspire our Sons / And Magnify Our Zeal / Grait God of Worship protect Our Cause / And we.ll Submit Thy Will
William H. Guthman

Simeon Smith served as a lieutenant in a company of minutemen under Captain Thomas Knowlton which marched from Ashford, Connecticut, on the Alarm of April, 1775. His term of Alarm service was six days. Although he is not shown on the existing records until 1776, it is obvious that Smith served at Camp No. 3 at Charlestown during the Siege of Boston at least until December 12, 1775. He is shown as a captain in Colonel Philip Burr Bradley's Battalion, General James Wadsworth's Connecticut Brigade, serving from the spring until November of 1776. That year his battalion was stationed in New Jersey until November, when it was sent across the river to New York to assist in defending Fort Washington. Unfortunately, the fort fell to the British and most of the garrison were taken prisoner. Although Smith was not listed as being among them, the records contain neither a notice of the termination of his service nor further mention of him.

This horn is almost identical to the George Morley horn (No. 81) except that it is slightly larger. The superb engraving is in the same copperplate style with identical flourishes and the same rococo borders. There are two two-masted ships flying British flags sailing along with a fish between them and with a smaller single-masted ship at the stern of the ship on the right. Instead of the tombstone cartouche seen on the Morley horn, the Smith example has a four-line rhyme within a snakelike oval cartouche. There are four detailed and impressive-looking buildings, three in a triangle and two facing the triangle. As on the Morley horn, there is a recessed portion at the plug end, here decorated with three buildings in a row, two birds on a tree, two birds outside the tree, and a faint line drawing of what was probably the camp and surrounding terrain, showing rivers and bodies of water.

Catalogue

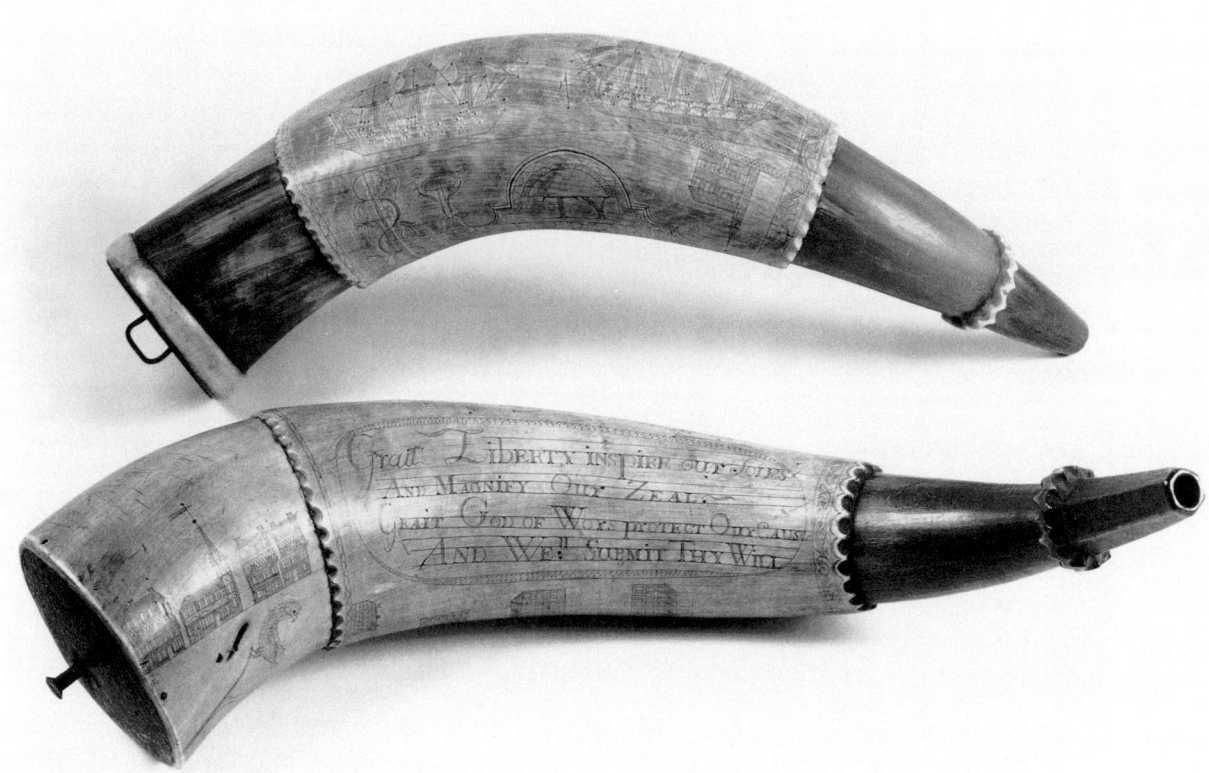

81 (Top, George Morley) and 82 (bottom, Simeon Smith): *the Morley-Smith Carver had a superb copperplate hand.*

The scroll border, identical to that on the Morley example, encircles the horn just before the scalloping at the beginning of the recessed portion 5¼ inches from the tip of the spout. A sawtooth-carved raised ring (again identical to Morley's) is 1⅞ inches from the tip of the spout; unlike the Morley example at this point, however, there is then 1⅞ inches of spout that is faceted into a hexagonal shape. The opposite recessed portion begins 2⅞ inches from the butt end; preceding it is the same engraved zigzag border as that on the Morley horn. This example has slightly finer scalloping at the beginning of the recessed portion at the plug end than at the spout end. The pine plug, unlike that of the Morley horn, was never painted; it is slightly rounded and is secured not by friction but by four small round hardware pins. There is a gun screw in the center of the plug, a hole where another screw might have been at one time, and two small square-cut nails that are driven into the plug next to the screw.

SIEGE OF BOSTON SCHOOL

83 *Frederick Robbins horn*

Roxbury, Massachusetts, September 7, 1775

Horn, pine, brass, iron, brown stain

OL: 12 in. (30 cm), W (of plug): 3 in. (7.6 cm)

Inscriptions: FREDERICK:ROBBINS:HIS / Horn: Made: September : The : 7th : 1775 : In : Roxbury / Camps / In: Defence: of: Liberty

The Connecticut Historical Society, Gift of Frederick Robbins Swan

Frederick Robbins (1756–1821) of Wethersfield, Connecticut, served in Captain John Chester's 9th Company, General Joseph Spencer's 2nd Connecticut Regiment, raised for the first call of troops in April and May of 1775. Marching by company, the regiment took post at Roxbury, where its members served out their enlistments. In Robbins's case, this was from May 12 to December 17, 1775. He also fought at the Battle of Long Island in 1776, at the Battle of White Plains, and served on a Wethersfield privateer, which was captured. He was later exchanged for Hessian prisoners.

The Robbins horn is competently carved with a

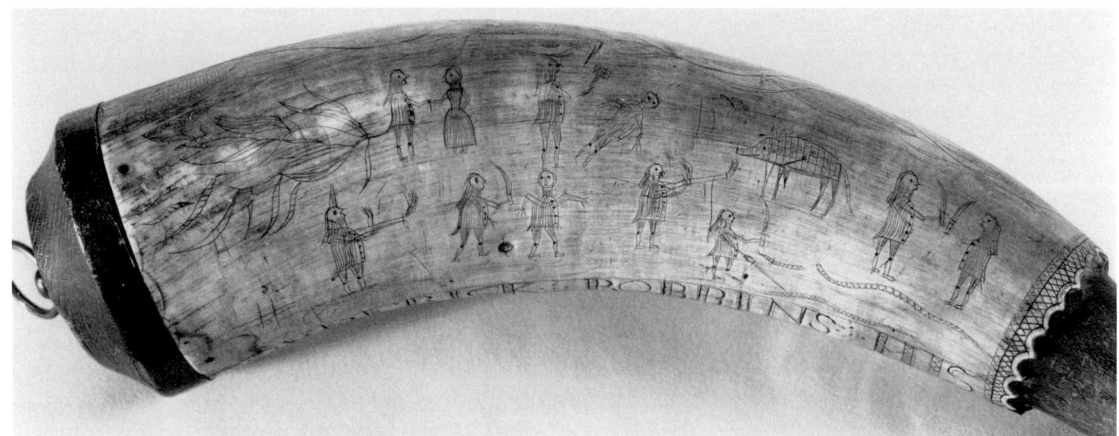

83. Soldiers engaged in a variety of activities from hunting to fighting to courting.

wealth of diversified vignettes and its calligraphy is deeply incised with cross-hatching on the uprights of the letters. Many vignettes portray the life of a soldier, with scenes that include firing a musket; chasing another, unarmed, soldier with a sword; shooting a pig; two soldiers fighting with swords; cutting up snakes with swords; knocking out another soldier with a mallet; and courting a lady. This last soldier sprouts an extraordinary multibranched tail that may be faintly suggestive. Two abstract rows of houses suggest Boston with Redcoats in the streets. Also in the streets are turtles, cats with wings, sunbursts, flowers on stylized stems, club-butted muskets, and a tavern sign on a post.

Neatly crosshatched borders precede the elegantly scalloped edge at the beginning of the recessed portion 2½ inches from the tip. There is similar scalloping on a raised ring 1¼ inches from the tip. The butt end has been trimmed slightly and fitted with a re-placement plug with a heavy beveled edge. The plug, stained dark brown, is held by four iron pins and has a brass drawer pull in its center.

BIBLIOGRAPHY: Grider, NYHS; Grancsay, No. 717, 66.

SIEGE OF BOSTON SCHOOL
    Signed by James Greenfield (w. 1775–1777)

84 *James Greenfield horn*

Roxbury, Massachusetts, July 17, 1775
Horn, pine
OL: 16¾ in. (42.5 cm), W (of plug): 2¾ in. (7 cm)
Inscriptions: JAMES GREENFIELD HIS HORN / HE MADE * IN ROXBURY * JULY * the 17th [A]D 1775
Private Collection

James Greenfield (born 1752) is listed as a private

Catalogue

84. *Two of James Greenfield's favorite motifs were Loreleis and lions with amusing facial expressions.*

from May 12, 1775, to December 19, 1775, in Captain Samuel Gale's 8th Company (Killingworth), Colonel Samuel Holden's 6th Connecticut Regiment. He resided in Lyme, but where he died is not known. Corporal Jonathan Murray, Jr., and Sergeant Abner Ely (Nos. 85 and 86) served in the same company, and Nathaniel Sunsimon (No. 87) served in the 3rd company of the same regiment. The horns of these three men are also attributed to Greenfield.

Greenfield may have been illiterate and had difficulty in forming his letters properly, for the calligraphy of all four of his horns, while carefully planned and deeply incised, is crudely done and is the poorest feature of his work. His figural carving is far superior. The two horns Greenfield carved in 1775 (this one and Murray's) depict one or two Loreleis and the British lion wearing a crown. The Lorelei on this horn is placed among five large sailing ships, possibly indicating that the British fleet is being lured onto the rocks of the Boston shoreline. Directly below the Lorelei is a panoramic view of Boston, at the far left is a small vignette of what appears to be Roxbury, and directly above is the British lion. Also included are a longboat with seven passengers, six of whom are rowing; schools of fish; and birds seated on perches, in flight, or strutting. The lion and the Lorelei, which stare directly at the viewer, have identical expressions.

All four horns are scalloped where the recessed portion begins; on this horn, that occurs 5 inches from the tip. The recessed area extends for 2 inches, and then the horn returns to normal size, although it is crudely faceted in an octagonal shape. There is no border at the plug end, although the edge has a neatly carved lip with two holes for the carrying strap. The flat plug is secured by three wooden pegs.

BIBLIOGRAPHY: *Bulletin*, Fort Ticonderoga Museum, XII, No. 3 (October, 1967), 189; Swayze, 124–126.

SIEGE OF BOSTON SCHOOL
Attributed to James Greenfield (w. 1775–1777)

85 *Jonathan Murray horn*

Roxbury, Massachusetts, July 31, 1775
Horn
OL: 15¼ in. (38.7 cm), W (of plug): 3 in. (7.6 cm)
Inscriptions: THIS BRIG : SALY / JONATHAN MURRY / HIS : HORN : MADE : IN : ROXBUR^Y / JULY : THE : 31 : IN : THE : YEAR 1775
Private Collection

Corporal Jonathan Murray, Jr. (1750–1785), of Killingworth, Connecticut, set out with that town's minutemen when the Lexington Alarm called for men to march for "The Relief of Boston" on April 19, 1775. He served on that occasion for six days under Captain Samuel Gale, who was also captain of the 8th Company in Colonel Samuel Holden Parsons's 6th Connecticut Regiment in which Murray served from May 6 to December 19, 1775.

The attribution of this horn to Greenfield (No. 84) is based on its large deeply incised but crudely executed block letters. The letter "x" in ROXBURY is almost identical on both horns. There are two Loreleis on this horn, a lion, and birds and fish, all of which are quite close to those on Greenfield's horn. An officer in a tricorn hat stands with arms akimbo in front of a church which is adjacent to a fort. The officer's face is identical to that of the Lorelei on the Greenfield horn. The brig *Sally*, which is very much like the ships on the Greenfield horn, may refer to the brig *Sally* that is listed in the Naval Records of the Revolution: a whaling ship of that name bound for Nantucket with a cargo of whale oil was captured by the British and taken to Boston between June and December of 1775. This horn also displays a large house, a fancy scroll border over the owner's name which functions as the water in which the brig floats, and several neatly carved geometric designs, including a compass star.

A simple looped border precedes the scalloping at the beginning of the recessed portion 3¾ inches from the tip. The recessed portion continues for 1¾ inches, after which the horn resumes its normal size but is

85

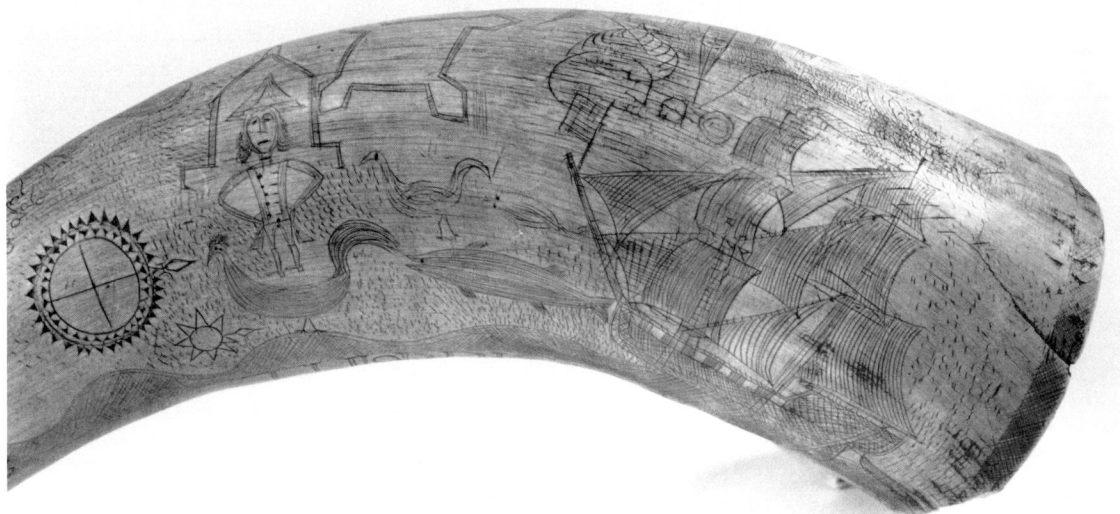

85. *Although his calligraphy was inferior, James Greenfield was an able engraver of such subjects as the compass, the soldier, and the ship seen here.*

crudely worked with facets and notched at both ends. The plug is lost and the butt end of the horn is chipped and broken.

SIEGE OF BOSTON SCHOOL
   Attributed to James Greenfield (w. 1775–1777)

86 *Abner Ely horn*

Probably Peekskill, New York, May 24, 1777
Horn, maple, brown stain
OL: 15 in. (38 cm), w (of plug): 3¼ in. (8.3 cm)
Inscriptions: THIS HORN BELON^{gs} / TO ABNER ELEY × A / ENSIGN × OF × CAPT × ELISHA ELYS / COMPANY × IN COLO × DUGLIS × RIGE / MENT × LIBERTY × ILE HAVE OR DEATH / 1777 / THE MAY 24 1777 / THE LION
Private Collection

Abner Ely (1749–1805) of Lyme was a sergeant in Captain Samuel Gale's 8th Company, Colonel Samuel Holden Parsons's 6th Connecticut Regiment, from May 8 to December 19, 1775. James Greenfield (No. 84) also served in this company. Ely then served as an ensign in Captain Elisha Ely's Company (Lyme), Colonel William Douglas's 6th Continental Regiment, from January 1, 1777, to May 4, 1778.

The calligraphy here is the deeply incised block capital letters seen on all Greenfield horns. The spelling is phonetic, and in one place Greenfield miscalculated the spacing of the letters. The imagery focuses

86. *Among James Greenfield's individualistic vignettes were these two—a house with a flag and a pond filled with fish, two rowboats, and an island.*

86. *Detailed fortifications are another of Greenfield's trademarks, as are the lion and Lorelei with stylized crowns.*

on the Lorelei and the lion, whose expressions, as they stare out at the viewer, differ somewhat from those seen on Greenfield's own horn. There is a house that is identical to that on the Murray horn (No. 85) and a large fortification with firing cannon representing Roxbury Fort, where both Ely and Nathaniel Sunsimon (No. 87) served in 1775. An uneven circle surrounds a group of motifs that includes fish, trees, and two longboats with oars. There are also two circles with chip-carved borders, one containing the day and month, the other the year.

A crudely shaded border precedes the scalloped edge at the beginning of the recessed portion 2¾ inches from the tip; the recessing extends for 1 inch and then recedes deeper for ⅞ of an inch, where it ends at a raised ring ¾ of an inch from the raised tip of the spout. The plug end has no border, and the slightly rounded plug, which is secured by six wooden pegs, extends ¼ inch beyond the horn; it is beautifully carved with a pinwheel, as is the Sunsimon plug.

Abner Ely's second cousin Elihu Ely (1737–1815) owned a horn dated May 20, 1776 (not included in the exhibition), that displays inscriptions stating that it was MADE IN LYME. That horn is also decorated with a house, a Masonic square and compass, and a fort diagram. It, too, is probably Greenfield's work.

BIBLIOGRAPHY: Grider, NYHS; Grancsay, No. 299, 51; *Bulletin*, Fort Ticonderoga Museum, 191; Swayze, 156–158, 192–195; A. E. Brooks, *Illustrated Catalogue of the A. E. Brooks Collection* (1899), 183–188; Moses E. Beach and William Ely, *The Ely Ancestry* (New York, 1902), 65–66.

SIEGE OF BOSTON SCHOOL
Attributed to James Greenfield (w. 1775–1777)

87 *Nathaniel Sunsimon horn*

Probably Peekskill, New York, February 24, 1777
Horn, maple, iron, brown stain
OL: 14¼ in. (36.2 cm), w (of plug): 2⅞ in. (7.3 cm)
Inscriptions: NATHANIEL × SUNSAMAN / HIS HORN MADE IN THE YEAR 1777 / IN FEBRUARY THE 24 / MY LIBERTY ILE HAVE OR DEATH / R F
Private Collection

Nathaniel Sunsimon was a Mohegan or Niantic Indian who served as a private in Captain Samuel Prentice's 3rd Company (Stonington), Colonel Samuel Holden Parsons's 6th Connecticut Regiment, from May 12 to December 17, 1775. He is thought to have been from Preston, Connecticut. He served in the same regiment as James Greenfield (No. 84), who carved his horn, and Abner Ely (No. 86), whose horn is nearly identical to his.

There are minor differences between the Sunsimon horn and the Ely horn: the date is contained within the inscription on this example, while on the Ely horn it is inscribed within two circles; the tail of the Lorelei on this horn is longer and curled and both the Lorelei and the lion have different headpieces. There are more birds and fish on this horn and the chip-carved circles are incised with shaded pinwheel devices. The house and flag, border around the name, and calligraphy are the same on both horns, as are the two forts, except that this one has the initials R F for Roxbury Fort. The carved plugs are almost identical.

There is scalloping at the beginning of the recessed portion 2½ inches from the tip; the recessing extends for only 1 inch before the horn resumes its normal size and becomes a hexagonal spout with a raised lip. Similar narrow recessed portions are seen on the Greenfield and Murray horns. The rounded dark-stained maple plug, secured by six oblong iron tacks, extends ⅜ of an inch beyond the horn and is neatly carved in a pinwheel design which has a ¼-inch incised border of double lines and circles. Like the Ely horn, this was probably carved as a memento, not as a functional accoutrement.

BIBLIOGRAPHY: *Bulletin*, Fort Ticonderoga Museum, Vol. XII (October, 1967), 191; Swayze, 152–154, 177, 192–195.

SIEGE OF BOSTON SCHOOL
Attributed to Jacob Gay (w. 1758–1787)

88 *Stephen Upson horn*

Winter Hill, Massachusetts, December 12, 1775
Horn, pine, hardwood, iron, paper
OL: 13½ in. (34 cm), w (of plug): 2¾ in. (7 cm)
Inscriptions: STEPHEN UPSON / SUCCESS TO AMERICA / I. POWDER— / WITH.MY / BROTHER.BA / LL.HERO. LIKE

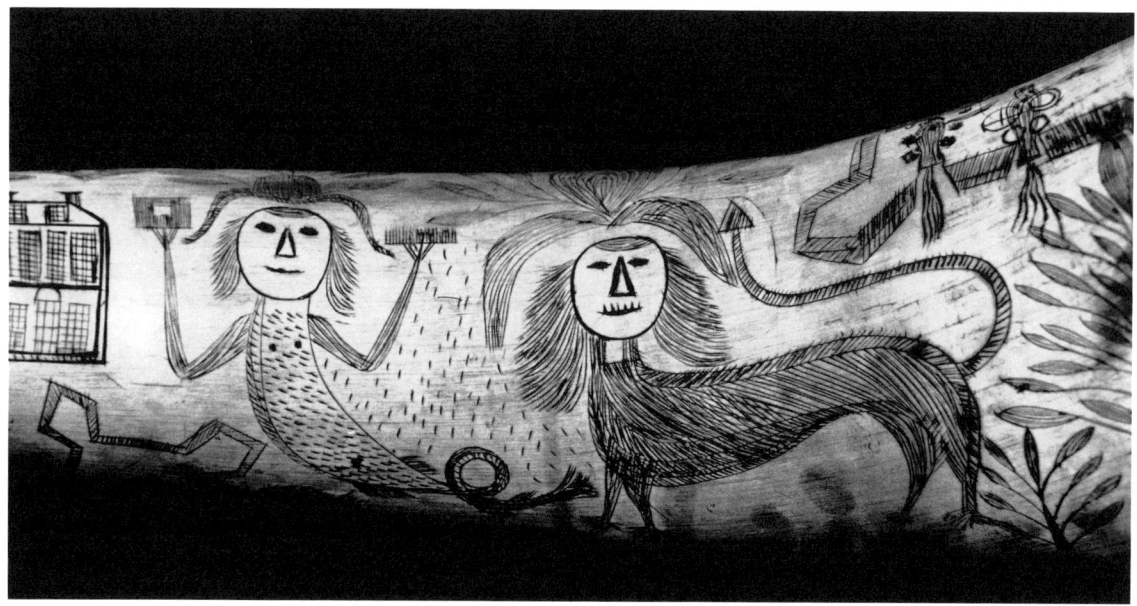
87

88

/ ALL DOTH CONCUER ALL / JGA fm 1775 his. horn. made.
AT. Winter Hill / Dec^m. y^e 12. 177^5 / TRY IT OUT
William H. Guthman

Stephen Upson is listed as a private serving from July 12 to December 20, 1775, in Captain Nathaniel Tuttle's 5th company, Colonel Webb's 7th Connecticut Regiment. He served at Winter Hill in 1775, and was killed at Harlem Heights during the New York campaign of September, 1776.

This is a fine Jacob Gay horn (although not so fine as Nos. 90 and 91), with elaborate architectural scrolls and pediments embellished with floral decoration bordering the inscriptions. There is a scene depicting two uniformed soldiers fighting a pistol duel that is captioned TRY IT OUT. There is also a profusion of wonderfully executed animals all labeled with the first letter of their names, such as "R" for rabbit.

A neat floral-swag border surrounds the area just preceding the recessed portion of the spout 3¼ inches from the tip. A raised ring occurs ½ inch from the tip, which contains what appears to be its original hardwood stopper. The flat pine plug, secured by friction alone, appears to be original as well. It has a small iron nail in the center and another iron nail, even smaller, one inch above. Incised block lettering, *DH 40*, appears on the spout, which also bears a paper label printed with the number *93*—very likely an exhibition label.

SIEGE OF BOSTON SCHOOL
Attributed to Jacob Gay (w. 1758–1787)

89 *Ephraim Moore horn*

Charlestown or Cambridge, Massachusetts, December 29, 1775
Horn, pine, iron
OL: 15 in. (38 cm), W (of plug): 2¾ in. (7 cm)
Inscriptions: EPHRAIM / MOORS his / Horn Made / At TEMPELS / Warf 29 1775 / Dec^r / IBW / Lech More / Plow Hill / Buncker Hill / Warf / Prospect / Winter Hill / Cambridg
Massachusetts Historical Society

An Ephraim Moor from Sudbury, Massachusetts, served as a minuteman in the Lexington Alarm company of Captain Isaac Locker, Sudbury Troop, Colonel James Brett's Massachusetts Regiment, which marched to Lexington on April 19, 1775. Another Ephraim Moore or Morse served in Colonel Moor's New Hampshire Regiment during the New York campaign in the fall of 1776. Temple's Wharf was probably located on Temple's Farm, between Cambridge and Charlestown near Prospect and Winter Hill. Captain Samuel Clarke, captain of the ship *Edith Warren*, gave this horn to the Massachusetts Historical Society in 1876.

Although not so finely carved as the Hull Curtis or Edward Sherburne horns (Nos. 90 and 91), both of which were carved in the same area in January of 1776, this fine example has similar features. Its block lettering is not embellished, save for two man-in-the-moon

89. Jacob Gay's depictions of fortifications and formations of soldiers represent the highest accomplishments of Siege of Boston carvers.

faces in the "O's" of the last name. The inscription is enclosed within an elaborate broken-scroll cartouche typical of Gay's work. The characters in the formation of five uniformed grenadiers marching with muskets shouldered and in the scene showing two large soldiers fencing with swords have Gay's typical monkey faces. The animals, which were a Gay specialty, include fish, a long-eared donkey, a rabbit, a running doe, a standing buck, and a dog.

Gay also engraved a detailed map of Boston and its surrounding fortifications, along with ships in the harbor and the river. This view of Boston is somewhat abstract, with the buildings formed into geometric designs. They are engraved with deep dark zigzag lines combined with squares outlining roof lines and building structures. This is reminiscent of professionally engraved depictions of Havana seen on Havana Expedition horns of the 1760s and also of the Samuel Selden horn (No. 103).

A grand scroll border of vines and leaves has an imposing man's head in profile at its center. This border, ⅜ of an inch wide, precedes a scalloped edge at the beginning of the recessed portion 4½ inches from the tip. From this point to a raised ring 2⅜ inches from the tip the spout is octagonally faceted, then becomes round again from the ring to the tip. There is no border surrounding the butt end, allowing the main designs to extend to the edge of the horn. The rounded pine plug, extending ½ inch beyond the horn, is secured by three round wooden pegs and an iron staple that is ⅜ of an inch wide and is driven through the horn into the plug at a right angle.

BIBLIOGRAPHY: Grider, NYHS; Grancsay, No. 601, 62; Stewart Culin, "Powder Horns of Men Who Made Our History," *The Philadelphia Press*, February 20, 1898, 33; *Minutes*, MHS, 1876.

90

90. *On Siege of Boston horns Jacob Gay satirized the British coat-of-arms, using patriotic American motifs and slogans.*

SIEGE OF BOSTON SCHOOL
   Signed by Jacob Gay (w. 1758–1787)

90 *Hull Curtis horn*

Charlestown or Cambridge, Massachusetts, January 1, 1776
Horn, pine, iron
OL: 18 in. (45.7 cm), W (of plug): 2¾ in. (6.1 cm)
Inscriptions: SUCCESS TO AMERICA / HULL / CURTISS / His Horn Made At Tem / Pels Farm Jeny 1 1776 J. Gay
William H. Guthman

Hull Curtis (born 1759) of Woodbury, Connecticut, is listed as having been detached to Lieutenant Colonel Thomas Knowlton's Rangers from Colonel Charles Webb's regiment. Knowlton's regiment, the 7th Connecticut, marched from Boston to New York and, in 1776, became the 19th Continental Regiment. After the defeat on Long Island a small body of select troops—mostly from Connecticut—was organized for special service behind enemy lines and put under Knowlton's command. He had served during the French and Indian War, had distinguished himself at Bunker Hill, and he led the Rangers to perform with valor on several occasions, during one of which he was mortally wounded. There was then a succession of officers chosen to succeed him, among them Nathan Hale. Under the command of Colonel Robert Magaw of Pennsylvania, the unit was captured while defending Fort Washington in Harlem during the Battle of New York on November 16, 1776. All of the Rangers were taken prisoner, and Hull Curtis does not appear in the records after that date. He is not recorded as having served during the Siege of Boston but he undoubtedly did, in Webb's regiment.

This is one of Gay's finest horns. Above the name, which is less elaborately engraved than those on other

Gay horns, is an imaginative coat-of-arms based on the British arms. A rampant lion and unicorn are supporting a crown over an oval cartouche containing a profile bust of a man who is sometimes identified as Washington, although he bears no resemblance to him. Surrounding the bust is the legend SUCCESS TO AMERICA. The lion has the animated cartoonlike face seen on most Gay lions. Another lion and another unicorn, one on either side of the coat-of-arms, appear to be attacking the beasts that support the arms. Above the coat-of-arms are several animals that have typical Gay expressions: the moose has a hooked snout, and a combination horse and mule has a laughing expression that Gay frequently employed. Other animals include a deer, a dog, a raccoon, a rabbit, and a bear.

The calligraphy of the owner's name is deeply shaded. The uprights of the letters are decorated with Gay's usual faces, trefoils, heart-shaped leaves, flowers, and geometric devices, but these are not employed so profusely as on other horns. The name is contained within a simple cartouche flanked by groups of three marching soldiers with monkeylike expressions. Behind the last soldier in each formation is a large and graceful floral design. Gay signed the horn in script, spelling his name correctly.

A simple line, lunette, and dot border precedes the scalloped edge at the beginning of the recessed portion 6 inches from the tip; there is a raised ring 2¾ inches from the tip and the end of the spout is turned in a ring. A simple scroll border decorates the plug end. The slightly rounded pine plug, extending ¼ inch beyond the horn and secured by eight wooden pins, has an iron screw eye in its center.

BIBLIOGRAPHY: Grancsay, No. 232, 49; duMont, cover, frontispiece, back cover, 76–77.

SIEGE OF BOSTON SCHOOL
    Attributed to Jacob Gay (w. 1758–1787)

91 *Edward Sherburne horn*

Cambridge, Massachusetts, January 23, 1776
Horn, pine, brass, paint, dark pigment
OL: 17 in. (43.2 cm) W (of plug): 3 in. (7.6 cm)
Inscriptions: EDWARD / SHERBURNE / HIS Horn / Janr:23: 1776 / SUCCESS / TO / Liberty / MADE AT CAMBRIDGE
William H. Guthman

Edward Sherburne was a major from Portsmouth, New Hampshire, who joined the army at Cambridge in 1775. He took part in the New York campaign and was made aide-de-camp to General Sullivan on October 9, 1776. He was killed on October 4, 1777, at the Battle of Germantown.

Despite its relatively plain calligraphy, this is possibly the most beautifully executed of all horns signed by or attributed to Jacob Gay. The name of the owner, engraved in prominent block letters darkened with pigment, is contained within a boldly carved and darkened rectangular cartouche. The rococo embellishment at the spout end of the cartouche becomes a border for nearly all of the carved section next to the

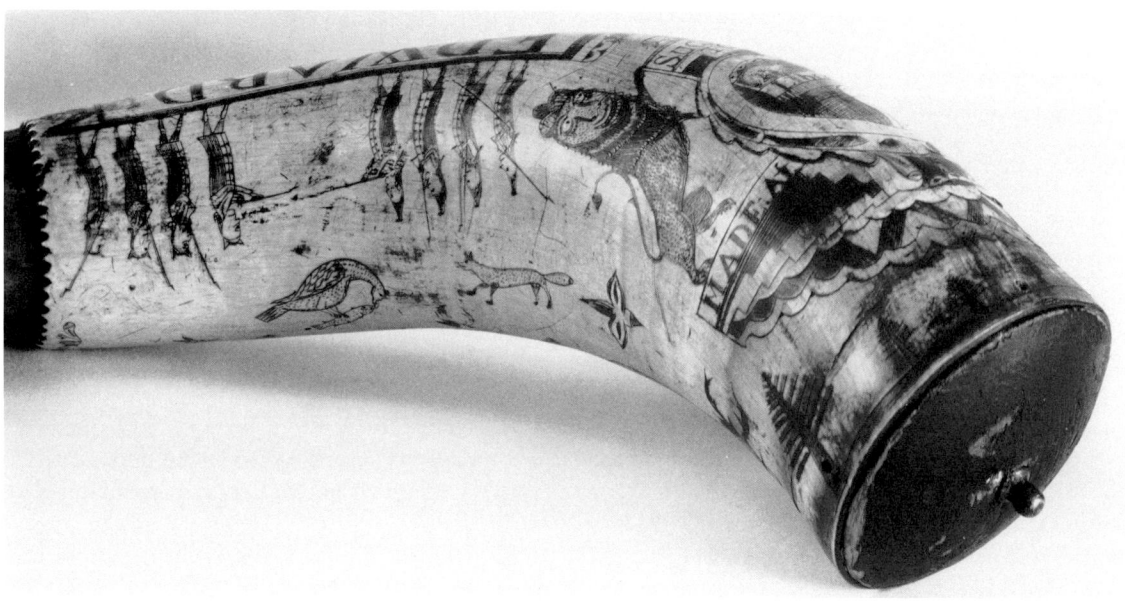

91

91. Jacob Gay's animals were always graceful and distinctive, as are the hook-nosed moose and long-eared fox seen here.

recessed portion. The entire plug end of the cartouche consists of a shaded geometric ornament.

Standing directly upon the top of the cartouche are eight soldiers divided into two opposing groups of four. The front soldier of each group is firing upon the other group, the second and third pairs are marching with their muskets upright, and the fourth pair are holding their muskets over their shoulders. Underneath the cartouche is a rococo design that separates into two sections, forming another cartouche for the date. Below this are three rows of animals, birds, and reptiles, each separated from the next by a tree or flower. A bird is devouring a fish, a moose is munching on a twig, and a beaver is about to enter a trap. Each boldly engraved creature has a typical Gay expression and form and is highlighted with pigment.

The most dramatic feature of the horn is its coat-of-arms, consisting of typical Gay lion and unicorn supports, a doe resting on top of the stepped cartouche that contains the words SUCCESS TO, and a heart below. The cartouche contains a bust that may be Washington. There is a rectangular cartouche above the head that contains the word *Liberty* and has two more animals standing on top. The lion and unicorn support both the heart and the cartouche; the lion stands upon a cartouche bearing the words MADE AT, and the unicorn stands on one that says CAMBRIDGE. The entire coat-of-arms is supported by an ambitious geometric platform that is deeply engraved and shaded with pigment.

The final and most unique feature of this horn is the face carved in relief at the tip of the spout, which seems to watch the powder as it is poured out. A collar of the most expertly carved scalloping this writer has seen on a Gay horn decorates the beginning of the

recessed portion 4¾ inches from the tip. There is a raised carved ring 2¼ inches from the tip that has a brass post with a ring bail mounted on it. The rounded pine plug, which is painted black, is secured by four wooden pegs. Another brass furniture knob is screwed into the center of the plug.

BIBLIOGRAPHY: Charles W. Brewster, *Rambles About Portsmouth* (C. W. Brewster & Son, 1859); Grider, NYHS; Grancsay, No. 773, 67; *Bulletin*, KRA, Fall, 1985, 7; *Bulletin*, ASAC, October, 1989, cover, No. 61, 49.

SIEGE OF BOSTON SCHOOL
Attributed to Jacob Gay (w. 1758–1787)

92 *James Martin horn*

Probably New York, New York, June 28, 1776
Horn, pine
OL: 7½ in. (19 cm) W (of plug): 1⅞ in. (4.8 cm)
Inscriptions: JAMES × MARTIN / HIS × HORN × MADE
In / June the 28 1776 you May See / this and [unfinished]
Henry Francis du Pont Winterthur Museum

A number of men named James Martin enlisted in New Hampshire and Massachusetts regiments during the Revolution. Because Jacob Gay was working in New York at this period, it seems logical to place this horn there.

This is the smallest Gay horn I've seen and could well have been a priming horn. The combination of large shaded block letters and competently engraved script is characteristic of Gay's work, as is the formation of soldiers. The faces of the soldiers, two of whom are firing at each other, do not have Gay's unmistak-

Catalogue

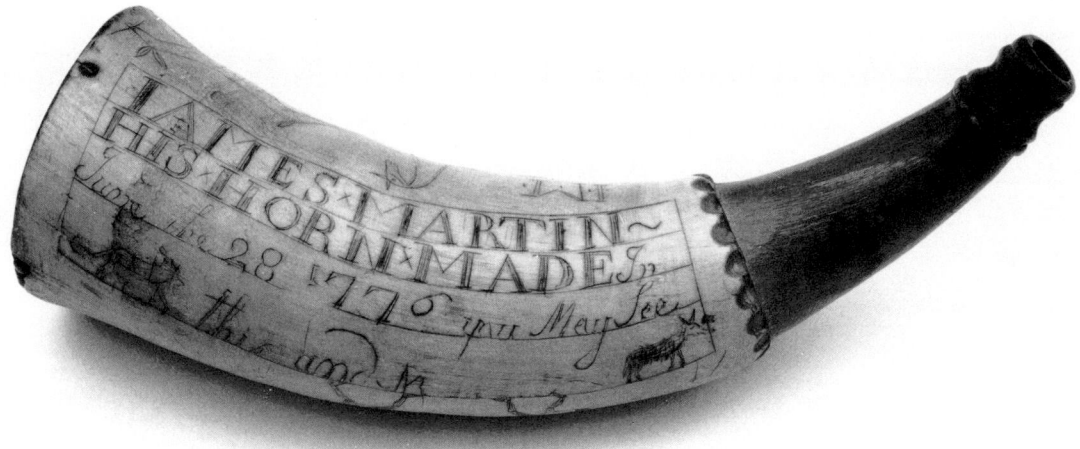

92

able monkeylike expressions but are nevertheless of a recognizable Gay type. Other vignettes, of two soldiers fencing and of a lion and a unicorn, are engraved on a far smaller scale than was usual for Gay. The lion and unicorn support a compass with a crown finial. Neatly carved figures of a horse and a running deer are nearby.

Scalloping precedes the beginning of the recessed portion 2½ inches from the tip. The end of the spout has a turned double ring and the flat pine plug is secured by four wooden pegs.

SIEGE OF BOSTON SCHOOL
  Attributed to Jacob Gay (w. 1758–1787)

93 *John Noyes horn*

Probably made in late 1775 or early 1776
Horn, pine, brass
OL: 12½ in. (31.8 cm) W (of plug): 2½ in. (6.4 cm)
Inscriptions: JOHN / NOYES / HIS . HORN / Liberty / SUCCESS TO AMERICA
New Hampshire Historical Society

Too many men of this name served in the Revolution to identify the owner with certainty. However, this

93

185

93. Jacob Gay's soldiers usually had whimsical expressions and odd-looking squashed headgear.

horn was carved for a soldier who served either in the Northern Campaign or in a camp around Boston.

Although the calligraphy is not embellished with ornaments, the block letters are boldly carved and accented with cross-hatching and the "O's" of the name have animated faces within them. There is a beautiful floral and geometric design on the spout side of the name and an interesting geometric design highlighted with shading that produces a curved checkerboard effect underneath. The plug end of the inscription is decorated with a coat-of-arms typical of Gay's work, consisting of an animated lion and a unicorn supporting a heart cartouche. Atop the heart is a grenadier with drawn sword and within it is a profile bust, possibly of Washington, with the word *Liberty* above the head.

This coat-of-arms is similar to several others in this exhibition that Gay engraved. The lion and unicorn are standing on a scroll that bears the motto SUCCESS TO AMERICA. Above the letter "L" carved within a broken arch are two groups of three soldiers each, confronting each other in battle formation. The lead soldier is firing, the second soldier is in the ready position, and the last soldier is at attention. Below this is a hunter, possibly an Indian, with his hound, firing his musket at a deer.

A neat scalloped edge precedes the recessed portion 4⅛ inches from the tip and there is a double raised ring 2¼ inches from the tip, which also has a double ring. The slightly raised plug extends ⅜ of an inch beyond the horn and is secured by four small round wooden pegs. A brass staple 1 inch long and ⅜ of an inch wide is driven into one end of the plug.

BIBLIOGRAPHY: Grider, NYHS; Grancsay, No. 623, 62.

SIEGE OF BOSTON SCHOOL
Attributed to Jacob Gay (w. 1758–1787)

94 *Colonel Samuel Connor horn*

Probably Fort No. 4, New Hampshire, July-December, 1776
Horn, cherry, iron, brass
OL: 15 in. (38 cm), W (of plug): 2½ in. (6.4 cm)
Inscriptions: COLONEL / SAMUEL / CONNOR / His Horn / SUCCESS TO THE AMERICAN ARMY / LIBERTY
William H. Guthman

Samuel Connor was the son of a prominent physician of Pembroke, New Hampshire. He served as captain of the 9th Company in December of 1775, when General Sullivan called for additional troops to replace the Connecticut troops whose enlistments had expired. During the New York campaign, from 1776 until he was discharged in 1777, Connor served as lieutenant colonel in Colonel Joshua Wingate's New Hampshire Militia Regiment. On September 29, 1777, he and three other former officers of the New Hampshire line volunteered to serve as privates under General Horatio Gates at Saratoga. Connor died of wounds on October 9, 1777.

This horn was undoubtedly carved by Jacob Gay during the Siege of Boston, probably early in 1776. The simple block lettering of the name is in the same style as that of the Sherburne horn (No. 91), but not so bold. The rectangular cartouche has an elaborate engraved rococo border consisting of floral, scroll, and geometric designs. The coat-of-arms at the plug end is supported by the typical Gay lion and unicorn, with an elaborate oval cartouche in the center. Within the cartouche is an American soldier with upraised sword derived from the engraving Paul Revere created in 1775 for Massachusetts paper currency and

94

bills of credit. Whereas the Revere engraving shows the soldier holding a rolled Magna Carta in his left hand, Gay has him holding the Scales of Justice. Gay also changed the wording from "Issued in defence of American Liberty" to "Success To The American Army—Liberty."

The coat-of-arms rests on a magnificent rococo scroll that encircles the plug end, serving as a border as well as a support for the arms. Among other motifs are a soldier with the typical Gay monkey face shooting at a deer and a rabbit and an expertly carved compass next to the lion.

A simple border precedes the scalloping at the beginning of the recessed portion 5 inches from the tip. There is then an elaborate series of raised rings ½-inch wide 2 inches from the tip, which itself has a double lip. A large brass staple, 2¼ inches long and ⅜ inch wide, is driven into the butt end of the horn just before the scalloped border. The beautifully turned cherry plug is secured by seven pins supplemented at a later date by five iron tacks. The plug, which has a central knob, extends ½ inch beyond the horn.

BIBLIOGRAPHY: *Bulletin*, KRA, Fall, 1985, 6.

94. Jacob Gay borrowed from Paul Revere for this patriotic design.

95

SIEGE OF BOSTON SCHOOL
Possibly by Jacob Gay (w. 1758–1787)

95 *Amos Bostwick horn*

New York, New York, May 21, 1776
Horn, hardwood, brass, iron, varnish, brown pigment
OL: 16½ in. (42 cm), W (of plug): 3½ in. (8.9 cm)
Inscriptions: AMOS / BOSTWI^{CK} / His Horn made may Y^c 21 / AD 1776 / SUCCESS TO AMARICA / MADE IN KNEW YORK
William H. Guthman

Amos Bostwick (1743–1829) of New Milford, Connecticut, served as ensign in Colonel Charles Webb's 19th Continental Regiment (formerly the 7th Connecticut Regiment) in 1776. That regiment took part in the Siege of Boston, then marched to New York but did not participate in either the Battle of Long Island or the Battle of White Plains. It was, however, engaged in the Battles of Trenton and Princeton. Amos Bostwick is recorded as having served in Captain Isaac Bostwick's Company, which crossed the Delaware River under the command of George Washington on December 25, 1776. Two other known horns belonged to men who crossed the Delaware that night (see No. 108). Bostwick also served as ensign in Colonel Bezaleel Beebe's Connecticut State Regiment in 1780.

Except for the distinctive animals and the format of the cartouche, this is not typical of Jacob Gay's work and may have been made by another carver who was copying Gay's style. The lettering is inferior to Gay's and the lion's eyes were enhanced at a later date, although none of the many other animals seems to have suffered that fate. The cartouche has a lion and a unicorn supporting a circle beneath a stylized crown as well as a profile bust of a man who probably represents Washington. The legend SUCCESS TO AMARICA surrounds the bust, and a scroll enhanced with floral branches underneath the cartouche bears the legend MADE IN KNEW YORK. A large moose with a hooked snout is munching on one branch of a tree while a bird perches on another.

A scroll border precedes the scalloped edge at the beginning of the faceted recessed portion 4½ inches from the tip and there is a double raised ring scalloped on all edges 1¼ inches from the tip. The wooden stopper may be original. There is also a scroll border at the butt end. The rounded plug, which extends ½ inch beyond the horn, is secured by eight wooden pins that have been supplemented by six iron nails. All of these fasteners may be original. The plug, which has an elaborate brass drawer pull in its center, was at some time given a heavy coat of ocher paint that has turned dark brown. An old coat of varnish remains on the horn.

Catalogue

96

SIEGE OF BOSTON SCHOOL
Attributed to Jacob Gay (w. 1758–1787)

96 *Samuel Webster horn*

Warwick, Rhode Island, September 26, 1777
Horn, pine, iron, reddish-brown pigment
OL: 14 in. (35.6 cm), W (of plug): 3 in. (7.6 cm)
Inscriptions: SAMUEL / WEBSTER / His Horn Made At Warwick In / Rhode Island Govement Sep^{tm} ye 26 Day 1777 / SUCCESS / TO THE / AMARICA^n / ARMS
William H. Guthman

Samuel Webster served in Captain Simon Marston's Company, Colonel Joseph Senter's New Hampshire Regiment, raised for the defence of Rhode Island on July 15, 1777. His company was discharged on January 7 and 8, 1778.

This horn and that of Colonel Samuel Connor (No. 94) are almost identical. Both have fine patriotic cartouches, and the one on this example has exceptionally fine and elaborate rococo scrolls that surround a creature that appears to be a cross between a rampant lion and a rampant horse—perhaps a conflation of the lion and the unicorn. The legend SUCCESS TO THE AMARICAN ARMS surrounds the cartouche and there is a flintlock musket in an open panel at its crest.

The expertly carved double-line block lettering with crosshatched shading resembles that of the Connor horn. The inscription is set within a rectangular cartouche bordered by wavy broken lines, a series of small dashes, and beautiful floral scrollwork. The remainder of the horn is sparsely decorated with the winged horse, Pegasus; with a long-necked bird holding a wriggling worm in its beak; and with a hunter aiming his musket at a bird in a tree.

There is neat scalloping at the beginning of the recessed portion 5½ inches from the tip, and then deeper recessing begins 2 inches from the tip. No raised ring is provided to secure the carrying strap, which suggests that this horn was never actively used. The plug is secured by four small wooden pegs and a 1½-inch-long iron staple that is driven into the side of the horn near the plug. The slightly rounded plug, extending ⅜ of an inch beyond the horn, is painted a heavy reddish-brown.

BIBLIOGRAPHY: *Antiques and the Arts Weekly*, August 21, 1987, advertisement.

SIEGE OF BOSTON SCHOOL
Attributed to Jacob Gay (w. 1758–1787)

97 *Elijah Bradbury horn*

West Point, New York, December 21, 1778
Horn, pine, brass, iron
OL: 12½ in. (31.8 cm), W (of plug): 2½ in. (6.4 cm)
Inscriptions: ELIJAH / B^{R}ADBARY / HIS:HORN:MADE:AT: WEST / point December ^{ye} 21th 1778 / DRINK:A:BOUT / DRINK:A:BOUT RM
Maine Historical Society

Elijah Bradbury was a corporal in Captain Stephen Jenkins's Company, Colonel Thomas Poor's Massachusetts Regiment, from May 25, 1778, to February

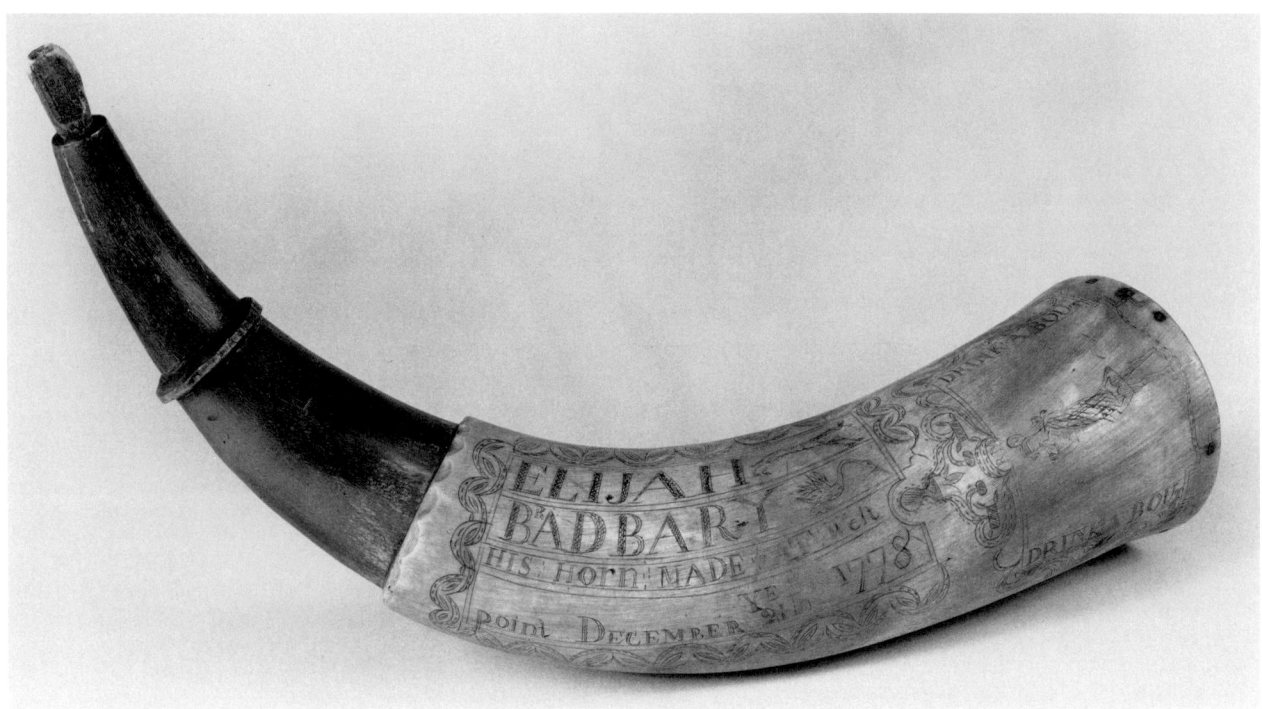

97

97. *Here Gay turns from patriotic concerns to social commentary.*

9, 1779. The regiment served along the Hudson River at Peekskill; Fort Clinton, one of the small forts guarding West Point that was next to Fort Montgomery on the west bank of the river; and King's Ferry.

Although the calligraphy of the name is boldly and neatly executed (save for the belated insertion of the first "R" in Bradbury's name) and accented with crosshatching, it lacks Jacob Gay's typical ornaments and faces. It is surrounded, however, by a graceful cartouche consisting of floral scrollwork. Its animated animals include a running fox after the name ELIJAH and a goose after the name BRADBURY; a mule with a typical Gay snout; another goose; two deer; and a fox. A second cartouche contains an unusual abstract coat-of-arms, the initials RM, and the figure of an Indian drinking from a cup. On either side of this figure are the words DRINK:A:BOUT, which may be a derogatory comment on American Indians' tendency toward bouts of drinking or may simply be a reference to a round—or bout—of drinks.

Crude scalloping precedes the recessed portion 5 inches from the tip. A shaped stopper closes the spout. The slightly rounded plug, extending ¼ inch beyond the horn, is secured by six round wooden pegs. In the plug's center is a brass screw ⅜ of an inch in diameter and next to it is an iron tack of later date.

BIBLIOGRAPHY: Grancsay, No. 94, 44.

Catalogue

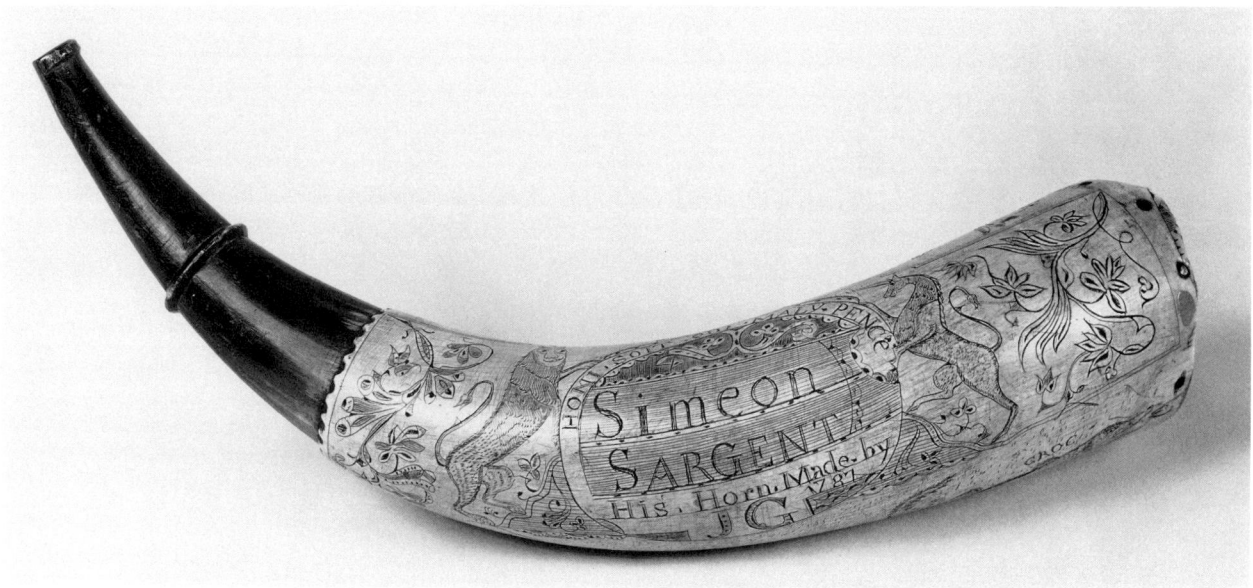

98

Catalogue only
SIEGE OF BOSTON SCHOOL
   Signed by Jacob Gay (w. 1758–1787)

98 *Simeon Sargent horn*

Possibly New Hampshire, 1787
Horn, pine, iron
OL: 16½ in. (42 cm), W (of plug): 2½ in. (6.4 cm)
Inscriptions: Honi Soit Qui Mal Y Pence / Simeon / Sargent / His Horn Made by / JG 1787 / Dolph / Sea Wolf / Croc
Carol and the late Thomas J. Segal

This horn, the latest Gay example known to me, is illustrated to emphasize the amazingly long time period during which this maker produced beautiful powder horns. Probably used for hunting, the horn exhibits a number of typical Gay features: a distinctive coat-of-arms, finely executed lettering, and imaginative animals, fish, trees, and the figure of a hunter. This horn and that of Amos Bostwick (No. 95), offer cause for further study of Jacob Gay's work and of the possibility that other carvers added to horns Gay may have begun and/or copied Gay's style.

The spout is 5¾ inches from the beginning of the nicely scalloped recessed portion and there is a raised ring 2⅜ inches from the scalloped edge and 3 inches from the tip. The slightly recessed pine plug is held by three wooden pins. There is evidence of a scalloped extension lobe on the rim of the plug end, but that is now missing; after it broke a heavy iron staple was added to hold the carrying strap.

SIEGE OF BOSTON SCHOOL
   Attributed to the Hosmer Carver
   (w. 1775–1776)

99 *Reuben Hosmer horn*

Concord, New Hampshire, May, 1775
Horn, maple, iron
OL: 13½ in. (34.3 cm), W (of plug): 3¼ in. (8.3 cm)
Inscriptions: Concord May 1775 / REUBEN HOSMER / HIS HORN / MASON / RH
Concord Museum

Reuben Hosmer first appears as a wagoner on the payroll of Captain Benjamin Mann's Company, Colonel James Reed's New Hampshire Regiment, from May 23, 1775, to August, 1775. He next turns up at Winter Hill on September 13, 1775, as a private in the same company and regiment. In 1777, listing Mason, New Hampshire as his home, he was engaged in the Continental Service of the United States in the 5th New Hampshire Regiment of Militia, Captain Frye's Company, Colonel Scammel's Regiment. Alexander Scammel was from Durham, New Hampshire, and Captain Isaac Frye was from Wilton, New Hampshire. Hosmer next appears in Captain Joseph Barret's Company, Colonel Moses Nichols's Regiment, which marched to reinforce General St. Clair at Ticonderoga in June of 1777. The pay roll, dated December 1777 at Hillsborough, New Hampshire, lists both Reuben and Nathaniel Hosmer in the same company. From April 23, 1777, to January, 1780, Reuben was again in Scammel's regiment, serving all

Drums A'beating, Trumpets Sounding

99

99. *The Hosmer Carver's decoration harks back to the formalized European tradition of the King George School.*

99. *A whimsical fish and the name of the Hosmer Carver's hometown, Mason, New Hampshire, join his more formal motifs.*

of this time as a private. From January 1, 1780, to January 1, 1781, he was in Captain Nicholas Gilman's company, Colonel Alexander Scammel's 3rd New Hampshire Regiment.

This is a fine Siege of Boston horn, identical in style to the Nathaniel Hosmer horn (No. 100). Its neatly executed calligraphy has large capitol letters and double-line lettering with accents of shading between the lines. The letter "O" is always dotted in the center and the lines of lettering are bordered by rectangular cartouches decorated with shaded lunettes almost identical to those on Nathaniel Hosmer's horn. The word MASON refers to Mason, New Hampshire, Reuben's hometown. Other motifs include a large crowned lunar-looking face; two large, carefully engraved geometric flowers; and a long fish that also appears on Nathaniel's horn. A double border with shaded lunettes appears before the nicely scalloped recessed portion that begins 5 inches from the tip of the spout; a raised ring appears 2¾ inches from the tip and the horn is hexagonal from that point to the tip. There is a wide border consisting of shaded lunettes at the plug end and the rounded maple plug, which extends less than ¼ inch from the end of the horn, is secured by six round wooden pins. Four small square-cut nails are driven into the center of the plug all the way to the heads; the letters RH are also neatly carved in the plug's center.

BIBLIOGRAPHY: Grancsay, No. 456, 57.

SIEGE OF BOSTON SCHOOL
Attributed to the Hosmer Carver
(w. 1775–1776)

100 *Nathaniel Hosmer horn*

Mason, New Hampshire, May 1, 1776
Horn, maple
OL: 18½ in. (47 cm), w (of plug): 3¼ in. (8.3 cm)
Inscriptions: MASON*MAY 1*1776 / NATHANIEL*HoSMER / HIS*HORN*ADoM / N H
Private Collection

Although Hosmer does not appear on the New Hampshire rolls until 1777, he undoubtedly marched to Boston and may have been part of the New Hampshire troops called up in December of 1775, when the Connecticut enlistments expired. His first official listing, however, occurred in 1777 when he signed up to serve in Captain Joseph Barrett's Company of militia raised in Hillsborough for the alarm to assist at the evacuation of Ticonderoga. He is listed as serving with a horse he supplied, and was paid for, in transporting provisions. Hosmer is also listed as serving in the Rhode Island campaign of 1778.

This unusually large, beautiful horn is similar in format to many Siege of Boston horns—its architecture, for example, resembles that of the George Morley and Gershom Mott horns (Nos. 81 and 101) in having recessed portions at both ends—but it is in many ways nearly identical to the Reuben Hosmer horn (No. 99). Its calligraphy is neatly executed in large capital letters shaded by lines, the words separated by a variety of geometric devices. The inscription is enclosed by rectangular cartouches decorated with shaded lunettes.

100

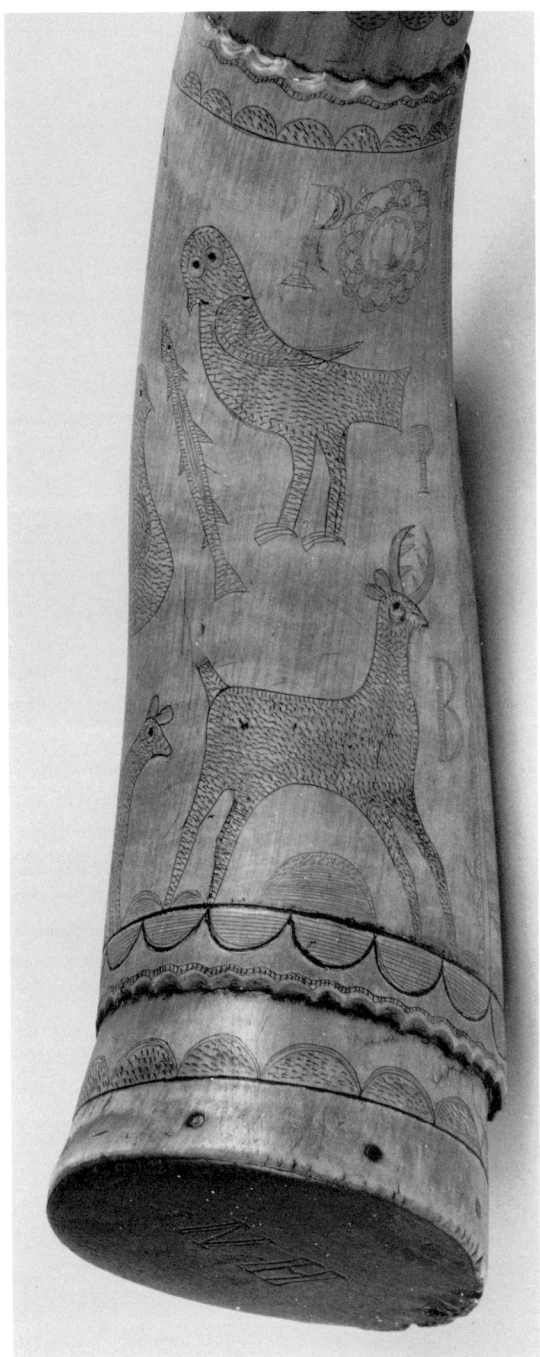

100. *The recessed plug end was an occasional Siege of Boston feature.*

Five creatures are engraved in an almost professional hand. Two are identified with a capital letter, "P" for parrot and "B" for buck. The unlettered animals are a dove, a rooster, and a large dog or wolf. All the wildlife is carefully drawn and intricately feathered or filled in with fur.

There is a double border next to a beautifully scalloped edge at the beginning of the recessed portion 6½ inches from the tip. There is another neat border ½ inch closer to the tip where deeper recessing begins and there is a scalloped and engraved raised ring 3 inches from the tip. The remainder of the recessed portion is hexagonal in section. Fine scalloping also precedes a recessed portion 1½ inches from the butt end and there is another border ¾ of an inch from the butt end. The flat maple plug, secured by eight round wooden pegs, has the initials *N H* incised in large double-lined letters at its center. Two holes ¼ inch apart and ⅛ inch from the edge secured the carrying strap.

SIEGE OF BOSTON SCHOOL

101 *Gershom Mott horn*

Roxbury, Massachusetts, March 11, 1776
Horn
OL: 20 in. (50.8 cm), W (of plug): 3¼ in. (8.3 cm)
Inscriptions: GEARSHOM : MOTT : HIS : HORN : MADE / IN * ROXBURY : CAMP * MARCH * THE * 11ᵀᴴ * A D 1776 / LIBERTY : OR : DEATH / MADE : IN : ROXBURY / CP / FORT / BOSTON
Private Collection

Gershom Mott served as a lieutenant in Colonel John Lamb's 2nd New York Artillery Regiment; as a captain in the 1st New York Regiment in June, 1775; as a captain in Colonel John Nicholson's Continental Regiment on March 8, 1776; and as a captain in Colonel John Lamb's 2nd Continental Artillery from January 1, 1777, to June of 1783. His name is correctly spelled "Gershom," and the small capital A inserted in his name is a later addition, as is the TH in the date.

Although he was not an artistically gifted calligrapher, the engraver of this extremely large horn did his best to be consistent. The carving and lettering are carefully planned and meticulously executed and the chip-carved border around the inscription is of high quality. The chip-carved sunburst with animated face at the beginning of the inscription and the deeply carved coat-of-arms with the inscriptions LIBERTY OR DEATH and MADE IN ROXBURY C[am]P are expertly done. Other decorations include a lone unicorn supporting the coat-of-arms, a man-of-war under full sail, a Lorelei, another unicorn, a deer, two whales, two birds, and a single-masted ship. A large alligator with a plaid design on its body and a passive lion on its back is walking above a four-sided fort which is probably Roxbury Fort. These motifs, along with a compass and an extremely fine view of Boston, cover the entire surface of the horn.

Catalogue

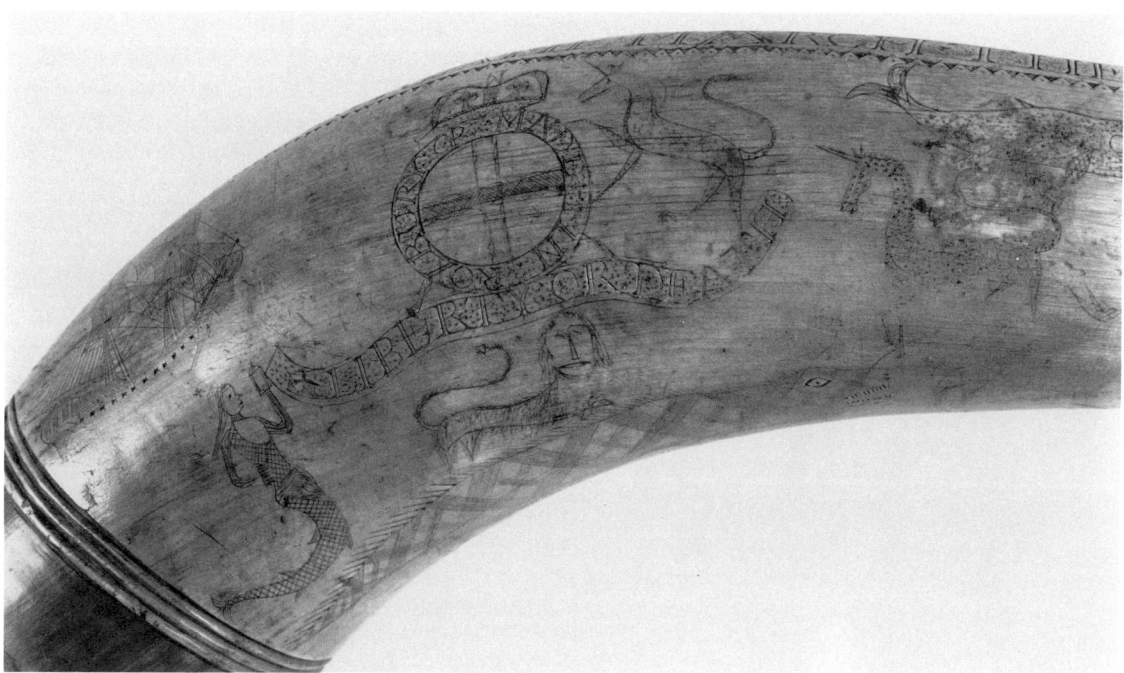

101. *Here is another characteristic Siege of Boston transposition of the British coat-of-arms into an American patriotic symbol.*

All of the above decorations are far more literal and realistic than those of other Siege of Boston horns. The engraver evidently consulted print sources and copied them closely. The alligator with a lion on its back could be interpreted as the British Army's trying to subdue America, of which the alligator could reasonably be thought to be a symbol.

A neat geometric border precedes the finely scalloped edge at the beginning of the recessed portion 4¼ inches from the tip. There is a double raised ring 2¼ inches from the tip and the horn is ten sided between these rings and the raised lip at the tip. Like the Morley and Nathaniel Hosmer horns (Nos. 81 and 100), the Mott horn has a recessed portion at the

195

Drums A'beating, Trumpets Sounding

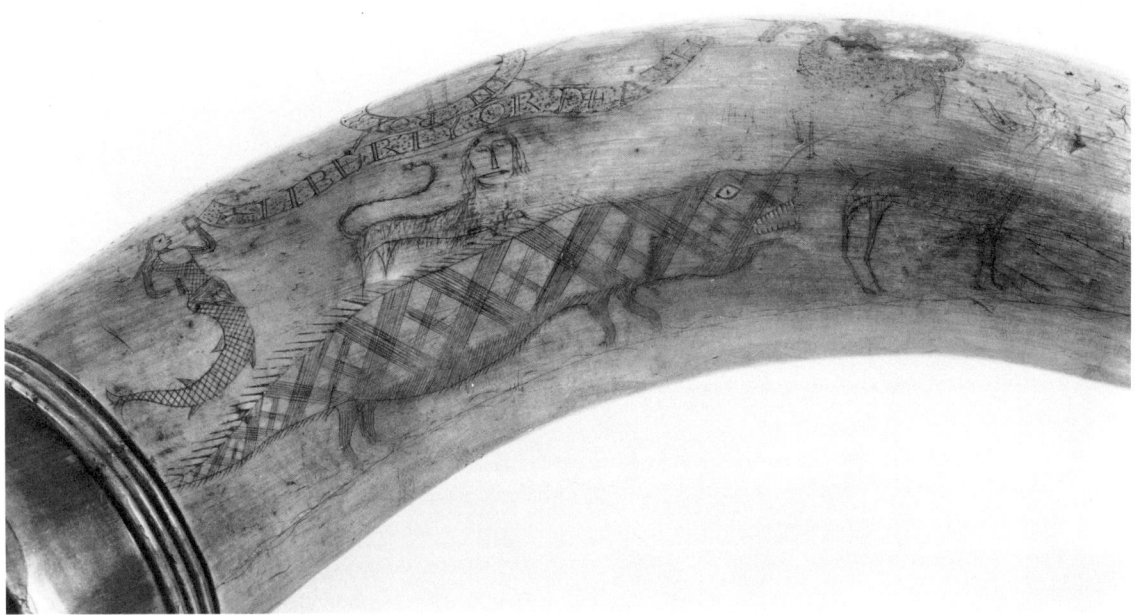

101. *This scene perhaps symbolizes the British repression of America: the alligator, or America, is forced to carry the lion, or Great Britain, on its back.*

butt end; preceding it is a deeply incised three-ring border. The plug is missing, and portions of the recessed butt end have broken off at the former locations of the pins.

BIBLIOGRAPHY: *The Providence Sunday Journal*, March 20, 1921, 5th section, 5 (story on an exhibition of powder horns at the Rhode Island Historical Society); Grancsay, No. 604, 62; Swayze, 121–123, 179.

SIEGE OF BOSTON SCHOOL

102 *Jabez Gooddel horn*

New York, New York, area, June 7, 1776
Horn, mahogany, iron
OL: 19¼ in. (49 cm), W (of plug): 2¾ in. (7 cm)
Inscriptions: Jabez Gooddel's Horn Made June Ye 7th / A.D. 1776 To Arms to arms O Free born Sons / Exert the Sword And Spear / Oppose the Tyrant and / bands; Defend your Rights most Dear. / Street Firing / Make / ready / Present / Fire / Eager the Soldier meet his desperate foe / With an Intent to give the Fatal Blow / The cause he fights For Animates him high / Namely Religion and dear Liberty / For These he Conquers or more bravely dies /And Yields himself A willing Sacrifice
James E. Routh, Jr.

102

Catalogue

102. *This scene showing formations of soldiers illustrates how closely Siege of Boston carvers' work resembled that of Lake George School carvers.*

Jabez Gooddel served in Captain Noadiah Hooker's 6th Company (Farmington), Colonel Joseph Spencer's 2nd Connecticut Regiment, from May 8 to December 18, 1775. The company was recruited mainly from Middlesex County and marched to the camps around Boston, taking post at Roxbury; detachments took part in the Battle of Bunker Hill and in Arnold's expedition to Canada. The regiment was reorganized in 1776 under Colonel Samuel Wyllys, marched to New York and took part in the New York, Long Island, and White Plains campaigns. Gooddel's record is not complete, but he undoubtedly enlisted for these campaigns. The date on his horn indicates that he helped with the New York fortifications prior to the Long Island campaign.

This is a beautifully carved horn in the style of, but not by, Jacob Gay. Its calligraphy is good but not exceptionally so. Formations of uniformed soldiers are facing each other, with officers holding halberds in the rear. The forward troops are shown in the three firing positions, while the rear troops are marching in formation. Nicely carved schools of fish swim in water indicated by hundreds of fine dots or bubbles and by floral vines floating gracefully in the current.

A nicely scalloped edge precedes the recessed portion 5¼ inches from the tip and there is a scalloped raised ring 1¾ inches from the tip followed by octagonal faceting of the spout. The butt end has no border and the extremely large lathe-turned mahogany plug is not original; it appears to be about one hundred years old. It is fastened to the horn by both wooden and iron pins.

BIBLIOGRAPHY: Grancsay, No. 386, 54; Catalogue, Walpole Galleries, January 10–11, 1924, and August 6, 1925; *Bulletin*, Fort Ticonderoga Museum, Vol. 12, No. 3 (October, 1967), 190; duMont, 84; *Bulletin*, ASAC, No. 42 (Spring, 1980), 2.

SIEGE OF BOSTON SCHOOL

103 *Major Samuel Selden horn*

Boston, Massachusetts, March 9, 1776
Horn, walnut, iron
OL: 19½ in. (49.5 cm), W (of plug): 2⅞ in. (7.3 cm)
Inscriptions: LYME, MARCH: THE 9th A.D. 1776 / MAJOR. SAMUEL SELDENS:P: HORN / MADE: FOR:THE DEFENCE:OF. LIBERTY: / The Regulars. Brst. Work / Ship. AMARICA. Mortar / The Redouts / Yankes. Brest. Work / Boston. Neck / Block / House
Massachusetts Historical Society

No record of Samuel Selden's participation in the Siege of Boston survives, but he obviously served there. He is first listed in 1775 as a militia major from

103

Hadlyme, Connecticut. In June of 1776 he was promoted to colonel in the Connecticut State Regiment and from June 20 he served in that capacity in the 4th Battalion, Brigadier General James Wadsworth's Connecticut State Brigade. His battalion was raised to reinforce Washington's forces in New York and was stationed on the East River when the British attacked on September 15, forcing the American troops to abandon New York City. Colonel Selden was taken prisoner on September 15 in the area of modern 34th Street and died in prison in New York City on October 11, 1776. This horn was given to the Massachusetts Historical Society in June of 1881 by James, Lord Bowes, of Liverpool, England, through his friend Thomas G. Frothingham of Boston.

The well-done block lettering consists of double-line engraving filled in with cross-hatching. The letters are neatly spaced, with punctuation between each word. There is an artistic and highly detailed plan of the British defenses in Boston, as well as of the American fortifications on Boston Neck. The buildings are laid out in an abstract geometric pattern, with the roofs and walls of the forts emphasized by heavy dark lines with cross-hatching. The staggered roof lines create an arresting geometric effect, as does the design of the masts and rigging of many ships packed tightly together, their hulls floating on a dark crosshatched sea. These ships form one massive motif. Interestingly, there is also a lion which is similar to the lion that decorates James Greenfield's horn (No. 84). On the reverse of the horn is a large man-of-war labeled AMARICA, carved in an exceedingly fine manner that demonstrates the engraver's familiarity with ships.

There is a simple linked lunette-and-dot border at the beginning of the recessed portion 5 inches from the tip. The recessed area is composed of three recessions, each successively smaller in diameter, and there is a notched raised ring 3¾ inches from the tip. The 1½-inch spout is raised, with a high base that diminishes toward the slightly flared lip. There is a recessed portion 1 inch wide at the butt end that is incorporated into the vignette of the fortifications on Boston Neck. The superb crown-shaped 1¼-inch-square extension lobe is engraved with a man-of-war on a crosshatched sea, with a watchful pair of eyes peering down over the masts; two holes are provided for the carrying strap. The walnut plug was elegantly turned on a lathe and at one time had a large metal fixture screwed into its center. It is now secured by twelve small iron pins.

BIBLIOGRAPHY: Grider, NYHS; Grancsay, No. 761, 67; *A History of the United States and Its People, from Their Earliest Records to the Present Time*, 7 vols. (The Burrows Company, 1904–1910), vol. 6, 23, illus.; J. L. Sticht, "Historical Military Powder Horns," *St. Nicholas Magazine*, 23 (1896), 993–997; *Minutes*, MHS, 1881.

104

SIEGE OF BOSTON SCHOOL

104 *John Abbott horn*

Canada, July, 1776
Horn, pine, paper, glass, tan putty, cork, brown ink
OL: 20½ in. (52 cm), W (of plug): 3 in. (7.6 cm)
Inscriptions: JOHN : ABBOT : H H 1776 / INDEPENDENCE : DE.ᴰ : July : 1776 / John / Abbot / his horne / made in The / Year / 1776 / LF
New Hampshire Historical Society

At the request of the Continental Congress, New Hampshire raised a regiment in June and July of 1776 to reinforce the army in Canada. John Abbott was mustered in Captain William Harper's Company, Colonel Isaac Wyman's New Hampshire Regiment, in July, and his horn is one of the few to make direct reference to the signing of the Declaration of Independence on July 2.

This horn is in some respects crudely engraved, but its calligraphy is neat, legible, and evenly spaced, and has an interesting rabbit motif after the date and a four-bastion fort following the two-line inscription. It is profusely carved with vignettes of Woodland warfare: soldiers are firing from behind trees or bushes, lying on their stomachs, or striking with tomahawks from points of ambush. Wild animals, including bear, foxes, ducks, and geese, are involved in some of the scenes and one shows a soldier wrestling with a bear. Since depictions of frontier-fighting tactics are extremely rare, this horn assumes great importance despite its mediocre carving.

Deep sawtooth carving precedes the recessed portion 4½ inches from the tip and there is a triple raised ring 2¾ inches from the tip. From the rings to the tip, the horn is faceted octagonally. A piece of cork remains lodged in the spout. The plug, probably of pine, is recessed into the horn ½ inch from the edge and secured by round wooden pegs. A paper inscription is affixed to the plug by a piece of glass secured by putty, much like the glass of a surveyor's compass. Two holes halfway between the plug and the edge of the horn once secured the carrying strap, but are now covered with putty, indicating that the glass has been reset. The glass itself appears to be original.

BIBLIOGRAPHY: Grider, NYHS; Grancsay, No. 3, 41.

SIEGE OF BOSTON SCHOOL

105 *Howland or Rowland horn*

Boston, Massachusetts, or vicinity, about October, 1775
Horn, pine, iron
OL: 12 in. (30.5 cm), W (of plug): 2¼ in. (5.7 cm)
Inscriptions: [RO or HO]WLAND:S:HORN:M:IN:THE / LIBERTY:OR:DEATH / OLD:THOMAS:GAGE:GOING:FROM: BOSTON / FOX / COCK
William H. Guthman

Because this horn was cut down before Rufus Grider saw it in 1891, the owner's name is lost. However, we can establish where and when the horn was made on

Drums A'beating, Trumpets Sounding

105. *This patriotic vignette possibly depicts Patrick Henry under the words* Liberty or Death.

105. *Here General Thomas Gage evacuates Boston holding onto his tricorn hat, his musket slung over his shoulder backward indicating that he's leaving in a rush.*

the basis of the surviving inscriptions and of the vignette showing General Thomas Gage, commander of British forces in North America. Gage was relieved of his command in October of 1775, whereupon he sailed for England. He is shown here in profile, with one hand holding a musket upside down over his shoulder and the other clapped onto his hat, to emphasize the swiftness of his departure. Another view of a man, perhaps Patrick Henry, is shown beneath a ray of light, holding the key of Liberty in his left hand and an upraised sword in his right, with the banner *Liberty or Death* framing his head. Above this vignette is a stylized crown.

This was the first horn that captured the interest of Rufus Grider and inspired him to begin his collection of powder-horn drawings. In an article written in 1895 he stated that when he first began to travel about the Mohawk Valley drawing pictures of historical relics, he was not interested in powder horns because he felt they had no historical value. Then when he was visiting his friend Robert M. Hartley of Amsterdam, New York, he saw this horn, which, he said, "taught me to see those objects differently.... It opened my eyes to see them as I do at present. [This horn] was used to drive the British out of Boston, and ... was copied and a separate collection [of drawings of powder horns] begun."

The drawing of this horn is now No. 35 of the Revolutionary series in the Grider drawings at the New-York Historical Society. The caption to the drawing states that the horn belonged to Robert M. Hartley and that prior to his obtaining it, it had been shortened, "making it more convenient for hunting purposes."

The horn could be attributed to Jacob Gay on the basis of several motifs: the patriotic vignettes with stylized crowns above, the faces of the characters in the vignettes, the depiction of a cock, the tricorn hat Gage wears, and the fox — all these resemble similar motifs on signed Gay examples. The labeling of the cock and fox here has many precedents on Gay horns; sometimes Gay put an initial on or under an animal, sometimes the entire name. The lettering is almost identical to Gay's simpler style of block letters, and the draperylike border at the spout end is similar to Gay borders. Too much of the horn is missing for a strong attribution to this artist, but the quality of the carving is good enough so that it could be one of his less important works. (In addition to the decorations mentioned, there are as well an abstract house and a vine with tulips. Grider also depicted a goose on his drawing, but he may have added it incorrectly from memory.)

There is fine scalloping at the beginning of the recessed area 4 inches from the tip. The horn resumes its normal dimensions 2¼ inches from the tip and becomes octagonally faceted, with neat scalloping at the edge. The flat pine plug — a replacement at the time the horn was cut down — is secured by two iron tacks and has an iron tack driven into its center.

BIBLIOGRAPHY: Grider watercolor, NYHS; "Forward," Grider drawings, 1895, NYHS; A. J. Wall, ed., Rufus Grider, "Powder Horns, Their History And Use," *Bulletin*, NYHS, 15 (1931–1932), 3.

SIEGE OF BOSTON SCHOOL

106 *Caleb Johnson Hall horn*

Possibly Boston, Massachusetts, area or the Northern Campaign, 1776

Horn, pine, iron

OL: 15 in. (38 cm), W (of plug): 3½ in. (8.9 cm)

Inscriptions: Caleb : Johnson Hall his / Horn. This you See I / For Liberty. With / Powder And Ball Kills / all / 1776 / I & He My Soard And / that fights Hart Shall Never / For Liberty Part:Shew Me / A Sign For Good: And let My hater / See: And be Ashamed: Because Lord / thou Dost help And Comfort / me

The Connecticut Historical Society, Gift of John C. Hall

Rufus Grider recorded when he drew this horn in 1890 that its then owner, John C. Hall of Berlin, thought that the horn was carried by his grandfather, who served in Colonel Herman Swift's Regiment in 1780; he also thought that the horn was made by a British deserter. The Connecticut records confirm that a Caleb Hall served in Swift's Regiment, as well as that a Caleb Hall served in General Wooster's 1st Connecticut Regiment in 1775. The only person named Caleb Johnson Hall who is recorded in Connecticut was born in Mansfield in 1763.

The format of this horn is typical of those produced early in the Revolution. Its carver was careful, but not highly skilled — the spacing of letters becomes cramped at the end of lines, the spelling is phonetic, and the doggerel verses or slogans are odd. Despite the crudeness of its workmanship, however, it is quite graphic. The creatures illustrated have a cartoonlike animation that lends the horn great charm. The rendition of a cat and a bird, for example, is pleasing and well proportioned. The sword- and tomahawk-wielding character is curious though, because soldiers were required to provide a sword, bayonet, *or* tomahawk as part of their equipment, and it is unlikely that a soldier would have carried both a sword *and* a tomahawk.

106

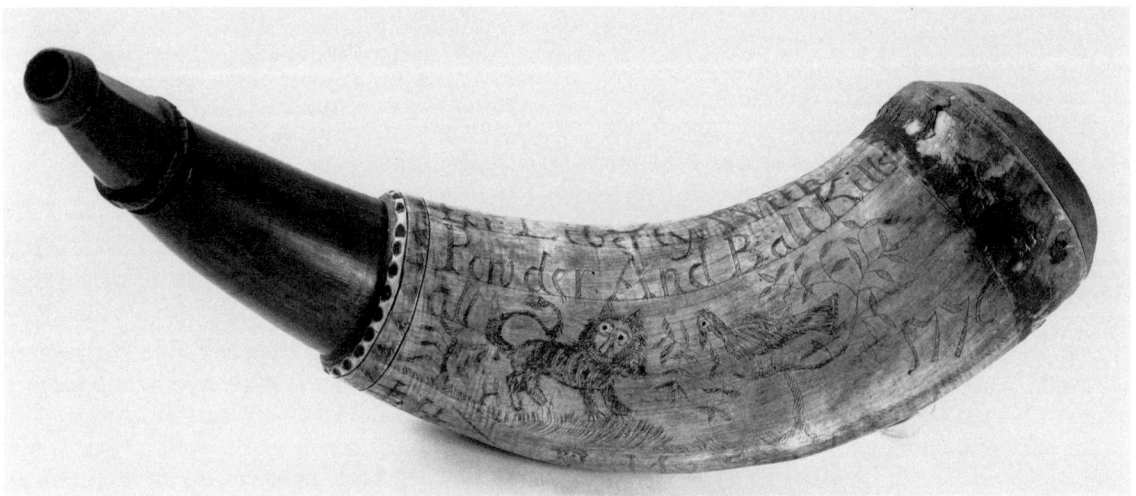

106. *This feline and bird express the whimsical nature of much American powder-horn decoration.*

A crudely scalloped or notched edge precedes the recessed portion 5¾ inches from the tip. There is a fine raised ring notched into a gadroon pattern 2¼ inches from the tip and the balance of the spout is nicely faceted into a chamfered octagonal shape. The lip is raised, neatly turned, and ½ inch wide. Both the plug end and the rounded plug have some water damage. The plug, which extends ½ inch beyond the horn, is secured by four large round wooden pegs and has a rosehead nail driven into an upper corner.

BIBLIOGRAPHY: Grider, NYHS; Grancsay, No. 408, 55.

SIEGE OF BOSTON SCHOOL

107 *Jonathan Gardner horn*

Probably Roxbury, Massachusetts, 1776
Horn, maple, brass
OL: 12 in. (30.5 cm), w (of plug): 3 in. (7.6 cm)
Inscriptions: Jonathan × Gardner × His × Horn 1776 / Liberty × Property × or × Death
Concord Museum

There are several Jonathan Gardners listed in the Massachusetts records and the most likely candidate for this horn is the Jonathan from Sunderland, Massachusetts, who was recorded at Roxbury in September

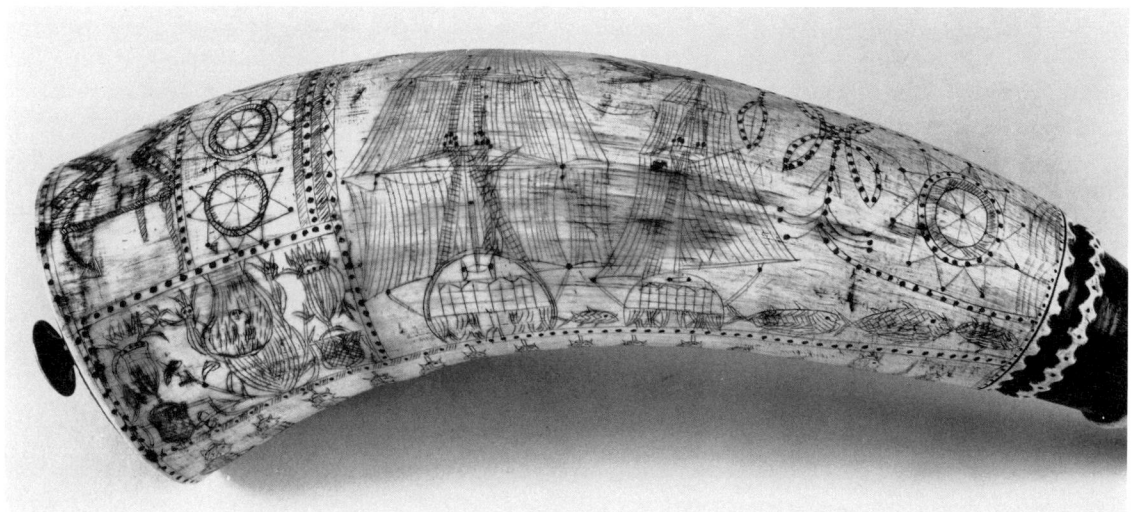

107. Compasses, animated flowers, and two ships' sterns illustrate the nautical motifs popular on Siege of Boston horns.

of 1776 and at Dorchester Heights in November of 1776 as a private in Captain Samuel Taylor's Company, Colonel Nicholas Dike's Regiment. He later served for three years in Colonel Wigglesworth's Massachusetts Regiment. The other Jonathans listed do not appear during this period, except for one who was the second lieutenant of an armed vessel in 1776, but this horn, with its formation of soldiers, does not appear to be nautical. Connecticut and New Hampshire do not list any Jonathan Gardners during this period.

This is an extremely visual horn. It is deeply engraved and accented with black pigment and its carver was meticulous in the placement of his geometric designs, trees, birds, and formations of uniformed soldiers. Rather than fighting facing each other, as is so often the case on horns of this period, this group of fourteen soldiers appears to be marching in single file. The two in the lead are officers, one with a fusil held horizontally at knee level, the other with upraised saber, and the two just behind them are a fifer and drummer. Below this formation are two other figures of uniformed men, apparently officers, one with upraised sword and the other with his sword held point downward. On the same plain but behind these figures is a hunter aiming his musket at a bird in a tree.

107. Formations of soldiers, birds, and trees were popular Siege of Boston motifs.

Above the formation of soldiers are two tall-masted ships under sail with carefully executed sterns; around them are fish, a bird, whimsical plants animated with human faces, and geometric devices, some of which appear to be in the form of ship's compasses. Although these nautical motifs might suggest that Jonathan was the seaman, many Siege of Boston horns have nautical views and fish, as well as compasslike geometric devices. The formation of soldiers precludes, to my mind, any nautical connection.

There is a finely carved sawtooth border at the beginning of the recessed portion 2½ inches from the tip of the spout, followed by a beautifully carved raised ring in a sawtooth pattern ¼ inch from the recessed border. There is a simple dotted border at the base of the horn. The slightly raised maple plug, secured by four small round wooden pins, is carved with an expertly executed pinwheel which has a brass desk-drawer pull in its center to secure the carrying strap.

BIBLIOGRAPHY: Grancsay, No. 92, 54.

SIEGE OF BOSTON SCHOOL

108 *Aaron Foot horn*

McKonkey's Ferry (Washington Crossing), Pennsylvania, December 27, 1776
Horn, pine, iron, brass, brown pigment
OL: 16¼ in. (41.3 cm), W (of plug): 3¼ in. (8.3 cm)
Inscriptions: AARON.FOOT. HIS. HORN. MADE / AT PENNSYLVANIA. Decemb-ʳ 27 AD 1776 / American / HESSIAN / R P
William H. Guthman

An Aaron Foot is recorded as serving in Captain Isaac Bostwick's Company (New Milford), Colonel Charles Webb's Regiment, in 1776. That regiment marched from Boston to New York and served in the New York campaign from April to December of 1776. Although it did not engage in the Battle of Long Island, it did participate in the Battles of White Plains and Trenton. Captain Bostwick's Company crossed the Delaware River with Washington on December 25, 1776. (At least two other horns that belonged to men who took part in the crossing are known, the Ensign Amos Bostwick horn, No. 95, and the Jonathan Cruttenden horn, which is in a private collection and is not in the exhibition.) There is no question that the Aaron Foot in Bostwick's Company was the owner of this horn, and of the several men of that name listed in records of the period, the most likely seems the one who came from Washington, Connecticut (his birth and death dates are unknown).

This is one of the very few genuine horns with phrases and imagery relating to a specific military event. Both the Foot and Cruttenden horns show American and Hessian soldiers and note the location PENNSYLVANIA, while the Cruttenden example also records the location White Plains. However, the notable feature of both horns is the elliptical way in which the victory at the Battle of Trenton is referred to: rather than a celebration of the victory, with the date and location of the battle, both horns note simply when and where they were carved. Only the satirical vignette showing an American infantryman with saber in hand chasing a retreating Hessian whose musket is slung over his shoulder refers to the battle.

What happened was this: the troops had crossed the Delaware River late in the evening of December 25, had successfully made a surprise attack and routed the defending Hessian troops, and had then crossed back over to the Pennsylvania side of the river on the

Catalogue

108. *The American soldier driving off the Hessian in this scene is a rare reference to a specific Revolutionary War battle.*

26th. The return was difficult—several soldiers froze to death in the boats on the return trip and it took others almost twenty-four hours to return to their camps. The troops then rested while their officers planned the maneuver against Princeton on January 3, 1777. It was only later that this battle was recognized as one of the turning points of the American Revolution. I have seen two other Battle of Trenton horns that I do not consider period.

One of the most beautifully carved horns of the Revolutionary period with carving in the manner of Jacob Gay, this was most likely carved as a memento, rather than as a functional accoutrement. The lettering is neat, shaded, adequately planned, and of excellent quality, although it doesn't approach the quality of the best carvers. The engraved animals are particularly fine and include a large moose munching on a tall plant with a bird perched on its back; two foxes, one of which is attempting to climb a tree while the other is running; two deer; and a peacock. There are also a house; a two-masted and a three-masted ship, both expertly carved; several trees and bushes; a horse;

and the vignette of the American chasing the Hessian. The incomplete Cruttenden horn, which shares some of the same motifs, was probably modeled on this example.

A neat geometric border precedes the finely carved scalloping at the beginning of the recessed portion 5¾ inches from the tip. Beginning 2½ inches from the tip are two raised rings with fine zigzag raised carving; they are ½ inch apart and the remaining portion of the spout is faceted octagonally. There is a plainer version of the spout-end border at the plug end. The flat brown-painted pine plug is secured by twelve square-headed iron nails, three of which are missing, and has a brass furniture knob in its center. There are also two nails driven all the way in and the carved initials *R P* (the initial "P" is partially covered by the brass knob) on the plug.

BIBLIOGRAPHY: Guthman, *U.S. Army Weapons*, 93; Swayze, 149–151; duMont, 88; *Bulletin*, KRA, Fall, 1985, 10; Suzanne Corlette, *The Pulse of the People: New Jersey 1763–1789* (New Jersey Historical Commission, et al., 1976), 99.

SIEGE OF BOSTON SCHOOL

109 *Stephen Newell horn*

Springfield, Massachusetts, September 24, 1777
Horn, brass, iron
OL: 15⅛ in. (38.4 cm), W (of plug): 6³⁄₁₆ in. (16.2 cm)
Inscriptions: STEPHEN NEWELL HIS H / HORN MADE AT Springfield / September the 24 AD 1777 / MAY THE BLOSSOM OF LIBERTY / NEVER FAIL AND THE KING AND TIRANTS NEVER / PREVAIL / Saratogue Stephen Newell his hand / United States of Liberty or death friend
The Abby Aldrich Rockefeller Folk Art Center

Stephen Newell, probably from Sturbridge, Massachusetts, served as a private in Captain Abel Mason's Company (Sturbridge), Colonel Job Cushing's Massachusetts Regiment, from August 16 to November 30, 1777. The company was drafted to serve in the Northern Army during the Saratoga campaign.

The calligraphy is good, but not expert, double-line block lettering with diagonal short dashes and deeply incised dots for serifs. Two formations consisting of

109

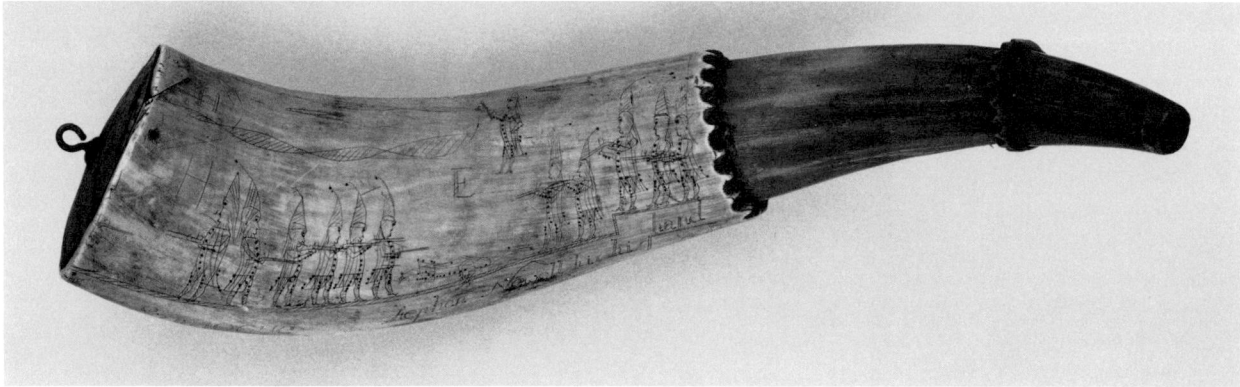

109. *Details of soldiers' uniforms are often accurate on Siege of Boston horns.*

a total of thirteen soldiers are shown firing muskets at each other; the uniform buttons and the soldiers' eyes are indicated with the same deep dots used as serifs on the lettering. Two soldiers are shown conversing and another lies dead near the *Liberty or death* motto. There are several fish and an officer carrying a sword that are oddly placed.

The edge of the horn is neatly scalloped at the beginning of the recessed portion 6 3/16 inches from the tip. The recessed portion is faceted with eleven sides up to a notched raised ring 2 3/8 inches from the tip. Just below the ring is a brass band that at one time held a brass ring for securing the carrying strap. The slightly rounded pine plug, secured by three square iron pins, has a brass-hook ring at its center.

BIBLIOGRAPHY: Grancsay, No. 619, 62.

REVOLUTIONARY WAR SCHOOL
Attributed to Joshua Kendall (w. 1779)

110 *Joshua Kendall horn*

West Point, New York, February 29, 1779
Horn, pine, pewter
OL: 14 in. (35.6 cm), W (of plug): 2 3/4 in. (7 cm)
Inscriptions: JOSHUA: KENDAL / HIS:HORN:MADE:AT / WESTPOINT:FEBRUARY / THE:TWENTYNINTH: 1779
William H. Guthman

Joshua Kendall served in Captain Daniel Allen's Company, Colonel Samuel Wyllys's 3rd Connecticut Regiment, which served under General Heath at the Battle of Stony Point on July 16, 1779.

The decoration and the precise thick-and-thin calligraphy of this horn are of the highest quality. The ornament is as profusely and carefully worked as that of the finest needlework of the same period, and in much the same format. By 1779, even though a great deal of activity was occurring in the West Point vicinity (Stony Point was taken by the British on May 31 and was recaptured by Anthony Wayne on July 16), it must be taken for granted that this horn was carved as a memento and not for use as a military accoutrement. Cartridge boxes were by this time widely used by American troops.

This example represents the nonmilitary themes of most horns carved during the latter years of the Revolution and later. It is one of the neatest examples of soldier's art this writer has seen, exhibiting an extremely talented but totally unsophisticated style of craftsmanship. The block letters, accented with thick-and-thin calligraphy and highlighted with upright lines within the letters, are precisely engraved within double-line borders accented with dots. The lettering is bold, ranging in height from 3/4 of an inch to 1 inch. Above the lettering but within the same border is a stylized floral vine, and above the vine, within another identical border, are plants with many leaves, partridgelike birds, and a snail. Below the lettering, amidst a profusion of shrubbery and trees abundant with leaves, birds, and a squirrel, are detailed views of two elegant houses joined by a fence and an arched gate. There are also two smaller houses, one sitting on top of and one fenced in by a decorative serpentine-pattern border with dotted leaflike devices between the curves. The trees and shrubs have beautifully ar-

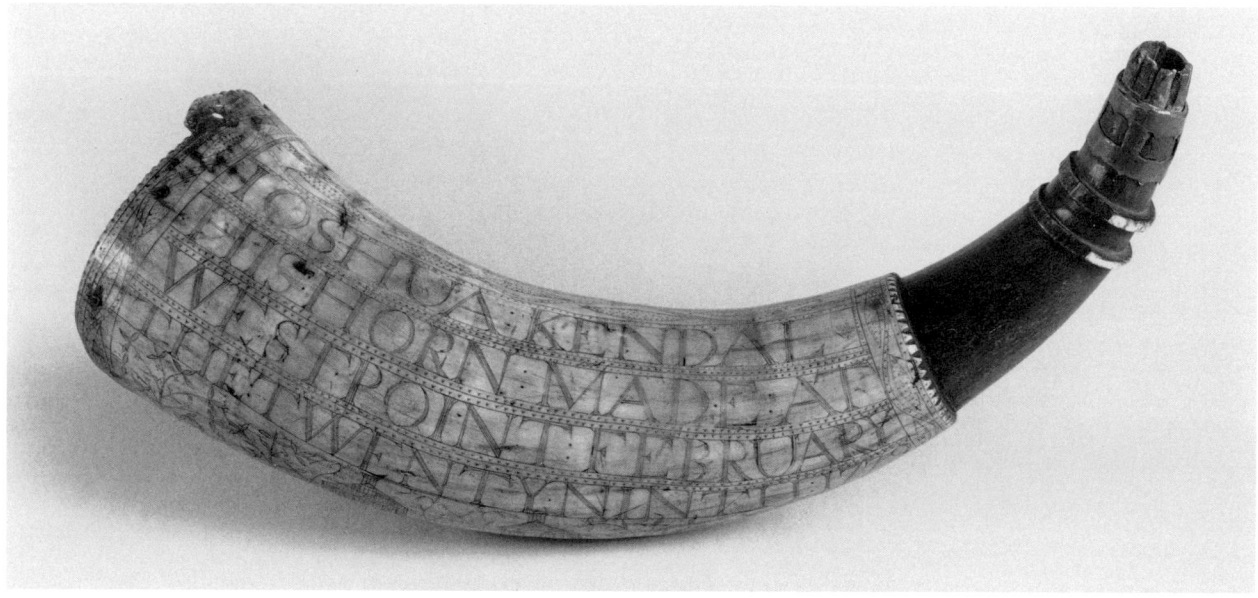

Drums A'beating, Trumpets Sounding

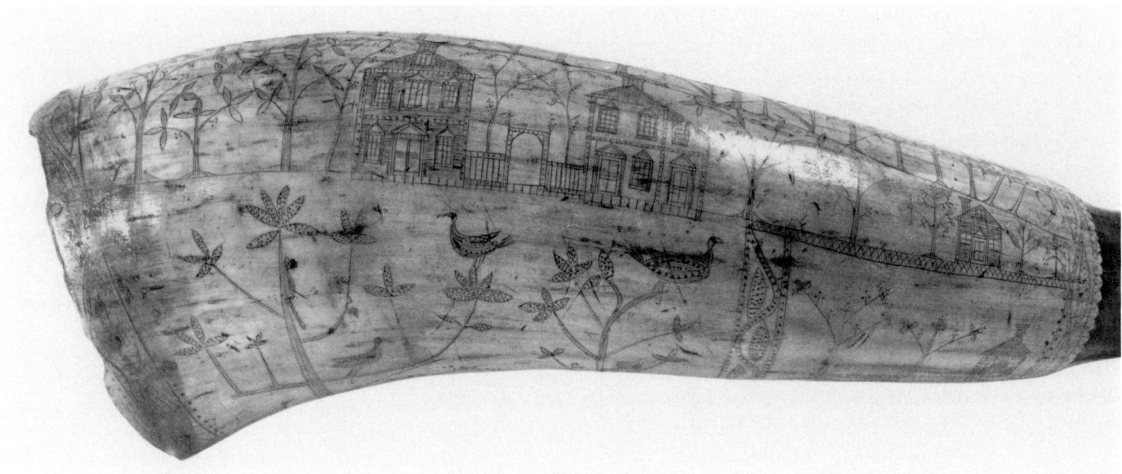

110. *The scene depicted on this horn resembles those often seen on needlework of this period.*

ranged branches, all extending upward, with finely shaped leaves.

Another serpentine border precedes the finely scalloped edge that introduces the recessed portion, which is 4 inches from the tip of the spout. There are two raised rings ½ inch apart, the closest to the tip being 1½ inches from the end of the horn. The spout is reinforced with two pewter bands linked by four reinforcing bars; the total width of this pewter fixture is ¾ inch. The plug end has a serpentine border identical to that of the spout end and a beautifully scalloped edge. The scalloping at both spout and plug ends is expertly and finely notched. The flat pine plug, which protrudes ⅛ inch beyond the horn, has a finely scalloped extension lobe 1⅜ inches long and ⅜ inch deep, which is pierced with two holes for the strap. The plug is secured to the horn by six round wooden pegs.

REVOLUTIONARY WAR SCHOOL
Attributed to the Bedford Carver
(w. 1778–1785)

111 *Jacob Forman horn*
Bedford, New York, 1779
Horn, pine, brass
OL: 17½ in. (44.5 cm), w (of plug): 2⅝ in. (6.8 cm)
Inscriptions: H✶F / Jacob ✶ Formons ✶ Horn / Made ✶ in ✶ Bed ✶ Ford / 1779
William H. Guthman

Although the horns associated with this carver have in the past been assigned to Bedford, Pennsylvania, they are far more likely to have been made in Bedford, New York. Not only does Jacob Forman's name appear on the rolls of a regiment from Westchester County, New York, but two other horns by this maker

111

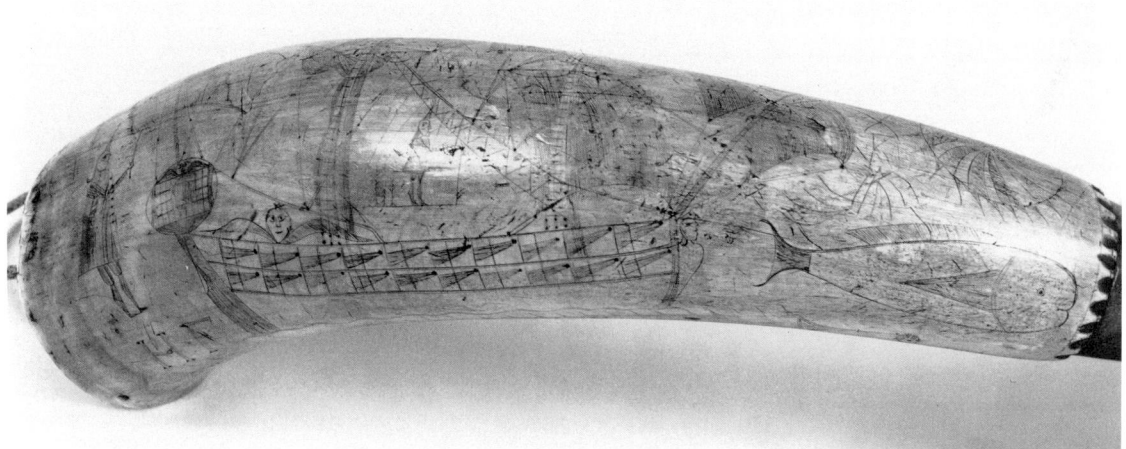

111. *The Bedford Carver employed amusing faces and figures within his ships' rigging and as figureheads.*

were made for the Conkling family of Westchester (private collection), and a fourth bears the name *Calman* (Library of Congress), also a Westchester County name. Bedford, New York, was burned by the British in 1779. A late friend of the writer purchased the Forman horn in 1952 in Beaver Dam, Wisconsin, from a family whose ancestors had taken it to the Midwest before the Civil War from "somewhere in the East."

The calligraphy is bold and carefully executed in large shaded block lettering. The inscription is contained within a rectangular cartouche 6½ inches long that has a shark-finlike decoration along the top and pointed extensions at both ends. The last numeral, "9," in the date has been accidentally or intentionally defaced to look like a "2." A mounted soldier with sword in hand stands beside the cartouche above two rows of marching soldiers—ten to a row—depicted holding muskets with bayonets fixed. There are two more soldiers below the second row.

Both faces and complete soldiers armed with swords peer out from the rigging and the deck of an elaborately rigged ship. Faces are also shown on the stern board and the figurehead. On the starboard side are twenty gunports with cannon firing. Among the many flags visible are long pennants flying from the mastheads and a large flag, which appears to be the Grand Union flag with thirteen vertical stripes in the canton in the upper left-hand corner, flying from the stern. A thirteen-stripe pennant flows from a mast. All the soldiers wear mitre hats and whimsical expressions. Around the ship are fish of various types and sizes and a large pinwheel.

There is crude scalloping at the beginning of the recessed portion 4⅞ inches from the tip, a raised ring 2½ inches from the tip, and hexagonal faceting from there to a raised lip at the spout. The plug end has no border. The flat pine plug, secured by four large round wooden pegs, has a large brass furniture knob in its center.

BIBLIOGRAPHY: Swayze, 107–109.

Catalogue only
REVOLUTIONARY WAR SCHOOL
    Attributed to the Bedford Carver
    (w. 1778–1785)

112 *James Konkun horn*

Bedford, New York, 1784
Horn, pine, iron
OL: 13½ in. (34 cm), W (of plug): 2¾ in. (7 cm)
Inscriptions: James∗Kon∗Kuns∗Horn / Made∗in∗Bed∗Ford ∗1784
William H. Guthman

This horn is in the same style as the Forman horn (No. 111), with identical calligraphy and similarly detailed ship, fish, faces, and pinwheel designs; instead of formations of soldiers, however, this example has only a single uniformed soldier standing facing the stern of the ship.

The horn is crudely scalloped at the beginning of the recessed portion 4¼ inches from the tip of the spout. There is a raised ring 2¼ inches from the beginning of the recessed portion and 2 inches from the tip, and another raised ring at the tip of the spout. The stopper in the spout is a late replacement. There is no border at the plug end. The rounded pine plug extends ⅜ of an inch beyond the end of the horn and has a large hole in its center that might have held a furniture knob similar to that on No. 111. Four iron screws secure the plug to the horn.

Catalogue only
REVOLUTIONARY WAR SCHOOL
Attributed to the Bedford Carver
(w. 1778–1785)

113 *John Conklin horn*

Bedford, New York, 1785
Horn, pine
OL: 9 in. (22.9 cm), w (of plug): 2¼ in. (5.7 cm)
Inscriptions: John∗ConKlins∗Horn / Mad∗ in∗Bedford / October∗ye 27∗1785
William H. Guthman

This priming horn is a miniature duplicate of Nos. 111 and 112, with ships, fish, and faces. There are no soldiers here, but there was little room available for such additional decoration.

There is no border at the beginning of the recessed portion, which begins 3⅜ inches from the tip of the spout. There is a raised ring 1¾ inches from the beginning of the recessed portion, 1½ inches from the tip of the spout, and the spout is faceted hexagonally from the raised ring to the tip, which also has a raised ring. The wooden stopper is a later addition. There is no border decoration at the plug end. The slightly rounded pine plug, secured by five wooden pegs, extends a fraction of an inch beyond the end of the horn and has an iron staple in its center.

BIBLIOGRAPHY: Grancsay, No. 207, 48.

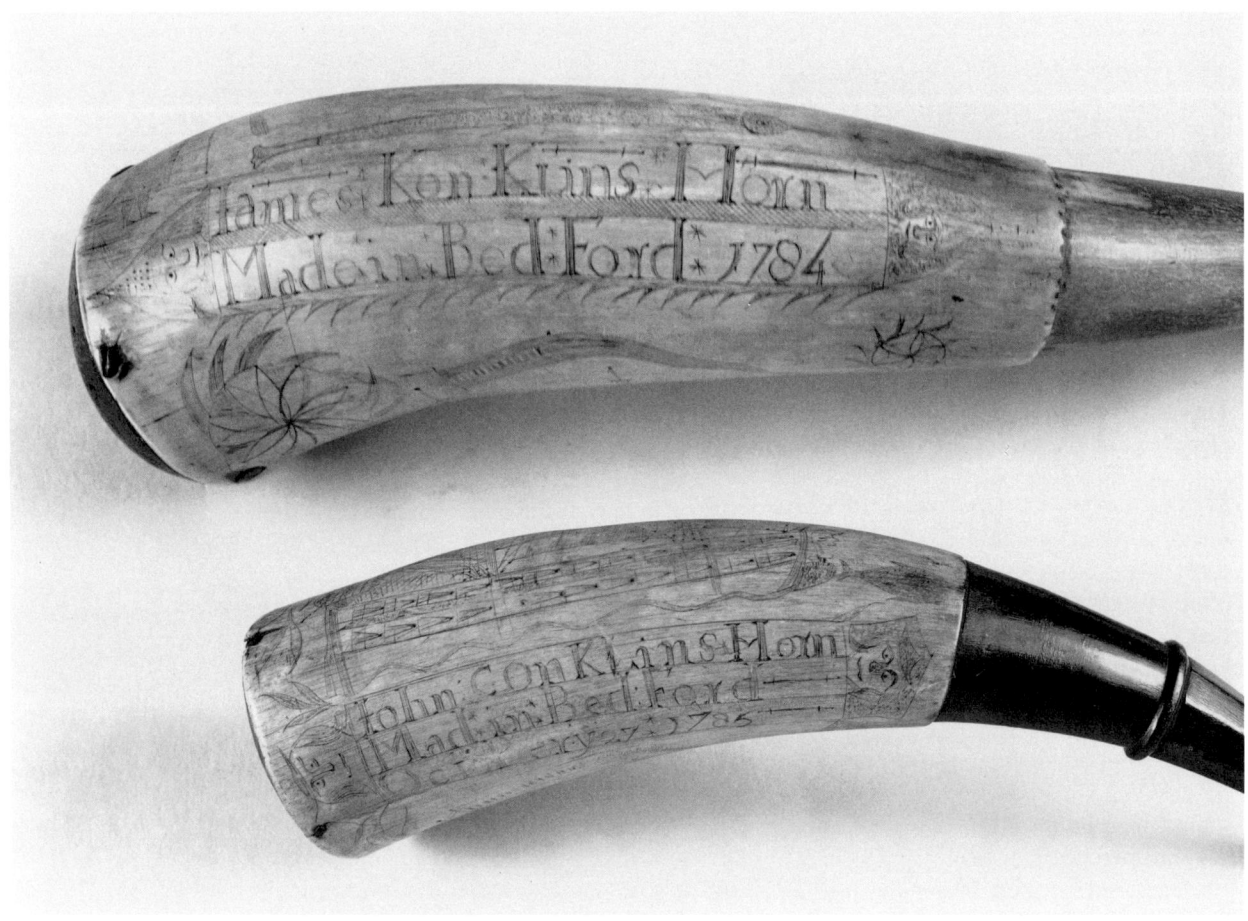

112 (Top James Konkun) and 113 (bottom, John Conklin), both catalogue only. *The Bedford Carver's bold, carefully executed calligraphy appears consistently on the five horns so far attributed to him.*

Catalogue

REVOLUTIONARY WAR SCHOOL

114 *Henry Turner horn*

Possibly the Southern Campaign, 1781
Horn, pine, iron, brass, red pigment, leather
OL: 16½ in. (41.9 cm), W (of plug): 3 in. (7.6 cm)
Inscriptions: HANRY . TURNER / His . HoRN . Made . Y^e 1781
Private Collection

There is no record of Henry Turner's service before 1778, but in January of that year he enlisted in Colonel Seth Warner's Vermont Regiment for the duration of the war. He was commissioned a lieutenant in 1781. Turner was from Pownal, Bennington County, Vermont, which is only a few miles above Pittsfield, Massachusetts, where the horn was sold at auction about 1974.

This is a delightful horn, displaying many features of the Siege of Boston School and sharing many characteristics of Jacob Gay's Amos Bostwick horn (No. 95). Among similar motifs are the animals, trees, vines, and the hunter dressed in detailed eighteenth-century attire; he is shown poised in front of his hound, aiming at a buck and a doe over the heads of two birds that may be chickens or pigeons. There is also a bird perched in a tree between the birds and deer, surmounted by a large horse and stylized tulips. Three ships under full sail resemble those carved by James Greenfield, with crews peeking out from the rigging and on deck. These comical crew members add another pleasing dimension to the horn.

The calligraphy is neatly carved, but not accomplished: the letters vary in size and do not conform to the guidelines—some even appear to be bouncing off them. The phonetic spelling, HANRY, is amusing.

A lunette border laid out with a compass precedes the evenly carved scalloping at the beginning of the faceted recessed portion 5½ inches from the tip. There is a neatly carved and scalloped raised ring 2⅜ inches from the tip, after which the spout is faceted octagonally. There is another lunette border at the plug end. The slightly rounded reddish-brown plug, extending ⅜ of an inch beyond the horn, is secured by eight round wooden pegs. Two rosehead nails and a brass nail attach a leather strap to the center of the plug and there are two larger brass nails that appear to be later additions on either side of the strap. A fragment of the original woven strap remains on the spout, covering some of the faceting and the raised carved ring.

BIBLIOGRAPHY: Swayze, 24, 168–170.

Bibliography

POWDER HORNS

Brockett, S. F., 1941, "Keep Your Powder Dry," *The Magazine Antiques*, Vol. XXXIX, No. 1, 22–24.

Coakley, William D., 1975, "A Horn from the Siege of Boston," *The Canadian Journal of Arms Collecting*, Vol. 13, No. 3, 92–97.

Comstock, Helen, Ed., 1959, "Powder Horn Inscribed Robert Rogers," *Portfolio of the Old Print Shop*, Vol. XVIII, No. 9, 211.

Dickinson, Thomas A., 1900, "Rufus Alexander Grider," *Worcester Society of Antiquity Proceedings*, March–April, 1900, 110–113.

duMont, John S., 1978, *American Engraved Powder Horns*, Canaan, New Hampshire, Phoenix Publishing.

Fort Ticonderoga Museum, Ed., 1936, "Powder Horns," *The Bulletin of the Fort Ticonderoga Museum*, Vol. IV, No. 1, 9–12.

Grancsay, Stephen V., 1945, "American Engraved Powder Horns," *The Magazine Antiques*, Vol. XLVII, No. 1, 26–28.

———, 1946, *American Engraved Powder Horns*, NY, The Metropolitan Museum of Art.

Grant, Madison, 1987, *Powder Horns and Their Architecture*, York, PA, Madison Grant.

Grider, Rufus A., 1931, "Powder Horns and Their Use," (Ed. by A. J. Wall), *The New-York Historical Society Quarterly Bulletin*, Vol. XV, No. 1, 3–24.

———, 1895, "Catalogue of a Collection of Powder Horns Embracing the World," Accompanying the Photographs of the Grider Collection of Powder Horn Drawings, Special courtesy of the New-York Historical Society, NYC.

Guthman, William H., 1978, "Powder Horns of the French and Indian War 1755–1763," *The Magazine Antiques*, Vol. CXIV, No. 2, 312–331.

———, Ed., 1979, *Guns and Other Arms*, 124–153, NY, Main Street Press.

———, 1980, "Fantasy Designs on Powder Horns," *Man at Arms*, Vol. II, No. 4, 43–45.

———, 1981, "Carved Powder Horns," *Bulletin of the American Society of Arms Collectors*, No. 44, 35–50.

———, 1989, "Powder Horns Carved in the Provincial Manner," *Bulletin of the American Society of Arms Collectors*, No. 61, 9–20.

Milliman, Crosby, 1967, "An Exhibition of American Engraved Powder Horns of the Colonial and Revolutionary Periods," *The Bulletin of the Fort Ticonderoga Museum*, Vol. XII, No. 3, cover, 175–192.

Routh, James E., Jr., 1980, "American Engraved Powder Horns," *Bulletin of the American Society of Arms Collectors*, No. 42, 2–19.

Sawyer, Charles Winthrop, 1929, "The Why and How of Engraved Powder Horns," *The Magazine Antiques*, Vol. XVI, No. 4, 283–285.

Segal, Thomas, 1983, "American Engraved Powder Horns," *Michigan Antique Arms Collecting*, Vol. 2, 16–21.

Swayze, Nathan L., 1978, *Engraved Powder Horns of the French and Indian War and the Revolutionary War Era*, Yazoo City, MS, Gunhill Publishing Co.

Vail, R. W. G., 1943, "The Grider Collection of Schoharie Valley Pictorial History," *Journal of the Schoharie County Historical Society*.

Williams, John S., 1843, "Ancient Relic," *American Pioneer*, Vol. II, frontispiece, 3–16.

KING GEORGE'S AND FRENCH AND INDIAN WARS

Bates, Albert C., 1903 & 1905, 2 Vols., *French-Indian War Rolls, 1755–1762*, Hartford, Connecticut Historical Society.

———, 1918 & 1920, 2 Vols., *The Fitch Papers 1754–1766*, Hartford, Connecticut Historical Society.

———, 1931, *The Two Putnams*, Hartford, Connecticut Historical Society.

Betts, C. Wyllys, 1894, *American Colonial History, Illustrated by Contemporary Medals*, NY, Scott Stamp & Coin Co.

Butler, B. C., 1868, *Lake George & Lake Champlain*, Albany, Weed, Parsons & Co.

Chapin, Howard M., 1918, *A List of Rhode Island Soldiers & Sailors in the Old French & Indian War 1755–62*, Providence, Rhode Island Historical Society.

———, 1920, *A List of Rhode Island Soldiers & Sailors in King George's War 1740–1748*, Providence, Rhode Island Historical Society.

Chichester, Henry Manners & Burges-Short, George, 1986, *The Records and Badges of Every Regiment and Corps. in the British Army*, London, Greenhill Books.

Clark, Delphina L. H., 1964, *Phineas Lyman*, Springfield, Connecticut Valley Historical Museum.

Cuneo, John R., 1959, *Robert Rogers of the Rangers*, NY, Oxford University Press.

Day, Richard E., Ed., 1909, *The Calendar of the Sir William Johnson Manuscripts*, Albany, New York State Library.

deForest, Louis Effingham, 1926, *The Journals and Papers of Seth Pomeroy*, NY, Society of Colonial Wars.

———, 1932, *Louisbourg Journals 1745*, NY, Society of Colonial Wars.

Drake, Samuel G., 1870, *Five Years French & Indian War, 1744–1749*, Albany, Joel Munsell.

Ford, Worthington C., 1894, *British Officers Serving in America, 1754–1774*, Boston.

———, Ed., 1898, *General Orders of 1757*, NY, Dodd Mead & Co.

Gipson, Lawrence Henry, 1936–1970, Vols. 1–15, *The British Empire Before the American Revolution*, NY, Alfred A. Knopf.

Gridley, Luke, 1906, *Luke Gridley's Diary, 1757*, Hartford, The Acorn Club.

Goss, David K., and Zarowin, David, Eds., 1985, *Massachusetts Officers and Soldiers in the French & Indian Wars, 1755–1756*, Boston, The Society of Colonial Wars in Massachusetts & The New England Historic & Genealogical Society.

Hastings, Hugh, State Historian, 1897 & 1898, 2 Vols., *Second & Third Annual Reports of the State Historian of the State of New York*, Albany, State of New York.

———, Ed., 1911, *Orderly Book and Journal of Major John Hawks*, NY, Society of Canadian Wars.

Hough, Franklin B., 1860, *Diary of the Siege of Detroit in the War with Pontiac*, Albany, J. Munsell.

Humphreys, David, 1794, *An Essay on the Life of Hon. Major-General Israel Putnam*, Middletown, Moses H. Woodward.

Jackson, H. M., 1953, *Rogers Rangers*, privately printed.

Johnson, Sir William, 1921–1965, *Sir William Johnson Papers*, Vols. 1–14, Albany, State University of NY, Division of Archives and History.

Knap, Isaac, 1895, *The Diary of Isaac Knap of Newbury*, Boston, Colonial Wars Society.

Kelby, William, 1892, *New York Muster Rolls, 1755–1764*, NY, New-York Historical Society.

Knox, J., 1769, 2 Vols., *An Historical Journal of the Campaigns in North America*, London, for the author.

Leach, Douglas Edward, 1973, *Arms for Empire*, NY, Macmillan Co.

Lincoln, Charles Henry, 1909, Vol. XI, *Manuscript Records of the French & Indian War*, Worcester, American Antiquarian Society.

———, 1912, 2 vols., *Correspondence of William Shirley*, NY, Macmillan Co.

Loescher, Burt G., 1945, *History of Rogers Rangers*, San Francisco.

———, 1957, *Genesis, Rogers Rangers*, Vol. III, Burlingame.

———, 1969, *Rogers Rangers*, San Mateo.

Mackay, Robert E., Ed., 1978, *Massachusetts Soldiers in the French & Indian Wars, 1744–1755*, Boston, Society of Colonial Wars in Massachusetts and the New England Historic Genealogical Society.

Mante, Thomas, 1772, *The History of the Late War in North America*, London, W. Strahan & T. Cadell.

Mason, Richard A., Ed., 1966, *The Diary of Jabez Fitch, Jr. 1757*, Rogers Island Historical Association.

Matloff, Maurice, Ed., 1969, *American Military History*, Washington, U.S. Army.

Millan, J., 1745, *The Succession of Colonels to All His Majesty's Land Forces From Their Rise to 1744*, London, Printed for J. Millan.

———, 1755, *A List of the General and Field Officers As They Rank in the Army*, London, Printed for J. Millan.

———, 1756–1765, 10 Vols., British Army Lists, *List of the General and Field Officers*, London, Printed for J. Millan.

Montresor, James, 1882, *Collections for the Year 1881*, NY, New-York Historical Society.

New York State Historical Association, 1911, *Proceedings of 12th Annual Meeting*, Vol. X, French and Indian War bibliography, 238–365, Black Watch at Ticonderoga, 367–464, Glens Falls, New York State Historical Association.

———, 1914, *Proceedings of 14th Annual Meeting*, Vol. XIII, Historic Oswego, Sir William Johnson & Pontiac, Defenses of Oswego, Fur Traders of Oswego, Old Trail From the Mohawk to Oswego, The Capture of Oswego in 1756, Lineage of

Col. George Monro, New York State Historical Association.

———, 1922, *Proceedings of 22nd Annual Meeting*, Vol. XX, Historic Spots Near Lake George, The History of Fort Ticonderoga, New York State Historical Association.

———, 1934, *Proceedings of 34th Annual Meeting*, Vol. XXXII, Amherst in 1759, Fort Niagara 1759–1763, "Major Robert Rogers, Trader" and "Archaeology of Ticonderoga," New York State Historical Association.

O'Callaghan, E. B., Ed., 1850, 1851, 4 Vols., *Documentary History of New York*, Albany, State of NY.

———, 1856–1861, Vols. 1–10 and Index, *Documents Relating to the Colonial History of the State of New York*, Albany, State of New York.

———, 1866, *Calendar of Historical Manuscripts 1644–1776*, Albany, State of NY.

———, 1929, *Calendar of New York Colonial Commissions, 1680–1770*, NY, New-York Historical Society.

Pargellis, Stanley, 1969, *Military Affairs in North America 1748–1765*, New Haven, Archon Books.

Peckham, Howard H., 1947, *Pontiac and the Indian Uprising*, Princeton, Princeton University Press.

———, 1964, *The Colonial Wars 1689–1762*, Chicago, The University of Chicago Press.

Potter, C. E., 1866 (Muster Rolls), *The Military History of the State of New Hampshire, 1623–1861*, Concord, McFarland & Jenks.

Pound, Arthur, 1930, *Johnson of the Mohawks*, NY, Macmillan Co.

Quaife, Milo Milton, Ed., 1958, *The Siege of Detroit in 1763*, Chicago, Lakeside Press.

Rawlyk, G. A., 1967, *Yankees at Louisbourg*, Orono, University of Maine.

Reid, Max W., 1910, *Lake George and Lake Champlain*, NY, G. P. Putnam's Sons.

Richards, Frederick B., 1911, *The Black Watch at Ticonderoga*, Glens Falls, New York State Historical Association.

Roberts, Kenneth, 1936, *Northwest Passage*, NY, Doubleday, Doran.

Rogers, Mary Cochrane, 1917, *A Battle Fought on Snow Shoes*, Derry, Published by the Author.

Rogers, Robert, 1765, *A Concise Account of North America*, London, J. Millan.

———, 1765, *Journals of Major Robert Rogers*, London, J. Millan.

Samuel, Sigmund, 1934, *The Seven Years War in Canada*, Toronto, Ryerson Press.

Schutz, John A., 1961, *William Shirley*, Chapel Hill, University of North Carolina Press.

Scull, G. D., 1881, *Journals of Capt. John Montresor*, NY, New-York Historical Society.

Sheldon, George, 1895–96, 2 Vols., *A History of Deerfield*, Pocumtuck Valley Memorial Association.

Sloane, William Milligan, 1901, *The French War and the Revolution*, New York, Charles Scribner's Sons.

Smith, J.E.A., 1869, *The History of Pittsfield, MA, 1734–1800*, Boston, Lee & Shepard.

Smith, Joseph Jenks, 1900, 1901 & 1907, 3 Vols., *Civil & Military List of Rhode Island, 1647–1800*, Providence, Preston & Rounds Co. (1900 & 1901) & Joseph H. Smith (1907).

Stewart, Charles H., 1964, *The Service of British Regiments in North America*, Ottawa, Dept. of National Defense.

Stark, Caleb, 1831, *Reminiscences of the French War*, Concord, Luther Roby.

Stone, William L., 1865, *The Life and Times of Sir William Johnson, Bart.*, Vols. 1 & 2, Albany, J. Munsell.

Valentine, Alan, 1970, 2 Vols., *The British Establishment, 1760–1784*, Norman, University of Oklahoma Press.

Voye, Nancy S., 1975, Ed., *Massachusetts Officers in the French & Indian Wars 1748–1763*, Boston, Society of Colonial Wars in Massachusetts, Massachusetts Archives and the New England Historic Genealogical Society.

Webster, J. Clarence, after 1908, *Journal of Abijah Willard 1755*, NB, New Brunswick Historical Society.

———, 1931, *The Journal of Jeffery Amherst*, Chicago, Ryerson Press.

Whitton, F. E., 1971, *Wolfe and North America*, Port Washington, NY, Kennikat Press.

Willett, William H., 1831, *Narrative of the Military Actions of Col. Marinus Willet*, NY, Carvill.

Williams, Stephen W., 1847, *The History of the Williams Family*, Greenfield, Merriam & Merick.

AMERICAN REVOLUTION

Allen, Robert S., 1983, *The Loyal Americans*, Ottawa, Canadian War Museum.

Adler, Mortimer J., 1976, *The Revolutionary Years*, Chicago & London, Encyclopedia Brittanica, Inc.

Berg, Fred Anderson, 1972, *Encyclopedia of Continental Army Units*, Harrisburg, Stackpole Books.

British Library, 1975, *The American War of Independence*, London, British Museum Publications, Ltd.

Boatner, Mark Mayo, III, 1966, *Encyclopedia of the American Revolution*, NY, David McKay.

———, 1973, *Landmarks of the American Revolution*, Harrisburg, Stackpole Books.

Bolton, Charles Knowles, 1902, *Letters of Hugh Earl Percy From Boston & New York 1774–1776*, Boston, Charles E. Goodspeed.

Bray, Robert & Bushnell, Paul, Eds., 1978, *Diary of a Common Soldier in the American Revolution*, Dekalb, Northern Illinois University Press.

Cappon, Lester J., 1976, *Atlas of Early American History, The Revolutionary Era 1760–1790*, Princeton, Princeton University Press.

Chittenden, Abraham, 1922, *Orderly Book of Lt. Abraham Chittenden, Aug. 1776–Sept. 29, 1776*, Hartford, privately printed.

Clark, William Bell, Ed., 1963–1969, Vols. 1–4; Morgan, William James, Ed., 1970–1976, Vols. 5–7, *Naval Documents of the American Revolution 1774–1777*, Washington, DC, Government Printing Office.

Connecticut Historical Society, 1899, Vol. II Collections, *Orderly Books & Journals Kept by Connecticut Men in the American Revolution, 1775–1778*, Hartford, Connecticut Historical Society.

Corlette, Suzanne, 1976, *The Pulse of the People: New Jersey 1763–1789*, Trenton, New Jersey State Museum.

Dana, Elizabeth Ellery, 1924, *The British in Boston*, Cambridge, Harvard University Press.

Dann, John C., 1977, *The Revolution Remembered*, Chicago, The University of Chicago Press.

Dawson, Henry B., 1858, 2 Vols., *Battles of the United States by Sea & Land*, NY, Johnson, Fry & Co.

Deerin, James B., 1976, *The Militia in the Revolutionary War*, Washington, Historical Society of the Militia and the National Guard.

Dickerson, Oliver Morton, 1936, *Boston Under Military Rule, 1768–1769*, Boston, Chapman & Grimes.

Doggett, John R., 1849, *Boston Massacre*, NY, John Doggett, Jr.

Drake, Samuel Adams, 1875, *Bunker Hill*, Boston, Nichol & Hall.

Ellis, George, 1875, *History of the Battle of Bunker's (Breed's) Hill, on June 17, 1775*, Boston, Lockwood, Brooks & Co.

Fairbanks, Jonathan L. & Cooper, Wendy A., 1975, *Paul Revere's Boston, 1735–1818*, Boston, Museum of Fine Arts.

Fernow, Berthold, 1887, Vol. XV, *Documents Relating to the Colonial History of the State of New York*, Albany, Weed Parsons & Co.

Force, Peter, 1837–1846, Vols. I–VI, *American Archives, Fourth Series*, Washington, DC, Authority of An Act of Congress.

———, 1848–1853, Vols. I–III, *American Archives, Fifth Series*, Washington, DC, Authority of An Act of Congress.

Ford, Worthington C., 1897, *British Officers Serving in America, 1774–1782*, Brooklyn, Historical Printing Club.

French, Allen, Ed., 1926, *A British Fusilier in Revolutionary Boston*, Cambridge, Harvard University Press.

———, 1934, *The First Year of the American Revolution*, Boston, Houghton Mifflin Co.

Frothingham, Richard, 1873, *History of the Siege of Boston*, Boston, Little Brown & Co.

Goodrich, John E., 1904, *State of Vermont Rolls of the Soldiers in the Revolutionary War*, Rutland, Vermont Legislature.

Hammond, Isaac W., 1885–1889, 4 Vols., *State of New Hampshire, Rolls of the Soldiers in the Revolutionary War*, Concord, Vols. 1 & 2, Manchester, Vols. 3 & 4, New Hampshire Legislature.

Hamilton, C. D., 1976, *1776*, London, Times Newspapers Ltd.

Harris, William W. and Allyn, Charles, 1882, *The Battle of Groton Heights*, New London, Charles Allyn.

Heath, William, 1798, *Memoirs of Major-General Heath*, Boston, I. Thomas & E. T. Andrews.

Heitman, Francis B., 1914, *Historical Register of Officers of the Continental Army*, Washington, DC, Rare Bookshop Publishing Co.

Hinman, Royal R., 1842, *The War of the American Revolution*, Hartford, E. Gleason.

Hoadly, Charles J., 1894 & 1895, 2 Vols., *Public Records of the State of Connecticut, Vol. I, 1776–1778, Vol. II, 1778–1780*, Hartford, State of Connecticut.

Bibliography

Hughes, Thomas, 1947, *A Journal by Thomas Hughes, 1778–1789*, Cambridge, England, University Press.

Hutchinson, Peter O., 1884, 2 Vols., *The Diary & Letters of His Excellency Thomas Hutchins, Esq.*, Boston, Houghton Mifflin Co.

Johnston, Henry P., Ed., 1889, Muster Rolls, *Connecticut Men in Military and Naval Service During the War of the Revolution*, Hartford, Adjutant General of Connecticut.

Kaplan, Sidney, 1973, *The Black Presence in the Era of the American Revolution*, National Portrait Gallery, Greenwich, New York Graphic Society.

Kegan, Elizabeth Hamer, 1975, *To Set a Country Free*, Washington, Library of Congress.

Kinnman, M. E., 1931, *Order Book Kept by Peter Kinnman, July 7–Sept. 4, 1776*, Princeton, Princeton University Press.

Kleeb, Arlene Phillips, 1975, *Lexington & Concord*, Ann Arbor, William L. Clements Library.

Knight, Erastus, Comp., and Mather, Frederick G., Ed., 1901, *New York in the Revolution* (Supplement), Albany, Oliver A. Quayle.

Labaree, Benjamin Woods, 1964, *The Boston Tea Party*, NY, Oxford University Press.

Library of Congress, 1906, *Naval Records of the American Revolution*, Washington, DC, Government Printing Office.

Lincoln, William, 1838, *The Journals of the Provincial Congress of Massachusetts*, Boston, Dutton & Wentworth.

Lossing, Benson, J., 1860, 2 Vols., *The Pictorial Field Book of the Revolution*, NY, Harper & Bros.

Marshall, Douglas W. & Peckham, Howard H., 1976, *Campaigns of the American Revolution, An Atlas of Manuscript Maps*, Ann Arbor, University of Michigan Press.

Martyn, Charles, 1921, *The Life of Artemas Ward*, NY, Artemas Ward.

Massachusetts, Secretary of the Commonwealth, Muster Rolls, 1896–1908, Vols. I–XVII, *Massachusetts Soldiers and Sailors of the Revolutionary War*, Boston, Wright & Potter Printing Co.

Melvin, Andrew A., 1902, *The Journal of James Melvin*, Portland, Hubbard W. Brant.

Middlebrook, Louis F., 1925, *Maritime Connecticut During the American Revolution*, Salem, Essex Institute.

Millan, J., 1766–1783, 18 Vols., British Army Lists, *List of the General and Field Officers*, London, J. Millan.

Miller, Lillian B., 1975, *The Dye is Now Cast*, Washington, National Portrait Gallery.

Montgomery, Charles F. and Kane, Patricia E., 1976, *American Art: 1750–1800 Towards Independence*, New Haven, Yale University Art Gallery and London, Victoria & Albert Museum.

Nebenzahl, Kenneth & Higginbotham, Don, 1974, *Atlas of the American Revolution*, Chicago, Rand McNally & Co.

New-York Historical Society, Collections 1914 & 1915, Published 1916, 2 Vols., *Revolutionary Muster Rolls 1775–1783*, NY, New-York Historical Society.

New York, Office of Secretary of State, 1868, Vols. I & III, *Calendar of Historical Manuscripts Relating to the War of the Revolution*, Albany, Weed, Parsons & Co.

Roberts, James A., 1897, Muster Rolls, *New York in the Revolution*, Albany, Weed, Parsons & Co.

———, 1898, Muster Rolls, *New York in the Revolution*, Albany, Brandon Printing.

Sabine, Lorenzo, 1847, *The American Loyalists*, Boston, Charles C. Little and James Brown.

Sadik, Marvin, 1974, *In the Minds & Hearts of the People*, Greenwich, New York Graphic Society.

Sawtell, Clement C., 1968, *The Nineteenth of April, 1775*, Lincoln, Sawtells of Somerset.

Scheer, George E., Ed., 1963, *Private Yankee Doodle, Joseph Plumb Martin*, NY, Popular Library.

Smith, Joseph Jenks, 1900, 1901, 1907, 3 Vols., *Civil and Military List of Rhode Island*, Providence, Preston & Rounds Co. (Vols. 1 & 2), Providence, Joseph J. Smith (Vol. 3).

Stevens, Paul L., 1987, *A King's Colonel at Niagara 1774–1776*, Youngstown, NY, Old Fort Niagara Association.

Stewart, Charles H., 1964, *The Service of British Regiments in North America*, Ottawa, National Defense Library.

Swett, S., 1826, *History of Bunker Hill Battle*, Boston, Monroe Francis.

Wade, Herbert T., 1953–54, Vols. 89 & 90, *A Massachusetts Soldier of the Revolution*, Salem, Essex Institute.

Walett, G. Francis, 1976, *Patriots, Loyalists & Printers*, Worcester, American Antiquarian Society.

Ware, Thomas A., 1965, *Revolution in the Highlands*, National Temple Hill Association, Cornwallville, Hope Farm Press.

Wheeler, Robert G., 1976, *The Struggle and the Glory*, Dearborn, Henry Ford Museum.

Who's Who in America, Eds. of, 1976, *Who Was Who During the American Revolution*, Indianapolis & NY, Bobbs-Merrill Co.

Willey, W. L., 1925, *The Order of Military Merit*, Society of the Cincinnati, Cambridge, Harvard University Press.

CALLIGRAPHY

Baltimore Museum of Art, Peabody Institute Library, Walters Art Gallery, 1965, *2,000 Years of Calligraphy*, Baltimore, Walters Art Gallery.

Bickham, George, 1741, *The Universal Penman*, London, Bickham.

———, 1747, *A Short Description of the American Colonies*, London, Bickham.

Bonacini, Claudio, 1953, *Biblioteca Bibliografica Italica, Arti Scrittorie e Della Calligrafia*, Florence, Sansoni Antiquariato.

Doede, Werner, 1958, *Bibliographic deutscher Schreibmeisterbücher*, Hamburg, Dr. Ernst Hauswedell & Co.

Drogin, Marc, 1983, *Yours Truly King Arthur*, NY, Taplinger Publishing Co.

Fairbank, Alfred, 1960, *A Book of Scripts*, Baltimore, Penguin Books.

——— & Wolpe, Berthold, 1960, *Renaissance Handwriting*, Cleveland & New York, The World Publishing Co.

Gill, Nathaniel & Whiteman, J. H., 1832, *Knight's New Book of Seven Hundred & Fifty Eight Plain Ornamental & Reverse Cyphers*, London, F. Knight.

Hector, L. C., 1958, *The Handwriting of English Documents*, London, Edward Arnold Ltd.

Morison, Stanley, 1962, *Calligraphy 1535–1885*, Milan, La Bibliofila.

Osley, A. S., 1980, *Scribes and Sources*, Boston, David R. Godine.

Parker, Muriel, 1986, *Illuminated Letter Designs in the Historic Style of the Middle Ages*, Owings Mills, MD, Stemmer House.

Randall, Lillian M. C., 1989, *Medieval and Renaissance Manuscripts in the Walters Art Gallery*, Baltimore, Johns Hopkins University Press.

Shepherd, Margaret, 1977, *Learning Calligraphy*, NY, Collier Books.

Wardrop, James, 1963, *The Script of Humanism*, Oxford, Clarendon Press.

DESIGN AND ORNAMENT

Allen, Charles Dexter, 1894, *American Book-Plates*, NY, MacMillan & Co.

Benes, Peter, 1977, *The Masks of Orthodoxy, Folk Gravestone Carving in Plymouth, Mass., 1689–1805*, Peter Benes & The American Council of Learned Societies.

———, 1981, *New England Prospect*, Boston, Boston University.

Blair, Claude; Boccia, Lionello G.; Fahy, Everett; Nickel, Helmut; Norman, A.V.B.; Phyrr, Stuart W.; and La Rocca, Donald J., 1992, *Studies in European Arms and Armor*, Philadelphia, Philadelphia Museum of Art.

Black, Mary and Lipman, Jean, 1966, *American Folk Painting*, NY, Clarkson N. Potter.

Brigham, Clarence, 1954, *Paul Revere's Engravings*, Worcester, American Antiquarian Society.

Bumgardner, Georgia B., 1971, *American Broadsides*, Barre, MA, Imprint Society.

Cooper, J. C., 1987, *An Illustrated Encyclopedia of Traditional Symbols*, London, Thames & Hudson Ltd.

Cresswell, Donald H., 1975, *The American Revolution in Drawings and Prints*, Washington, DC, Library of Congress.

Dolmetsch, Joan D., 1976, *Rebellion and Reconciliation, Satirical Prints on the Revolution at Williamsburg*, Williamsburg, VA, The Colonial Williamsburg Foundation.

Dublin Seminar, 1978, *Puritan Gravestone Art II*, Boston, Boston University.

Dufty, Arthur Richard, 1974, *European Swords and Daggers in the Tower of London*, London, Her Majesty's Stationery Office.

Duval, Francis Y. and Rigby, Ivan B., 1978, *Early American Gravestone Art in Photographs*, NY, Dover Publications.

Fales, Martha Gandy, 1970, *Early American Silver*, NY, Funk & Wagnalls.

Fairbanks, Jonathan L. & Trent, Robert F., 1982, 3 Vols., *New England Begins: The Seventeenth Century*, Boston, Museum of Fine Arts.

Ferguson, George, 1961, *Signs & Symbols in Christian Art*, London, Oxford University Press.

Fleming, William, 1956, *Arts and Ideas*, NY, Henry Holt & Co.

Fowble, E. McSherry, 1987, *Two Centuries of Prints in America 1680–1880*, Charlottesville, VA, University Press of Virginia for Winterthur Museum.

Franco, Barbara, 1976, *Masonic Symbols in American Decorative Arts*, Lexington, MA, Museum of Our National Heritage.

Frederickson, N. Jayne and Gibb, Sandra, 1980, *The Covenant Chain, Indian Ceremonial & Trade Silver*, Ottawa, National Museums of Canada.

Friedlaender, Walter, 1965, *Mannerism & Anti-Mannerism in Italian Painting*, NY, Schocken Books.

BIBLIOGRAPHY

Garratt, John G., 1985, *The Four Indian Kings*, Ottawa, Public Archives, Canada.

George, Diana Hume and Nelson, Malcolm A., 1983, *Epitaph and Icon*, Orleans, MA, Parnassus Imprints.

Gillon, Edmund Vincent, Jr., 1966, *Early New England Gravestone Rubbings*, NY, Dover Publications.

Grancsay, Stephen V., 1950, *Master French Gunsmiths' Designs of the Mid-Seventeenth Century*, NY, Greenberg Publisher.

———, 1970, *Master French Gunsmiths' Designs*, NY, Winchester Press.

Hamilton, Sinclair, 1968, *Early American Book Illustrators and Wood Engravers 1670–1870*, Princeton, Princeton University Press.

Harlow, Thompson R., 1958, *Morgan B. Brainard's Tavern Signs*, Hartford, The Connecticut Historical Society.

Heal, Ambrose, 1925, *London Tradesmen's Cards of the XVIII Century*, NY, Charles Scribners Sons.

Heckscher, Morrison H. and Bowman, Leslie Greene, 1992, *American Rococo 1750–1775*, NY, Harry N. Abrams, Inc.

Honour, Hugh, 1975, *The New Golden Land, European Images of America*, NY, Pantheon Books.

Hubbard, R. H., 1972, *Thomas Davies in Early Canada*, Canada, Oberon Press.

Kindig, Joe, Jr., 1960, *Thoughts on the Kentucky Rifle in Its Golden Age*, Wilmington, DE, George N. Hyatt.

Kindig, Joe K., III, 1989, *Artistic Ingredients of the Long Rifle*, York, The Kentucky Rifle Association.

Landauer, Bella C., 1927, *Early American Trade Cards*, NY, William Edwin Rudge.

Lowance, Mason I., Jr., and Bumgardner, Georgia B., 1976, *Massachusetts Broadsides of the American Revolution*, Amherst, MA, University of Massachusetts Press.

Ludwig, Allan I., 1966, *Graven Images*, Middletown, CT, Wesleyan University Press.

Moure, Nancy Wall and Donelson, F. Hoopes, 1974, *American Narrative Painting*, Los Angeles, Los Angeles County Museum of Art.

Mugridge, Donald H., 1947, *An Album of American Battle Art*, Washington, DC, Library of Congress.

Newman, Eric P., 1976, *The Early Paper Money of America*, Racine, Wisconsin, Western Publishing Co.

Nickel, Helmut, Pyhrr, Stuart W., and Tarassok, Leonid, 1982, *The Art of Chivalry*, NY, Metropolitan Museum of Art.

North, Anthony, 1982, *Victoria & Albert Museum, European Swords*, London, Her Majesty's Stationery Office.

Lewis, Philippa, & Darley, Gillian, 1986, *Dictionary of Ornament*, NY, Pantheon Books.

Rabson, Carolyn, 1974, *Songbook of the American Revolution*, Peaks Island, Maine, NEO Press.

Reilly, Elizabeth Carroll, 1975, *A Dictionary of Colonial American Printers' Ornaments & Illustrations*, Worcester, MA, American Antiquarian Society.

Reilly, Bernard F., Jr., 1991, *American Political Prints, 1766–1876*, Boston, G. K. Hall & Co.

Shadwell, Wendy J., 1969, *American Printmaking, the First 150 Years*, Washington, DC, Smithsonian Institute Press.

Shumway, George, 1980, *Rifles of Colonial America*, 2 Vols., York, George Shumway Publisher.

Speltz, Alexander, 1959, *The Styles of Ornament*, NY, Dover Publications.

St. George, Robert Blair, 1979, *The Wrought Covenant*, Brockton, MA, Brockton Art Center.

Stokes, I. N. Phelps and Haskell, Daniel C., 1933, *American Historical Prints, Early Views of American Cities, Etc.*, NY, New York Public Library.

Swain, Margaret, 1973, *The Needlework of Mary Queen of Scots*, NY, Van Nostrand Reinhold Co.

Swan, Susan Burrows, 1977, *Plain and Fancy, American Women and Their Needlework, 1700–1850*, NY, Holt, Rinehart and Winston.

Whitehill, Walter Muir, Ed., 1973, *Boston Prints and Printmakers, 1760–1775*, Boston, The Colonial Society of Massachusetts.

Winslow, Ola Elizabeth, 1930, *American Broadside Verse*, New Haven, Yale University Press.

Windham, William and Townsend, George E., 1768, *A Plan of Discipline for the Use of the Norfolk Militia*, London, J. Millan.

FORTS

Armour, David A., Ed., 1971, *Attack at Michilimackinac*, Mackinac Island, Mackinac Island State Park Commission.

———, 1972, revised ed., *Treason? At Michilimackinac*, Mackinac Island, Mackinac Island State Park Commission.

Cleaveland, Rev. John, 1959, "Journal of Rev. John Cleaveland, June 14, 1758–Oct. 25, 1758," *The Bulletin of The Fort Ticonderoga Museum*, Vol. 10, No. 3, 192–236.

DeCosta, B. F., 1871, *Notes on the History of Fort George*, NY, J. Sabin & Sons.

Desandrouins, Capt., 1972, "Memoir on the Defense of the Fort of Carillon," *The Bulletin of the Fort*

Ticonderoga Museum, Vol. XIII, No. 3, 197–226.

Dunnigan, Brian Leigh, 1987, *Glorious Relic, The French Castle and Old Fort Niagara*, Youngstown, NY, Old Fort Niagara Association.

———, 1989, *Forts Within a Fort—Niagara's Redoubts*, Youngstown, NY, Old Fort Niagara Association.

Emery, Frank B., 1931, *Old Michigan Forts*, Detroit, Mutual Liability Co.

French, Allen, 1928, *The Taking of Ticonderoga in 1775: The British Story*, Cambridge, Harvard University Press.

Fry, Bruce W., 1984, *"An Appearance of Strength," The Fortifications of Louisbourg*, 2 Vols., Ottawa, Parks Canada.

Fucron, Thomas B. and Boyle, Elizabeth Ann, 1955, *Fort Ticonderoga 1755–1955*, Lake George, Fort Ticonderoga.

Fuller, Archelaus, 1970, "Journal of Archelaus Fuller, May–Nov., 1758," *The Bulletin of the Fort Ticonderoga Museum*, Vol. XIII, No. 1, 5–17.

Goodrich, Calvin, 1940, *The First Michigan Frontier*, Ann Arbor, University of Michigan Press.

Grant, Bruce, 1965, *American Forts*, NY, E. P. Dutton & Co.

Hamilton, Edward P., 1959, *Lake Champlain and the Upper Hudson Valley*, Ticonderoga, NY, Fort Ticonderoga Association.

Hammond, John Martin, 1915, *Quaint and Historic Forts of North America*, Philadelphia, J. B. Lippincott Co.

Hill, William H., 1929, *Old Fort Edward*, Fort Edward, NY, Bullard Press.

Hultzen, Claude H., Sr., 1939, *Old Fort Niagara*, Buffalo, Old Fort Niagara Association.

Lewis, Theodore Burnham, Jr., 1970, "The Crown Point Campaign 1755," *The Bulletin of the Fort Ticonderoga Museum*, Vol. XII, No. 6, 393–426 & Vol. XIII, No. 1, 19–88.

Luzader, John F., Torres, Louis, and Carroll, Orville W., 1976, *Fort Stanwix*, Washington, DC, National Park Service.

McLennan, J. S., 1967, *Louisbourg*, Sydney, Fortress Press.

Muller, John, 1746, *A Treatise Containing the Elementary Part of Fortification*, London, J. Nourse.

Pell, S.H.P., 1951, *Fort Ticonderoga—A Short History*, Lake George, Fort Ticonderoga Museum.

Peterson, Harold L., 1964, *Forts in America*, NY, Charles Scribner's Sons.

Reid, Max W., 1906, *The Story of Old Fort Johnson*, NY, G. P. Putnam's Sons.

Roberts, Robert B., 1980, *New York's Forts in the Revolution*, Rutherford, Madison, Teaneck, NJ, Fairleigh Dickinson Press.

Scott, Albert John, 1927, *Fort Stanwix and Oriskany*, Rome, NY, Rome Sentinel Co.

Stone, Lyle M., 1974, *Fort Michilimackinac 1715–1781*, East Lansing, Michigan State University.

Voorhis, Ernest, 1930, *Historic Forts and Trading Posts of the French Regime and of the English Trading Companies*, Ottawa, Dept. of the Interior.

Wheelwright, Nathaniel, 1960, "Nathaniel Wheelwright's Canadian Journey, 1753–54," 1960, *The Bulletin of the Fort Ticonderoga Museum*, Vol. 10, No. 4, 260–296.

WEAPONS & ACCOUTREMENTS

Calver, Wiliam L. and Bolton, Reginald P., 1950, *History Written with a Pick and Shovel*, NY, New-York Historical Society.

Darling, Anthony D., 1970, *Red Coat and Brown Bess*, Ottawa, Museum Restoration Service.

Guthman, William H., 1970, *U.S. Army Weapons, 1784–1791*, Meriden, American Society of Arms Collectors.

Huddleston, Joe D., 1978, *Colonial Riflemen in the American Revolution*, York, George Shumway Publisher.

Lindsay, Merrill, 1975, *The New England Gun*, New Haven, The New Haven Colony Historical Society.

Moore, Warren, 1967, *Weapons of the American Revolution and Accoutrements*, NY, Funk & Wagnalls.

Neumann, George C., 1967, *History of the Weapons of the American Revolution*, NY, Harper & Row.

———, and Kravic, Frank J., 1975, *Collector's Illustrated Encyclopedia of the American Revolution*, Harrisburg, Stackpole Books.

Peterson, Harold L., 1956, *Arms and Armor in Colonial America, 1526–1783*, NY, Bramhall House.

———, 1968, *The Book of the Continental Soldier*, Harrisburg, Stackpole Books.

Index

NOTE: *a bracketed number [ ] indicates the catalogue-entry number of a specific powder horn*

A

Abatis (fortification term), defined, 63
Abbott, John, powder-horn owner [104], 199; horn illustrated, 199
Abel, Captain Joshua, 129
Abercrombie, Captain James, powder-horn owner [38]; horn illustrated, 122
Adam and Eve (decorative motif), 148
Albany, New York, 26, 38, 40, 63, 64, 131; County Militia, 9th Regiment, 151
Allen, Captain Daniel, 207; General Ethan, 66
Allenstown, New Hampshire, 42, 44, 49, 156, 158
American Revolution, 20, 21, 22, 25, 34, 44, 48, 49, 50, 51, 52, 55, 57, 59, 60, 61, 66, 69, 110, 147, 151, 156, 159, 160, 166, 184, 185, 201, 205, 207; faked dates on powder horns, 20; school of carvers, 59. *See also* Revolutionary War
Amherst, General Sir Jeffrey, 64, 66
Amsterdam, New York, 201
Andrus, Christopher, powder-horn owner [76], 59, 166–167; horn illustrated, 167
Angel of death (decorative motif), 123
Angels, winged (decorative motif), 129, 131
Animals (decorative motif): alligator, 194, 195; bear, 122, 183, 199; beaver, 184; boar, 154; buck, 126, 181, 194, 211; camel, 154; cat, 174, 201; deer, 101, 120, 122, 128, 156, 181, 199, 205; doe and fawn, 37, 75, 156, 157, 174, 184, 211; dog, 114, 120, 122, 124, 127, 181, 183, 186, 194; donkey, 156, 181, 184; elk, 154, 156; fox, 101, 120, 126, 159, 190, 199, 201, 205; horse, 126, 136, 137, 138, 156, 185, 211, (and rider), 78, (winged), 189; horse/mule, 183; lion, 57, 58, 104, 117, 120, 154, 176, 178, 179, 194; moose, 101; mouse, 101; mule, 156, 190; pig, 174; rabbit, 180, 181, 183, 187; raccoon, 183; snail, 207; squirrel, 207; turtle, 159, 174; unicorn, 57, 58, 104; whale, 194; wolf, 194; zebra, 156
Architectural device (decorative motif), 35; pediment, 180; scroll, 180
Arnold, General Benedict, 51, 66, expedition to Canada, 197; Jabez, powder-horn owner [74], 58, 164–165; (horn illustrated), 164, 165; John, powder-horn owner [78], 168, (horn illustrated), 168
Ashford, Connecticut, 172
Augsburg, League of (1686), 24
Austin, Jesse, powder-horn owner [13], 90, 91; horn illustrated, 91

B

Bagley, Colonel Jonathan, 89, 112, 116, 131
Baird, Robert, powder-horn owner [32], 115, 116; horn illustrated, 115, 117, 118, detail illustrated, 23, 43
Baldwin, David, powder-horn owner [14], 91, 92, 93, 107, 150; horn illustrated, 92, detail illustrated, 40, 41
Banner (decorative motif), 167
Bannerman (catalogue), 76
Barns, Giles, powder-horn owner [64]; horn described, 48, 49, 152, 153; horn illustrated, 153
Barret, Captain Joseph, 191
Batterson, George, powder-horn owner [54], 140, 141; horn illustrated, 140, 141
Battery (fortification term), defined, 63
Bedel, Colonel Timothy, 58, 160
Bedford, New York, 60, 208, 209, 210; Pennsylvania, 208
Bedford Carver [111, 112, 113], 25, 208, 209, 210

Beebe, Colonel Bezaleel, 188
Bemus, Jotham, powder-horn owner [41], 126; horn illustrated, 125, 126
Bemus Heights, New York, 126
Bennet, Tilton, powder-horn owner [66], 155, 156; horn illustrated, 155, detail illustrated, 155
Bennett, Thaddeus, powder-horn owner [19], 100; detail illustrated, 21, 100
Berlin, Connecticut, 201
Berm (fortification term), defined, 63
Bible, as design source, 34
Birds (decorative motif), 36, 38, 40, 41, 42, 47, 55, 56, 59, 60, 73, 79, 80, 86, 99, 115, 117, 120, 124, 128, 137, 140, 156, 157, 161, 166, 167, 172, 176, 179, 184, 188, 201, 203, 211; in flight, 40; in trees, 114; on perches, 176; with worm, 189. Specific types: dove, 154; flamingo, 131; parrot, caged, 144; peacock, 73, 205; pigeons, 211
Blockhouse (decorative motif), 95
Blodget, Samuel, *Perspective Plan of the Battle of Lake George*, by, 140
Body (powder-horn part), defined, 31; scalloping on, 90
Border (decorative motif), abstract, 111; animal, 156; bird, 84, 156; chevron, 40, 41, 47, 93, 123; chip-carved, 194; cross-hatched, 113, 174; cloud, 87; C-scroll, 99, 104; demilune, 82, 136, 158; diamond, 75, 96; dogtooth, 40; dotted-vine, 131; double-line-square, 166, 193; elongated curves on, 13; floral, 41, 84, 99, 110, 120, 131, 135, 150; floral-scroll, 42, 75, 86, 96; floral-swag, 180; foliated-scroll, 75; geometric, 40, 80, 86, 110, 120, 158, 195, 205; intertwined-circle, 161; incised, 151; linked-oval, 136; looped, 148; lunette, 113, 138, 140, 152, 156, 166, 183, 193, 211; recessed, 194; rococo, 172; sawtooth-scalloped, 93; scalloped, 80, 99, 103,

221

Index

126, 131, 138, 140, 145, 149, 153; scroll, 37, 40, 58, 80, 96, 99, 104, 122, 137, 156, 172, 173, 176; S-scroll, 96; shell, 40, 87; shield-shape, 40; single-line with dots, 79, 80; triangular-device, 40, 131; vine, meander-and-leaf, 37, 135, 166; wave, 156; wriggle-work, 154; weedlike, 38; zigzag, 40, 75, 86, 147, 149, 172, 173

Boston, Massachusetts, 201, 204; Massacre, 49, (illustrated), 50, 158; plan of British defenses, 198; profile view of, 57; shoreline, 176; view of, 162, 163, 164, 166, 167, 176, 194

Boston Harbor encampments, 53, 167

Boston Neck, American fortifications at, 198

Boston Port Act (1774), 48, 94

Boston, Pre-Siege of, school of powder-horn carving, 51

Boston, Siege of, school of powder-horn carving, 20, 22, 25, 50, 51, 52, 53, 57, 58, 61, 86, 94, 157, 159, 162, 163, 164, 166, 169, 172, 176, 182, 185, 186, 188, 189, 196, 197

Boston Tea Party (1773), 48

Bostwick, Amos, powder-horn owner [95], 188, (horn illustrated), 188; Captain Isaac, 188, 204

Boylston, Massachusetts, 38. *See also* Shrewsbury North Parrish

Bradbury, Elijah, powder-horn owner [97], 189, 190; horn illustrated, 190

Bradstreet, [Colonel] John, 93

Bradley, Captain Philip Burr, 170, 172; Colonel, 172

Brass, band around powder horns, 80; charger, 153; screw, 84

Brattleboro, Vermont, 37

Brett, Colonel James, 180

British: coat-of-arms on carved powder horns, 49; flag, 86, 99, 151, 172; lion, 101, 104, 123, 127, 136, 137, 144, 176; regulars, 27th Regiment, 103; soldiers, 174

Brooks, A. E. (powder-horn collector), 76; Daniel, statement by, 158

Brown, Captain Isaiah, 91

Buildings (decorative motif), 78, 172

Bunker, Philip, powder-horn owner [42], 126–127; horn illustrated, 127

Bunker Hill, Fort, 165

Bush, John, powder-horn carver [11, A, B, 12, 13, 14, 15], 36, 37, 38–39, 41, 85, 86, 87, 88, 90, 91, 93

C

Calligraphy, examples of, 27, 28, 29, 35, 43, (from books), 58, 78, 87, 110, 111, 116, 117, 127, 151, 174, 191, 199, 211; illustrated, 27, 28, 29, 36, 38, 39, 40, 41, 42, 43, 44, 45, 49, 55–56, 169, 171, 190; with baroque floral designs, 93, 95; block-lettered, 37, 40, 79, 80, 82, 87, 111, 126, 136, 137, 142, 167, 206, 207, 209; in capital letters, 193; checkerboard, 186; copperplate, 40, 41, 42, 47, 48, 90, 92, 98, 100, 103, 107, 111, 114, 115, 128, 131, 133, 138, 142, 145, 149, 152, 172; with crossed muskets, 115, 150; crosshatched, 113, 126; with curled finials, 78; with diagonal shading, 82; with diamond-shaped device, 82; with double-line lettering, 73, 75, 78, 79, 193; with effigies, 123; European, 22; in an exercise book, 28; with faces and flowers, 127; with featherlike devices, 90, 93, 100; with finials on letters, 40; with floral and vine designs, 106; with floral devices, 124, 158; with geometric devices, 90, 93, 98, 100, 120, 122, 124, 158; German Renaissance, 40, 41; Gothic, 28, (illustrated), 40, 41, 79, 80; with heart-shaped leaves, 183; illuminated, 40, 159; with incised block lettering, 180, 184, 186, 189, 207; with intaglio lettering, (illustrated), 21; in King George's War style, (illustrated), 35, [1]; in Lake George School style, 40; with leaf devices, 100, 122, 123; lettering of, 42, 90, 95, 99, 103, 104, 122, 154; with lozenge, 82; with man-in-the-moon faces, 180–181; motto in, 106; pictorial devices in, 40, 86; with profile faces, 123; with rabbit motif, 199; rococo, 172; in sawtooth design, 106, 113, 199; scalloped, 172; scrollwork, 156, 157; C-scrolled, 75, 78, 84, 86, 87, 104, 115, 120, 122, 128, 137; serifs on, 73, 84, 98, 138, 152, 156, 206, 207; decorative, 98, 128; with shell scrolls, 86, (illustrated), 87; in Siege of Boston School style, 53; with single-line letters, 79; sources of, 28, (illustrated), 29; stylized, 78; with stylized flowers, 100, 122, 123; with stylized vegetation, 100; trefoils of, 183; uppercase letters of, 84; with "V" on ends of letters and dates, 35; in vinelike script, 126; wing motifs of, 115; winglike serifs of, 41, 47, 100, 115, 120, 132, 143

Cambridge, Massachusetts, 53, 94, 154, 166, 167, 168, 169, 180, 182, 183

Camp Winter Hill, Massachusetts, 111

Canada, 19, 53, 64, 67, 93, 95, 199; French fortifications in, 24

Canajoharie, New York, 32

Canfield, Captain Joseph, 139; Lieutenant Colonel Samuel, 170

Cannon, 113; (decorative motif), 178

Canoe (decorative motif), 161

Canterbury, Connecticut, 136, 165

Carillon, Fort. *See* Fort Ticonderoga

Carril, John, powder-horn owner [50], 137; horn illustrated, 137

Carrying Place, The (Fort Edward, New York), 41, 63, 84

Carrying strap (on powder horns), defined, 31

Cartridge box, described, 24; unavailability of, 24; use of, 24, 207

Cartouche (decorative motif), arched, 172; with broken scroll, 181; with floral edges, 112, 144; with floral scrollwork, 190; gravestone shaped, 59; irregular, 73; line, 164; with lunettes, 161, 193; oval, 172, 186; patriotic, 189; rectangular, 73, 74, 104, 154, 161, 184, 188, 209; with rococo border, 186; with rococo scrolls, 189; with scrolled edges, 123, 159

Case, Elijah, powder-horn owner [79], 169–170; horn illustrated, 169, 170

Castle, The (French fort), described, 68; history, 68, 69

Cavalry (decorative motif), 40, 107, 134, 135; illustrated, 134, 135

Cave, Titus, powder-horn owner [67], 156–157; horn illustrated, 156

Champion, Captain Henry, 110

Charles River, Massachusetts, 53

Charlestown, Massachusetts, map of, 163; view of, 43, 53, 161, 163, 167, 180, 182

Charlestown Neck, Massachusetts, 53, 158, 163, 164, 172; detail illustrated, 164

Chaussegros de Lery, Gaspard (French engineer), 64, 68

Cheonderoga. *See* Fort Ticonderoga

Index

Cherub (decorative motif), 40, 86, 93
Chester, New Hampshire, 49, 50, 155, 156, 157
Chevron (decorative motif), 44, 47, 102
Chip-carved work (decorative motif), 57; borders, 178, 194; circles, 179; sunburst, 194
Church (decorative motif), 154, 172, 176
Circle (decorative motif), 178, 179
Civil War, 209
Clark, the Reverend De Witt, 91; Captain Joel, 145; the Reverend Jonas, reminiscences of, 160
Clarke, Captain Samuel, 180
Clay pipe (decorative motif), 112
Cleaveland, Aaron, powder-horn owner [49], 136–137; horn illustrated, 136
Cleaveland-Carril Carver [49, 50, 51], work of analyzed, 46; 136–138
Clinton, General James, 66
Clock, tall-case (decorative motif), 167, 169
Coat-of-arms (decorative motif), 101; illustrated, 102, 104, 110, 111, 112; British, 104, 123, 136, 137, 148, 156, 157, 158, 167, 184, 186, 188, 190, 191, 194
Cock (decorative motif), 201; illustrated, 81
Coit, Captain, 136
Colchester, Connecticut, 110
Colerain, Massachusetts, 159
Committees of Correspondence, 48
Compass (decorative motif), 124, 148, 167, 185, 187, 194; with crown finial, 185; ship's, 204
Compasswork, chip-carved (decorative motif), 57
Concord, Massachusetts, Battle of (1775), 51, 52, 53; New Hampshire, 191, (illustrated), 192
Conklin, John, powder-horn owner [113], 210; horn illustrated, 210
Connecticut (State of), 38, 112, 116, 170, 182, 201, 203; coat-of-arms, 49, 64, 152
Connecticut River, 20, 22, 65, 85, 86
Connecticut River Valley, 21, 85
Connoisseurship, 22, 27, 28; importance of, 60
Connor, Samuel, powder-horn owner [94], 186, 187; horn illustrated, 187; detail illustrated, 12
Continental Congress, 48, 53, 199
Continental line, 1st Regiment, 162; 3d Regiment, 162; 15th Regiment infantry, 163

Continental motifs (decorative), 22, 34, 35
Cook, Jonathan, powder-horn owner [6], 37, 80; horn illustrated 79, detail illustrated, 36, 80
Cooke Captain James, 56, 161
Cooper, Enoch, powder-horn owner [35], 119–120; horn illustrated, 119
Cotton, William, powder-horn owner [51], 137, 138; horn illustrated, 138
Courtney, Edward, powder-horn owner [20], 101–103; horn illustrated, 101, 102, detail illustrated, 69
Coventry, Connecticut, 160
Cow's horn, as powder carrier, 24
Crain, Elisha, powder-horn owner [77], 167, 168; horn illustrated, 168
Crain-Arnold Carver [77, 78], 167–169
Crosby, Samuel, powder-horn owner [7], 80–81; horn illustrated, 81, detail illustrated, 18, 37
Crosses (decorative motif), 98
Crown Point, New York (fort), 42, 46, 47, 53, 63, 64, 86, 99, 120, 127, 133, 135, 137, 139, 146, 147, 148, 150
Cruttenden, Jonathan, 204, 206
Cuckold (decorative motif), illustrated, 46, 114, 144
Culch horns, defined, 26
Cumberland, Duke of. *See* William Augustus, Duke of Cumberland
Curlicues (decorative motif), 120
Currency Act (1764), 48
Curtain (fortification term), defined, 63
Curtis, Hull, powder-horn owner [90], 182, 183; horn illustrated, 182, detail illustrated, 55, 58
Cushing, Colonel Job, 206

D

Danbury, Connecticut, 162
Davidson, Hamilton, powder-horn owner [69], 44, 49–50, 158–159; horn illustrated, 158, 159, detail illustrated, 50
Death's-head effigy (decorative motif), 86, 93; illustrated, 86
Decanter and wine glass (decorative motif), 113
Declaration of Independence, 199
Decorative motifs: Adam and Eve, 148; alligator, 194; angel of death, 123; animals, 73, 74, 75, 78, 79, 101, 118, 122, 126, 127,

154, 155, 157, 158, 167, 180, 181, 184, 211, (large), 159, (on pedestal), 156; architectural scrolls, 180; arms, 137, (of Connecticut), 152; axe, felling, 151; banner, 167; battle formation, 186; bayonet, 151; bear, 183; beaver, 184; bird(s), 73, 74, 79, 115, 117, 122, 137, 140, 157, 172, 179, 184, 194, 204, 207, 211, (and cat), 201, (flamingo), 131, (and tree), 188, (in tree), 114, 156, 158, (and worm), 189; boar, 154, 155; boats, 114, (long), 178; Boston, (churches), 162, (map of), 181, (Massacre), 158, (plan of defenses), 197, (Siege of), 162, (view of), 194; bow and arrow, 151; brazen serpent, 83; British flag, 86; broken-scroll cartouche, 181; buck and doe, 211; building(s), 78, 102, 172; bushes, 205; busts, 184, 186, (of man), 188; calligraphy, illustrated, 35, 39, 40, 41, 75, 86, 87, 88, 90, 91; camels, 154, 155; cannon, 113; canoe, Indian war, 161; cartouches, 5, 73, 74, 113, (broken-scroll), 181, (and lunettes), 161, (rococo), 158; cat(s), (and bird), 201, (winged), 174; chariot and two mules, 156; chickens, 211; children, 154; chip-carved (border), 194, (sunburst), 194; church(es), 154, (Boston), 162; circle and crown, 188; clock, tall-case, 167, 169; cloud scrolls, 87; coat-of-arms, 101, (illustrated), 102, 104, 110, 131, (British), 123, 125, 126, 127, 137, 147, 148, 158, 167, 184, 186, 187, (British abstract), 156, 157, 169, (parody of), 110, 111, (and unicorn), 194, (with crown finial), 185, Masonic square, 178; Connecticut arms, 152; crew members, 211; cross, Maltese, 84; crown and circle, 188; Cumberland, Duke of, 73, 74; dancing couple, illustrated, 104, 122; death's-head, 86; deer, 120, 125, 127, 128, 136, 137, 183, 190, 194, 205, (prancing), 138, (running), 185, 187; devil, 83, 151, (holding serpent), 167; diamonds, 75; doe, 125, 184, (and fawn), 37, 74, 75, 76, (and buck), 211; dog(s), 120, 125, 127, 128, 183, (barking), 114, (and musket), 154; donkey, 156; dots, 131; dragon or sea monster, 102; drum and sticks, 151; ducks in

223

Index

flight, 105; dueling scene, 151; dwelling, soldiers', 180; elk, 154, 155; face(s), 154, 184, 209, 210; fantastic animals, 27, (illustrated), 36, 104; fence, 207; fencing soldiers, 181, 184; fiddle, 156; finial, compass-and-crown, 185; fish, 75, 78, 84, 122, 134, 142, 149, 158, 170, 171, 172, 178, 179, 197, 204, 207, 210; flags, 86, 113, 135, 209, (British on ship), 172, (Grand Union), 209; flamingo, 131; flintlock musket, 167, 189; floral, (borders), 102, 135, 150, (branches), 188, (designs), 107, 115, 183, 186, 189, (rococo scrolls), 190, (vines), 131, 144, 197, 207; flowers, 79, 86, (and vines), 73, 108, 109, 113, 120; formation, (battle), 186, (of soldiers), 43, 107; fort, 126, 144, 146, 166, 167, (British), 162, 164, (with church view), 154, (diagram), 113, 178, (four-sided), 194, (profile view), 154, (Roxbury), 178, (Edward, plan of), 101, (illustrated), 102; fortifications, 163, (American), 198; fox(es), 159, 190, 205; G.W. profile, 183; Gage, Tom, 167; Grand Union (flag), 209; geese in flight, 114; geometric, 37, (designs), 35, 75, 76, 77, 79, 86, 107, 122, 127, 139, 140, (devices), 23, 77, 115, 145, 147, 149, (figures), 184, 186, (horse), 156, (motif), 198, 203, 204; goose, 190; grenadier and sword, 186; griffin, 136, 137; hart, 184; heraldic device, 157; Hessian, (American chasing a), 204, 206, (soldiers), 204; hog, 151; horse(s), 137, 183, 185, 205, 211, (heads and necks), 163, (geometric), 156, (prancing), 135, (rampant lion-), 189, (rearing), 138, (riders), 136; horseman, 78, 110, 122, (mounted), 114; hound and hunter, 211; house(s), 114, 170, 174, 178, 205, (and fence), 207; hunter, 122, 127, 128, 136, (and dogs), 105, (and hound), 211, (and musket), 189, (shooting bird), 158, (and tree), 203; incised decoration, 92; Indian(s), 126, 146, (American), 161, (drinking), 190, (head of an), 86, (naked), 101, 102, (and soldiers fighting), 140, (war canoe of), 161, (with bow), 114; intaglio lettering, illustrated, 21; lake, abstract, 140; leaves, and vines, 181; lion(s), 117, 120, 137, 154, 155, 178, 188, 198, (rampant -horse), 189, (and unicorn), 110, 123, 126, 127, 144, 148, 152, 167, 183, 185, 186, 188; long boats, 174; Lorelei(s), illustrated, 104, 157, 167, 174, 178, 194; man, (bust of), 188, (-in-the-moon faces), 180, 181, (-of-war), 198, (and soldier), 166, (with horns), 114; man's head, profile, 181; map, 125, 126, 127, 148, 163, (of Boston), 181; marching soldiers, 183, 184; Masonic, square and compass, 178; mounted troops, 134; moose, 183, 184, 188, (eating plant), 205; mule(s), 156, 190; musket(s), crossed, 170, 171, (flintlock), 167, 189, (and hunter), 189, (non-firing), 150; New Haven Green, 148, 152; officer and soldiers, 197, 203, (with sword), 207; palisade and defenders, 148; parrot, 144; partridges, 201; peacock, 205; pediment, 180; Pegasus, 189; pennants, 209; pigeons, 211; pillars, 167; pinwheel, 209; pine tree, 144; plants, 84, 167, (moose eating), 205, leafy, 207; Polynesian, 161; profile head(s), 144, 149; rabbit, 183, 187; raccoon, 183; redcoats, 174; reptile(s), 157, 184; river(s), 144, 181, (and water), 172; rococo, 161, (floral scrolls), 190, (scrolls), 189; Roxbury Fort, 178; sailing ship(s), 78, 81, 97, 104, 115, 144, 146, 149, 174, (and oarsmen), 97, 154; scales of justice, 18; scalloped border, 131, 157; scalping scene, 122; scroll(s), 142, (rococo), 189, (rococo floral), 190; scrolled branches, 35; C-scrolls, 75, 78, 86, 87, 104, 115, 128, (with vine), 137; S-scroll border, 96; semicircles, 113; serpent, brazen, 83, (held by devil), 167; serpentine border, 208; shells, 86; ship(s), 76, 134, 161, 163, 167, 181, 198, 209, 210, 211, (British), 162, (with British flag), 172, (with cannon firing), 209, (single-masted), 194, (three-masted), 205, (two-masted), 205, (under sail), 204, (with uniformed soldier facing stern), 209; shrubs and trees, 207, 208; Siege of Boston, 162; sign, tavern, 174; sloop, 102, 120; snails, 207; snake, 83, 156, (with turtle), 157; soldier(s), 123, 125, 126, 127, 131, 134, 135, 180, (conversing), 207, (fencing), 181, 184, (firing), 105, 107, 184, 186, (on foot and mounted), 107, 147, (Hessian), 204, (holding weapons), 169, 170, (and Indians fighting), 140, (and man-of-war), 166, (marching), 183, 184, 209, (mounted), 209, (and officers), 197, 203, (in uniform, illustrated), [18], 21, 81, 169, 170, (in uniform facing ship's stern), 209; soldier's life, 174; spike tomahawk, 151; squirrel, 207; sun, 120; sunbursts, 84, 174, (chip-carved), 194; sword(s), 151, (and grenadier), 186, (officer with), 207, (and tomahawk wielder), 201; tall-case clock, 167, 169; tomahawk, (spike), 151, (and sword), 201; town view, 165; tree(s), 101, 136, 161, 164, 167, 170, 172, 178, 211, (and bird), 188, 203, (and hunter), 203, (of knowledge), 148, (and shrubs), 207, 208; triangle, 172; troops, mounted, 134; trumpets, 107, 113; tulips, 103, 104, 138, 139, 142, 211; turtle(s), 156, 159, 174, (with snake), 157; Union Jack, 97, 99, 113, 151; unicorn, 137, (and coat-of-arms), 194; vines, 154, 161, 211, (floral), 197, 207, (and leaves), 181, (and meanders), 37, 87, 115; war club, ball-headed and spiked, 151; Warren, Admiral, 74; water and rivers, 172; weapons, 140; whales, 194; whimsical characters, illustrated, 37, (in plants), 204; woodsman, 122; woman, 75; worm and bird, 189; wriggle-work border, 154; zebra, 154, 155; zigzag lines, 86

Deer (decorative motif), 101, 122, 128, 136, 137, 156, 183, 187, 194, 211; and doe, 124; prancing, 138; running, 120; miniature, 101

Deerfield, Massachusetts, 76, 86

Delaware River, 59, 188, 204

Demilune (fortification term), defined, 63

Demon/devil, 120, 151, 167; winged, 120

Denison, Captain John, 142

Designs, powder-horn, origins and sources of, 20–21, 34–35; frontier influence on, 20. *See also* Decorative motifs

DeVenter, Nathaniel. *See* Porter, Captain Nathaniel

Diamond, Thomas Smith, powder

Index

horn of [17], 98, 99; horn illustrated, 98, detail illustrated, 62
Dieskau, Baron, (death of), 86
Dike, Colonel Nicholas, 203
Dodge, John, horn of, 110
Dorchester Heights, Massachusetts, 203
Douglas, Colonel William, 177
Dragon (decorative motif), 102; rampant, 101
Drum and drumsticks (decorative motif), 113, 129, 151
Dueling scene (decorative motif), 151, 158
Durham, New Hampshire, 191
Durkee, Colonel John, 103, 162
Dyar, Colonel John, 136
Dyer, Colonel Eliphalet, 129, 142, 144

E

East Haddam, Connecticut, 133
East River, New York, 198
Edward Augustus, Duke of York, 64
Ely, Abner, powder-horn owner [86], 177, 178; horn illustrated, 177, 178
Ely, Elihu, horn, mentioned, 178
Ely, Captain Elisha, 177
Emplacement (fortification term), definition, 63
Encampments (decorative motif), 53
Entrenching tools (decorative motif), 40
Essex County, Massachusetts, 159
European traditions, influence of on American designs, 22, 23
Evert, Captain Nathaniel, 149
Extension lobe (powder-horn term), defined, 31; examples: 32, 35, 99, 126, 137, 158, 161, 191, 207, 208; crown shaped, 198
Eyre, William, British chief engineer, 64

F

Faces, 156, 209, 210; (decorative motif) human, 112, 154, 204; grotesque, 40; lunarlike, crowned, 193; mysterious, 47, (illustrated), 44, 144, 184; illustrated with wings, 48; with wings and feathers, 47
Fairfield, Connecticut, 104, 116, 131, 140
Fakes and fakers 19, 20; reasons for, 20
Feathers (decorative motif), 161
Ferris, Captain Reuben, 103, 108; 27th Regiment British regulars, 106

Ferrule, brass, 96, 142
Fifth Company (Fairfield), 116
Fifteenth Continental infantry, 163
Fifth New Hampshire Regiment of Militia, 191
Fifty-first American Provincial Regiment, 85
First Connecticut Regiment, 93, 103; 2nd Company, 103; 4th Company, 112; 7th Company, 91
First New York Artillery Regiment, 194
Fish (decorative motif), 36, 41, 47, 55, 75, 78, 99, 100, 122, 124, 131, 134, 142, 149, 158, 170, 171, 172, 178, 181, 184, 191, 193, 204, 207, 209, 211; schools of, 38, 40, 84, 86, 176, 179, 197
Fitch, Colonel Eleazor, 128, 142; 3rd Connecticut Regiment, 136, 142; 4th Connecticut Regiment, 128
Flag, 170, 172, 179; British, 96
Fletcher, David, powder-horn owner [2], 74, 75; horn illustrated, 74, 75
Flintlock, definition, 32; 166
Floral-and-vine design (decorative motif), 73, 75, 107, 131, 180, 183, 186, 197, 207; scrollwork, 189
Floral designs (decorative motif), 41, 46, 59, 75, 84, 105, 107, 108, 110, 112, 115, 116, 117, 140, 144, 148
Flowers (decorative motif), 79, 120, 184; eight-petaled, 112
Foliate designs (decorative motif), 147
Foot, Aaron, powder-horn owner [108], 204, 205, 206; horn illustrated, 205
Forman, Jacob, powder-horn owner [111], 208, 209; horn illustrated, 208, 209
Forerunners of Lake George School (of powder-horn carvers), 36–38
Fort(s): (decorative motif), 154, 161, 176, 178, 194; design of, 61; diagrams of, 40, 53, 95, 112, 113, 114, 126, 164, 166, 167, 178, 198; four-bastion, 199; French and Indian, Siege of Boston, and Revolutionary War: (Bunker Hill), 165, (Conti), 68, (Crown Point), 64–65, 66, 68, (Dummer), 37, 86, (Edward), 41, 44, 47, 63–64, 83, 88, 89, 95, 99, 101, 103, 112, 116, 122, 124, 131, 137, 142, 144, 146, 148, (brick kiln at), 102, 113, 147, (George [Oswego]), 67, 68, (La Gallette), 65, 138, (La Présen-

tation [see La Gallette]), (Lydius), 63, 64, (Lyman), 63, (Massachusetts), 86, (Montgomery), 190, (Niagara), 68–69, (Nook Hill), 167, (No. 1), 165, (No. 2), 165, (No. 3), 165, (No. 4), 20, 42, 58, 65, 100, 103, 104, 106, 116, 118, (Ontario), 67, (Oswego [Chueguen]), 66–67, (Ploughed Hill), 165, (Prospect Hill), 168, (St. Frédéric), 64, 66, 68, (Ticonderoga [Cheonderoga]), 53, 65–66, 144, (Vaudreuil [see also Ticonderoga]), 65, (William Henry), 37, 38, 41, 67, 68, 89, 91, 93, 95, 96, 98–99, 115, 116, 131, 136; horn carving at, 61; naming of, 61; profile view, 154
Fort Edward Creek, 63, 113
Fort Washington, fall of, 170, 172, 182
Fortifications, importance of, 61
Forty-second Royal Highland Regiment (British), 147
Fourth Connecticut Regiment, 116
French, Sergeant Ichabod, powder-horn owner [21], 103, 104; horn illustrated 103, 104, detail illustrated, 71
French and Indian War, 21, 22, 23, 34, 44, 48, 51, 61, 64, 65, 66, 68, 82, 94, 95, 96, 111, 127, 140, 182
Frontier, influence on horn carvers, 20, 24
Frye, Captain Isaac, 191
Funnel, definition, 32

G

Gadroon (decorative motif), 202
Gage, General Thomas, 29; depicted on powder horn, (illustrated), 60, 167, 200; described, 201
Gale, Captain Samuel, 8th Company (Killingworth, Connecticut), 175
Gardner, Jonathan, powder-horn owner [107], 202, 203, 204; horn illustrated 203, 204
Gargoyles (decorative motif), 40
Gates, General Horatio, 166, 186
Gay (Guay), Jacob, powder-horn carver [39, 40, 41, 42, 43, 69, 70, 88, 89, 90, 91, 92, 93, 94, 95, 96, 97, 98], 25, 36, 42–44, 46, 47, 49, 50, 51, 53, 57–58, 74, 80, 101, 122, 124, 125, 126, 127, 128, 134, 140, 148, 149, 155, 156, 158, 159, 179, 180, 182, 183, 184, 185, 186, 187, 188, 189, 190, 191, 201, 205, 211; calligraphy of, illustrated, 45; creativity, 50; possible attributions, 201;

Index

powder horns signed by, 44; signature of, illustrated, 44, 45; tradition concerning, 42, 43
Gentleman's Magazine, The (periodical), powder-horn design source, 56, 161
General Wolfe Tavern, Pomfret, Connecticut, 94
Geometric devices (decorative motif), 35, 37, 41, 42, 80, 84, 104, 107, 110, 115, 116, 117, 120, 123, 139, 140, 144, 145, 147, 149, 150, 170, 171, 176, 183, 184, 186, 193, 198, 203, 204, 206; flowers of, 193; houses of, 181, (illustrated), 23
George II, 101
Germantown, Pennsylvania, 32; Battle of, 183
Gilbert, J. H. Grenville (powder-horn collector), 33
Gilman, Captain Nicholas, 193
Glacis (fortification term), definition, 63
Glover, Colonel John, 154
Goding, William, powder-horn owner [40], 124, 125; horn illustrated, 124, 125
Goff, Jonathan, powder-horn owner [72], 162, 163; horn illustrated, 162, detail illustrated, 162
Goldthwait, Michael B., powder-horn owner [16], 96; horn illustrated, 97
Goldthwait-Smith-Diamond Carver [16, 17], 48, 96, 98
Gooddel, Jabez, powder-horn owner [102], 196, 197; horn illustrated 196, 197
Goodrich, Colonel Elizur, 88; Second Connecticut Regiment, 146
Gordon, Royal Engineer Harry, quoted, 67–68
Grand Union flag, 209
Grancsay, Stephen, 30, 33, 76, 146
Grant, Captain Noah, 7th Company (Windsor, Connecticut), 110, 131
Great Lakes (eastern), 19, 68
Greenfield, James, powder-horn carver and owner [84, 85, 86, 87], 56, 174, 175, 176, 177, 179, 211; Greenfield's horn illustrated, 175
Greenwich, Connecticut, 115, 131
Greenwood, Isaac J., 32, 33
Grenadiers, uniformed, marching (decorative motif), 181
Grider, Rufus Alexander, 19, 34, 59, 91, 92, 95, 110, 146, 151, 201; biographical sketch of, 32–34; drawings by, 25, 26, (illustrated), 33; papers of, described, 32, 33, 34
Griffin (decorative motif), 136, 137
Grotesque (decorative motif): figures, 38; faces, 40
Groton, Massachusetts, 159
Guay. *See* Gay

H

H. T., initials of Hugh Tolford, carver, 155
Haddam, Connecticut, 164
Hadlyme, Connecticut, 198
Hait, Colonel Jonathan, 115
Hale, Nathan, 182
Hall, Caleb Johnson, powder-horn owner [106], 201, 202, (horn illustrated), 202; John C., 201; Captain Street, 111
Hamilton, Alexander, 66; David, powder-horn owner [26], 108, 109, (horn illustrated), 108, 109
Hapgood, Asa, powder-horn owner [8], described, 37; 82; horn illustrated, 82
Hardy, John, drummer, 158
Harlem Heights, New York, 180, 182; Battle of, 170
Harper, Captain William, 199
Harris, Captain James, Jr., 5th Company, 128
Hart, Colonel John, New Hampshire Regiment, 137
Hartford, Connecticut, 112
Hartley, Robert M., 201
Hatfield, Massachusetts, 86
Havana, Cuba, 181; Siege of (1762), 48, 93
Havana Expedition (1762), 26, 94, 131, 181
Haviland, Lieutenant Colonel William (British officer), 103, 107, 108
Heath, General William, 207
Henry, Patrick, 201; possible portrait, (illustrated), 200, (described), 201
Heraldic device (decorative motif), 157
Hill, Rufus, powder-horn owner [10], 84, 85; horn illustrated, 84, detail illustrated, 85
Hill-Tyler Carver [10], 41, 84
Hillsborough, New Hampshire, 191
Hitchcock, Captain Amos, 6th Company (New Haven, Connecticut), 134; Ebenezer, powder-horn owner [48], 134, 135, (horn illustrated), 135
Hobart, Jonathan, powder horn of, inscription quoted, 20, 101; significance of, 20, 101
Hobby, Captain Thomas, 4th Company, 149; 5th Company (Greenwich, Connecticut), 116, 131; 6th Company, 115
Hodges, William (artist), 161
Hollis, New Hampshire, 76
Holt, Thomas, powder-horn owner [68], 157, 158; horn illustrated, 157
Hooker, Captain Noadiah, 6th Company (Farmington, Connecticut), 197
Hoosick River, 86
Horse and rider (decorative motif), 78, 126, 136, 137, 138, 185, 211
Horseman (decorative motif), 110, 114, 122, 135, 136; with winged horse, 189
Hosmer, Nathaniel, powder-horn owner [100], 59, 193, 194, (horn illustrated), 193, 194; Reuben, powder-horn owner [99], 191, 192, 193, (horn illustrated), 192
Hosmer Carver [99, 100], 191–193
Houses (decorative motif), 114, 154, 170, 174, 176, 178, 179, 198, 201, 205, 207
Howe, General William, 59
Howland, or Rowland, powder-horn owner [105], 199, 200, 201; horn illustrated, 200, detail illustrated, 21, 200
Hubbell, Captain Samuel, 5th Company (Fairfield, Connecticut), 116, 131
Hudson River, 22, 63, 94, 113, 162, 190
Humphrey, Captain Elihu, 4th Company (Simsbury, Connecticut), 169
Humphreys, David (author), 94
Hunting scene (decorative motif), 60, 105, 122, 124, 127; with hunter, 128, 136, 158, 191; showing hunter with musket, 154, 186; showing hunter with musket and dog, 189, 203, 211
Huntington, Colonel Jedediah, 8th Connecticut Regiment, 169
Hyatt, Captain J., 158

I

Indians, American (decorative motif), 40, 144, 147, 168; fighting, 140; naked, 101; Woodland, 161
Indian (decorative motif): canoe, 161; head, 86, 128, 149, (illustrated), 149; with bow, 114; war canoe, (illustrated), 56
Iroquois Confederacy, 65

Index

J

J.W. Carver [19, 32, 33, 34, 35, 36, 37], 25, 39, 42, 46, 47, 80, 90, 95, 100, 103, 110, 114, 115, 116, 119, 120, 128, 131, 134, 135, 140, 143, 145, 146, 149, 150, 152; calligraphy and designs described, (illustrated), 43; signature, (illustrated), 43

Johnson, Obadiah, powder-horn owner [75], 165, 166, (horn illustrated), 166, (detail illustrated), 165; Sir William, 64, 67, 68

Justice, scales of (decorative motif), 187

K

Kendall, Joshua, powder horn attributed to [110], 207, 208; horn illustrated 207, 208

King George's War, 21, 22, 29, 34, 41, 44, 73, 85; decorative styles of 34, 35; design changes in, 22; importance of, 22; powder horns of, 22, 80

King George's War School (of powder horn carvers), 34, 35, 36, 53, 59; motifs on Lake George School horns, 80; sylistic features of, 35

King William's War (1689–1697), 24

King's Ferry, New York, 190

Knowlton, Captain Thomas, 172; Lieutenant Colonel Thomas, 7th Connecticut Rangers Regiment, 182; 19th Continental Regiment, 182

Konkun, James, powder-horn owner [112], 209, 210; horn illustrated, 210

L

Lacey, Captain David, 116
Lake Champlain, 20, 32, 40, 44, 64, 131
Lake George, New York, 20, 21, 22, 32, 42, 44, 50, 63, 64, 65, 68, 84, 85, 103, 108, 110, 111, 116, 119, 120, 137, 140, 149, 151; Battle of, 68, 85, 86, 146

Lake George School (of powder-horn carvers), 40–48, 49, 50, 51, 59, 60, 73, 80, 101, 112, 115, 131, 133, 138, 140, 142, 143, 144, 145, 146, 152; characteristics of, 51, 52, 53, 144, 145, 147, 152; lesser hands of, 48

Lake Ontario, 66, 68, 114
Lake Ontario/Niagara region, 21, 22
Lamb, Colonel John, 194
Lathrop, Captain Ebenezer, 166

Latimer, Colonel Jonathan, 166
Lebanon, Connecticut, 146
Lechmere Point Redoubt, 165, 168, 180
Le Prestre de Vauban, Sébastien, 65
Lewis, Captain Eldad, 145; Jonathan Clark, powder-horn owner [70], 159–60, (described), 50–51, (political cartoon on), 49–50, (horn and detail illustrated), 51, 160

Lexington Alarm, 52, 111, 136, 164, 166, 170, 171, 180; school of powder-horn carvers, 51–52, 160–161

Lexington and Concord, Massachusetts, Battles of (1775), 52, 53, 94, 160

"Liberty" (motto—decorative motif), 57

"Liberty or death" (motto—decorative motif), 201, 207

Lion and unicorn (decorative motif), 44, 49, 110, 126, 127, 148, 152, 156, 157, 167, 175, 176, 178, 179, 183, 184, 188, 194, 198; with British lion, 75, 176, (crowned), 183, 188

Lititz, Pennsylvania, 32
Lively (brig), 49, 152
Locker, Captain Isaac, 180
Long Island, 182; Battle of, 170, 189, 204
Long Island Campaign, 197
Long Island Sound, 169
Longboat (decorative motif), and passengers, 176; and oars, 176

Lorelei (decorative motif), 40, 55, 57, 104, 157, 167, 175, 176, 178, 179, 194

Lotbinière, Michel Alain de, 65
Louis XIII, 34
Louis XIV, 24, 34
Louisburg, Expedition of (1745), 20, 75, 85, 160

Lounsbury, Samuel, powder-horn owner [23], 41, 105, 106; horn illustrated, 106

Lovejoy, John H. (artist), 95
Lunettes (decorative motif), 44, 84, 140, 156, 161, 193, 198; in circle, 80

Lydius, John Henry, 64
Lyman, Phineas, Colonel, 103, 104, 106, 112, 114, 116; Major General, 64, 90, 103, 104, 106, 112, 116, 128, 131

Lyme, Connecticut, 47, 57, 128, 175, 177

Lyme Carver. *See* Miller-Tribble Carver

M

Machicolation (fortification term), 63
McKonkey's Ferry, Pennsylvania, 204
McNeal, Captain Archibald, 108
Magaw, Colonel Robert, 182
Maltese cross (decorative motif), 84
Man (decorative motif), 151; profile bust, 114, 181; and woman [31], (detail illustrated), 71; dancing, 104, 122; firing musket, 150

Mann, Captain Benjamin, 191
Mansfield, Connecticut, 201
Map horns, defined, 26; author's preference, 26; faked, 26

Maps (decorative motif), 26, 126, 127, 148; of route from Fort Edward to Fort Henry, 112

Marblehead, Massachusetts, view of, 154

Martin, James, powder-horn owner [92], 184, 185; horn illustrated, 185

Marston, Captain Simon, 139
Mason, Captain Abel, 206
Mason, New Hampshire, 191, (illustrated), 192, 193

Massachusetts, 52, 67, 88, 116, 131, 184
Massachusetts Bay, 38
Massachusetts militia, 159
Massachusetts: Regiment, 112, Revolutionary War rolls, 160

Materials: brass, 79, 82, 91, 93, 96, 100, 101, 112, 116, 122, 124, 131, 136, 142, 152, 156, 157, 167, 170, 173, 183, 185, 186, 189, 202, 204, 206, 208; buff leather, 166; cork, 140, 199; deer hide, 126; dye, (brown), 30, (orange), 30, (vermillion), 30, (yellow), 30, 31; glass, 199; hemp, 146; horn, cow's, 24, 73, 75, 76, 78, 79, 80, 82, 84, 85, 87, 88, 90, 91, 93, 96, 98, 99, 100, 101, 103, 104, 105, 106, 108, 110, 111, 112, 114, 115, 116, 119, 120, 122, 124, 126, 127, 128, 129, 131, 133, 134, 136, 137, 138, 139, 140, 142, 143, 144, 145, 146, 147, 149, 150, 151, 152, 154, 155, 156, 157, 158, 159, 160, 162, 163, 164, 165, 166, 167, 168, 169, 170, 172, 173, 174, 175, 177, 179, 180, 182, 183, 184, 185, 186, 188, 189, 191, 193, 194, 196, 197, 199, 201, 202, 204, 206, 207, 208, 209, 210, 211; ink, brown, 199; iron, 73, 74, 75, 80, 85, 87, 88, 90, 96, 98, 99, 100, 103, 105, 106, 108, 110, 112, 115, 116,

Index

119, 120, 122, 124, 126, 129, 133, 136, 138, 139, 140, 142, 149, 151, 152, 154, 155, 158, 159, 164, 165, 166, 167, 168, 169, 170, 172, 173, 177, 179, 180, 182, 186, 188, 189, 191, 196, 197; leather, 98, 165, 211; paint, 183, (black), 149, (brown), 145, (red), 93, (reddish brown), 172, (vermillion), 101; paper, 84, 157, 179, 199; pewter, 79, 108, 154, 157, 180, 207; pigment, 73, 79, 85, 169, 170, (black), 90, 91, 93, 131, 136, 137, 143, 154, 167, 168, (brown), 188, 204, (dark), 99, 120, 126, 183, (red), 211, (dark blue), 119, (reddish brown), 119, 189, (vermillion), 112, 114; putty, tan, 199; rope, 166; sealing wax, 126, (red), 140; stain, brown, 164, 173, 177, 179, 183; string, 166; twine, 79; varnish, 146, 152, 160, 188, (red), 87; wood, 84, 88, 101, 108, 112, 114, 143, 149, (specific types): cherry, 98, 110, 122, 154, 185; hardwood, (pegs), 82, 110, (pins), 172, 188; mahogany, 196; maple, 99, 177, 179, 191, 193, 202; oak, 129; pine, 73, 74, 75, 76, 78, 79, 80, 82, 85, 87, 88, 90, 91, 93, 96, 100, 103, 104, 105, 106, 108, 110, 115, 116, 119, 120, 124, 126, 127, 128, 131, 133, 134, 138, 139, 140, 142, 144, 145, 146, 150, 152, 155, 157, 159, 160, 162, 163, 164, 165, 166, 167, 168, 169, 170, 172, 173, 174, 179, 180, 182, 183, 184, 185, 189, 191, 201, 204, 207, 208, 209, 210, 211; rosewood, 90; walnut, 137, 197; wool yarn, 140
Mather, Captain Timothy, 128
Maurepas, Frédéric, 64
Maynard, John, 65
Mead, James, 114
Meigs, Captain Return, 162
Meldrum, James, powder-horn owner [60], 147, 148; horn illustrated, 147, 148
Melvin, James (additional name on powder horn), 36, 37
Memento Mori Carver [29, 30, 31], 25, 39, 44–46, 112, 114, 144, 149; cartouche by, (illustrated), 45; motifs of, (illustrated), 44, 46; horn attributed to, 112–114
Men fencing (decorative motif), 158
Mermaids (decorative motif), 40, 55, 104

Merrimack River, 65
Middlesex County, Connecticut, 197
Middletown, Connecticut, 170
Miles, John, powder-horn owner [58], 145, 146; horn illustrated, 145
Milford, Connecticut, 145, 150
Miller, Thomas, powder-horn owner [44], 128, 129, 130; horn illustrated, 129, 130; detail illustrated, 47
Miller-Tribble, or Lyme, Carver [44, 45], 47, 128, 129, 131, 140; calligraphy of, 47; face motif of, (illustrated), 47; characteristic motifs of, 47, 131, 134, 140, 147
Mills, John, powder-horn owner [43], 127, 128; horn illustrated, 128; detail illustrated, 45
Minor, Richardson, powder-horn carver and owner [52, 53], 47–48, 115, 138–39; horns illustrated, 139
Mohawk River, 22, 26
Mohawk Valley, 32, 201
Mohegan Indian (Nathaniel Sunsimon), 179
Monster (decorative motif), 44, 113
Montcalm, General Louis Joseph, 39, 66
Montreal, Canada, 26, 64, 138
Moon, new (decorative motif), profile, 99
Moore, Ephraim, powder-horn owner [89], 57–58, 180–181; horn illustrated, 181
Morley, George, powder-horn owner [81], 172, 173; horn illustrated, 173
Morley-Smith Carver [81, 82], 172; illustrated, 173
Mott, Gershom, powder-horn owner [101], 194, 195, 196; horn illustrated, 195; detail illustrated, 196
Mottoes, patriotic (decorative motif), 49, 123, 143, 167
Mouse (decorative motif), 101
Murray, Jonathan, powder-horn owner [85], 176, 177; horn illustrated, 176, 177
Musicians (decorative motif): drummer, 203; fifer, 203
Musket (decorative motif), 174
Musket ramrod pipe (ferrule), 142
Mysterious face (decorative motif), illustrated, 47

N

Neptune with trident (decorative motif), 172

New England, 34, 52, 160; northern, 19
New Hampshire, 55, 57, 58, 89, 112, 116, 131, 158, 184, 191, 203
New Hampshire Colony, Regiment of Rangers, 160, 199
New Haven, Connecticut, 88, 116, 134, 152; Green (decorative motif), 49, 148, 152
New Jersey, 172
New Milford, Connecticut, 188
New York (city), 26, 44, 94, 159, 160, 182, 184, 188, 198, 204; (state), 19, 34, 58, 89, 112, 116, 131
New York Campaign (1776), 53, 137, 166, 182, 188, 197
New Zealand natives (in canoe—decorative motif), 56; war-canoe drawing described, 161
Newell, Stephen, powder-horn owner [109], 206, 207; horn illustrated, 206
Newton, Massachusetts, 86
Newtown, Connecticut, 38
Niagara River, 68
Nichols, Colonel Moses, 191
Nicholson, Colonel Francis, 63; Colonel John, 194
Ninth Connecticut Regiment, 118
Nook Hill, Fort, 167
North America, 19, 20, 24, 34, 59, 67, 73, 96, 101, 201; engraved powder horns of, 20
Northern Campaign (1776), 58, 201
Nova Scotia, French fortifications in, 24, 44
Noyes (Noyce), John, powder-horn owner [93], described, 58; inscription, 58; 185, 186; horn illustrated, 185, 186
No. 1, Fort, 165
No. 2, Fort, 165
No. 3, Fort, 165
No. 4, Fort, 20, 42, 58, 65, 100, 103, 104, 106, 116, 118

O

Oarsmen (decorative motif), 97
Octopuslike design (decorative motif), 84
Officer with tricorn hat (decorative motif), 176
Oswego, New York, 26, 44; fortifications at (carved on powder horn), illustrated, 114
Oswego River, 66, 114
Ottawa Indians, depiction of, 101
Oviatt, Alexander, powder-horn owner [62], 150; horn illustrated, 150

Index

P

Page, Aaron, powder-horn owner [25], 42, 108, 109; horn illustrated, 108, 109; described, 143, 144
Palisade (fortification term), definition, 63; and defenders (decorative motif), 148
Palmer, Lieutenant Christopher, powder-horn owner [55], 48, 142, 146, 149; horn illustrated, 142
Parapet (fortification term), definition, 63
Parker, John, powder-horn carver and owner [71], 55, 56, 160, 161; horn illustrated, 161
Parkinson, Sydney (artist-engraver), 161
Parsons, Colonel Samuel Holden, 175, 179
Patriotic slogans, on powder horns, 54
Patterson, Colonel John, 163; William, powder-horn owner [61], 149; horn illustrated, 149
Payson, Nathan, powder-horn owner [29], 112; horn illustrated, 113
Peck, Stephen, powder-horn owner [36], 120; horn illustrated, 121
Peekskill, New York, 57, 177, 178, 190
Pegasus (decorative motif), 189
Pelham, Henry (printmaker), 158
Pemberton, John, powder-horn owner [39], 44, 122, 123, 124, 128; horn illustrated 123, 125, detail illustrated, 45
Pembroke, New Hampshire, 186
Pennoyer, Captain John, powder-horn owner [73], 163, 164; horn illustrated, 163, detail illustrated, 164
Pennsylvania, 182, 204, 205
Penobscot Indian designs, [55], 142
Pepperrell, Sir William, 85
Pettibone, Captain Abel, 170
Pickering, Captain John, 137
Pickett, Nicholas Edgecomb, powder-horn owner [65], 154–55; horn illustrated, 154
Pinwheels (decorative motif), 178, 179, 204, 209
Pitkin, Lieutenant Colonel John, 103
Pittsfield, Massachusetts, 60, 85, 211
Plains of Abraham, Canada, 101
Platform (fortification term), definition, 63
Plaque, raised (decorative motif), 152
Ploughed Hill, Fort, 165
Political cartoon (decorative motif) 50–51; detail illustrated, 51
Polynesian motif (decorative motif), 161
Pomfret, Connecticut, 94
Pontiac (Indian chief), 93
Pontiac's War, 93
Pontiac's Rebellion (1763–64), 48
Porter, Captain Nathaniel (formerly mistakenly known as De Venter), powder-horn owner [59], 146, 147; horn illustrated, 146
Portsmouth, New Hampshire, 183
Pouchot, Captain Pierre, 68, 69
Powder Hill, Fort, 165
Powder horns, American carved, 30–31; as art form, 19, 20; characteristics, 21; selection of for exhibition and organization of into schools, 22–29; color described, 27; damaged, 27; decorative styles, 34–35; design styles as identification, 19; patina of, 27; pigment of, 27, 31; regional types of, 21; varnish on, 27, 31
Powder-horn carvers, schools of: King George's War, 35–36; Forerunners of Lake George, 36–38; Lake George, 38–48; Pre-Siege of Boston, 48–51; Lexington Alarm, 51–52; Siege of Boston, 51, 52–59; Revolutionary War, 59–60
Pownal, Bennington County, Vermont, 211
Pownall, Governor, 38
Pre-Siege of Boston School (of powder-horn carvers), 48–51, 154, 155, 156
Prentice, Captain Samuel, 3rd Company (Stonington, Connecticut), 179
Prescott, Brigadier General Oliver, 50, 159
Prescott, Ontario, Canada, 65
Preston, Captain Ephraim, 13th Company, 108
Prideau, General John, 69
Princeton, New Jersey, Battle of, 188, 205
Prospect, Massachusetts, 180
Prospect Hill, Fort, 168
Providence (ship), 159
Punchbowl, and goblet (decorative motif), 116; and pipes, 113
Putnam, Israel, powder-horn owner [15], 30, 93, 94, 95, 96, (horn illustrated), 94, 95; 3rd Connecticut Regiment, 166, 167; 4th Company, 166; Major Israel, 3rd Company, 136

Q

Quartering Act (1765), 48
Quebec, Canada, 39, 66; Campaign of 1759, 160
Queen Anne's War (1702–1713), 24

R

Rampart, (fortification term), definition, 63
Ravelin (fortification term), definition, 63
Redout (fortification term), definition, 63
Reed, Colonel James, New Hampshire Regiment, 191
Reptile (decorative motif): snake, 38, 82, 112, 148, 157, 174, 184; alligator, 194, 195
Revolutionary War School (of powder-horn carvers), 59–60. *See also* American Revolution
Revere, Paul, 49, 50; engraver, 158
Rhode Island, 89, 112, 116, 131, 189
Robbins, Frederick, powder-horn owner [83], 173, 174; horn illustrated, 174; detail illustrated, 54
Robie, Ichabod, powder horn of, 80
Rococo design (decorative motif), 161, 172, 183, 184; scroll, 187
Rockwell, John, powder-horn owner [31], 113, 114; horn illustrated, 113, 114; detail illustrated, 45, 46, 114
Rocky Gulch, New York, 146
Rogers, Major Robert, powder-horn owner [B], 65, 88, 103, 146, (drawing illustrated), 88; Island, New York, 64
Roman numerals (decorative motif), 163
Rowboat and oarsmen (decorative motif), 96
Roxbury, Massachusetts, 53, 56, 162, 163, 164, 167, 169, 170, 173, 175, 194, 196, 202; fort, 194; view of (decorative motif) 163, 166, 167, 176, 178
Royal Welsh volunteers (British regiment), 151
Ruggles, Brigadier General Timothy, 86

S

St. Claire, General Arthur, 66, 191
St. Francis Indian settlement, 65
St. Johnsville, New York, 32

Index

St. Lawrence River, 26, 65, 138; Valley, 151
Salem, Massachusetts, 91
Salisbury, Connecticut, 86
Sally (brig—decorative motif), 176
Sandyhook, New York, 122
Saratoga, New York, 66, 126, 186; Battle of, 166; Campaign of, 206
Sargent, Simeon, powder-horn owner [98], 191; horn illustrated, 191
Sawtooth designs (decorative motif), 112, 115
Sawyer, Charles Winthrop, quoted, 26
Scammel, Colonel Alexander, 3rd New Hampshire Regiment, 191, 193
Schoolboys' exercise books, illustrated, 28
Scrolls (decorative motif), 35, 84, 104, 124; with baroque flourishes, 48; copperplate, 142, 172, 186; floral, 188; shaded geometric-border, 38; rococo, 57; shell-like, 86
Sea monster (decorative motif), 40
Second Connecticut Regiment, 47, 90; 1st Company, 89, 90; 2nd Company, 88, 89, 138; 4th Company, 91; 6th Company, 115; 7th Company, 138
Selden, Major Samuel, powder-horn owner [103], 197, 198; horn illustrated, 198
Selkrig, Nathaniel, Corporal, powder-horn owner [24], 42, 106, 107, 108, 109, 144, 150; horn illustrated, 107, 109; detail illustrated: title page, 43
Selkrig-Page Carver [23, 24, 25, 26, 27], 25, 39, 41–42, 47, 49, 50, 73, 80, 99, 103, 105, 106, 108, 110, 130, 131, 134, 143, 147, 150
Seneca Indians, 68
Senter, Colonel Joseph, New Hampshire regiment, 189
Seventh Company (Fairfield, Connecticut), 116; Connecticut Regiment, 111
Sharon, Connecticut, 163; Massachusetts, 163
Sheffield, Massachusetts, 86
Sherburne, Edward, powder-horn owner [91], 183, 184; horn illustrated, 183, 184; detail illustrated, 54, 57
Ship(s) (decorative motif), 102, 161, 163, 166, 169, 172, 181, 194, 198; figureheads of, 37, 40, 57, 81, 201, 211; fully-rigged sailing, 78, 80, 104; in harbor, 53, 167, 198; single-masted, 166, 172, 194; tall-masked, 204; three-masted, 161, 205; two-masted, 172, 205; under sail, 115, 147, 149, 175, 211; in form of weathervane, 37; with cannon, 153
Shirley, Governor William, 67
Shrewsbury, Massachusetts, 20, 37, 38, 39; carving style of, 39; North Parrish of, 38
Siege of Boston: horns 52–55, 57, 163, 164, 172, 193, 195, 204; index features, 53–55; motifs described, 53; school of powder-horn carvers, 49, 51, 52–60, 109, 162–206, 186, 194, 204, 211
Siege of Havana. *See* Havana, Cuba, Siege of
Simsbury, Connecticut, 37
Simsbury Carver [79, 80], 169, 170, 171; horns by, illustrated, 170, 171
Smedley, Lieutenant Colonel James, 2nd Company (Fairfield, Connecticut), 131
Smith, Lieutenant Joseph, powder-horn owner [56], 48, 143, 144, (horn illustrated), 143; Captain Seth, New Haven Company of Minutemen, 170; Simeon, powder-horn owner [82], 172, 173, (horn illustrated), 173; William, powder-horn owner [1], 36, 44, 53, 54, 55, 73, 74, (horn illustrated), 74, (detail illustrated), 14, 35, 73
Smith-Fletcher Carver [1, 2, 3], 36, 37, 39, 73–76
Snake (decorative motif), 38, 82, 112, 148, 157, 174, 184
Soldier(s) (decorative motif), 40, 44, 105, 123, 127; and Indians, 126; chasing a Hessian, 59; courting a lady, 174; duelling, 180; fencing, 181, 185; fighting, 140, 174; firing, 107, 111, 197, 199; formation of, 43, (illustrated), [24], 46, 55, 58, 59, 110, 124, 126, 127, 131, 135, 170, 171, 197, 203; formation of for battle, 186, 204, 207; groups of, 184; Hessian, 204; in ambush, 199; line of, 42; marching, 107, 147, 183, 209; mounted, 107, 110, 134, 135, 147; on foot, 107, 108, 135, 147, 148; shooting, 187; trumpeting, 107; with flags, 135; with tomahawks, 101, 140, 141, 202
Sons of liberty, 94
Spaulding, Asa, powder-horn owner [22], 104, 105; horn illustrated, 105
Spear and tomahawk (decorative motif), 102
Spencer, Hobart, powder-horn owner [47], 46, 133, 134, (horn illustrated), 133; Lieutenant Colonel Joseph 131, General Joseph, 2nd Connecticut Regiment, 162, 164, 166, 170, 197
Spencer's Brigade, 169
Spencer-Hitchcock Carver [46, 47, 48], quality of work by, 46–47; 131–135, 145
Springfield, Massachusetts, 59, 206
Stamp Act (1765), 48
Star, eight-pointed (decorative motif), 99
Stiles, John, powder-horn owner [57], 144, 145; horn illustrated, 144, 145
Stillwater, New York, 44, 126
Stonington, Connecticut, 142
Stony Point, New York, 162; Battle of, 207
Storrs, Captain Experience, 2nd Company (Mansfield, Connecticut), 167, 169
Stratford, Connecticut, 37, 47, 82, 91, 133, 149
Stripping of powder horns, 27
Sturbridge, Massachusetts, 206
Stylistic periods of powder-horn decoration: King George's War (1744–1748), 35–39; Lake George (1755–1763), 40–48; Between the Wars (1763–1774), 48–51; Siege of Boston (1775–1778), 51–59; Revolutionary War (1776–1781), 59–60
Sudbury, Massachusetts, 65, 180
Suffield, Connecticut, 89
Sugar Act (1764), 48
Sullivan, General John, 183, 186
Sun (decorative motif), 38, 84, 145; chip-carved, 194
Sunburst (decorative motif), 174
Sunderland, Massachusetts, 202
Sunflower design (decorative motif), 111
Sunsimon, Nathaniel, powder-horn owner [87], 179; detail illustrated, 179
Swift, Colonel Heman, 201

T

Talus (fortification term), definition, 63
Tambling, Stephen, powder-horn owner [53], 47, 48, 139, 140; horn illustrated, 139

Index

Tavern sign (decorative motif), 174
Taylor, Meshach, powder-horn owner [3], 36, 54, 55, 56, 75, 76, (horn illustrated), 75, 76; Captain Samuel, 203
Telescope (decorative motif), 161
Temple Wharf, Temple's Farm, Massachusetts, 180
Third Connecticut Regiment, 1st Company, 89; 4th Company, 115
Third New Hampshire Regiment, 193
Ticonderoga, New York, 63, 66, 191
Tolford, Hugh, powder-horn carver [66, 67, 68], calligraphy of, 49; decorative features of, 49–50, 156; carving of, 49–50, 155, 156, 157
Tools (decorative motif): felling ax, 151; shovel, 151
Towns (decorative motif), scenes of, 40, 53
Townshend, Colonel George, 101
Tree(s) (decorative motif), 101, 136, 140, 148, 164, 166, 167, 170, 172, 178, 184, 188, 191, 203, 211; abstract, 161; and bushes, 205; of knowledge, 148; pine, 144
Trenton, New Jersey, Battle of, 59, 188, 204–205; Campaign of, 166, (described), 204–205
Tribble, John, powder-horn owner [45], mentioned, 110, 129–31; horn illustrated, 129, 130
Triangles (decorative motif), 44, 112
Tricorn hat (decorative motif), 201
Tuttle, Captain Nathaniel, 180
Trumpets (decorative motif), 107, 113
Tulip (decorative motif), stylized with leaves, 80; 82, 84, 104, 138, 139, 142, 144, 211; and vine 201
Turner, Henry, powder-horn owner [114], 211; horn illustrated, 211
Turtle (decorative motif), 157, 159
Tyler, Sol, powder horn of, 41

U

Unicorn (decorative motif), illustrated, 57
Union Jack (decorative motif), 96, 99; illustrated, 97
Upson, Stephen, powder-horn owner [88], 179, 180; horn illustrated, 180

V

Vail, Robert William Glenrole, quoted, 32
Valcour Island, New York, 66
Van Ness, Colonel Peter, New York Regiment, 151
Vaughan, Lieutenant Colonel John, British 46th Regiment, Royal Welsh volunteers, powder-horn owner [63], 151, 152; horn illustrated, 151; detail illustrated, 152
Vawn, Edward, 151; Richard, 151
Verelst, John (artist), 101
Vernon, Vermont, 37

W

Wadsworth, Brigadier General James Connecticut State Brigade (Durham), 170, 172; 4th Battalion, 198
Walker, Josiah, powder-horn owner [46], 46, 131, 132; horn illustrated, 132
Wallingford, Connecticut, 111
War, implements of (decorative motif), 40; canoe (decorative motif), Indian, 56
War of 1812, 69
Ward, Captain Andrew, 103; Colonel Andrew, 116; Colonel Andrew, Connecticut Regiment, 166; General Artemas, 52
Warner, Captain Robert, 162; Colonel Seth, 211
Warren, Admiral (decorative motif), 75; portrait on powder horn, [54]
Warwick, Rhode Island, 189
Washington, General George (decorative motif), 53, 66, 94, 188, 198; bust of, 58, 183, 184; portrait of on powder horns, 53; profile bust of, 188
Washington, Connecticut, 204
Waterbury, Captain David, 114, 116
Waterman, Zebulon, powder-horn owner [27], 110, 111, 131; horn illustrated, 110; detail illustrated, 70
Watertown, Massachusetts, 42
Wayne, General Anthony, 207
Weapons and accoutrements (decorative motif): 55, 59, 113, 151; ball-headed war club, 151; bayonet, 151, 201; bow and arrow, 151; flintlock musket, 166, 167, 189; halberd, 107, 161, 170; musket, 115, 140, 169, 170, 171, 174, 181; spike, 151; spike tomahawk, 140, 151; sword, 140, 151, 170, 174, 201, 203; tomahawk, 140, 201
Webb, Colonel Charles, 7th Connecticut Regiment, 111, 180, 188; 19th Continental Regiment, 182, 188, 204

Webster, [Captain] Samuel, powder-horn owner [96], 57, 189; horn illustrated, 189
West Point, New York, 58, 170, 190, 207
Westchester County, New York, 208, 209
Westminster, Massachusetts, 38, 82
Weston, Massachusetts, 86
Wethersfield, Connecticut, 88
Wheeler, David, powder-horn owner [34], 116, 117; horn illustrated, 117, 118
Whelpley, Isaac, powder-horn owner [33], 116; illustrated 117, 118
Whitaker, Samuel, powder-horn owner [37], 42, 120; horn illustrated, 121
White, Captain Samuel, 115
White Plains, New York, 204; Battle of, 166, 170, 188, 204; Campaign of, 197
Whiting, Colonel Nathan, powder-horn owner [12], 88, 89, 90, (horn illustrated), 89, 90; 46, 66, 106, 107, 110, 112, 115, 116, 131, 133, 134, 135, 144, 145, 146, 149, 150; Captain Samuel, 116
Whitney, Levi, powder-horn owner [9], 37, 38, 82, 84 (horn illustrated), 83; Lieutenant Samuel, 37, (powder horn of, described), 38; Samuel, 82; Sarah, 38
Wilcox. See Willcocks
Wigglesworth, Colonel, 203
Willcocks, Asa, powder-horn owner [80], 170–71; horn illustrated, 171
William Augustus, Duke of Cumberland, 73; portrait on powder horn, 73; illustrated, 73
Williams, Ephraim, 86; Esther, 86; Thomas, powder-horn owner [A], 41, 85, 86, (drawing), 87; Colonel William "Billy," biographical sketch of, 85, 86; William, powder-horn owner [11], 85, (horn illustrated), 85, (detail illustrated), 39, 86
Willson, David, powder-horn owner [4], 76, (horn illustrated), 77; George, powder-horn owner [18], 99, (horn illustrated), 99; Nathaniel, powder-horn owner [5], 78, (horn illustrated), 78
Wilton, New Hampshire, 191
Windham, Connecticut, 139; New Hampshire, 158
Windmills (decorative motif), 166

Index

Winglike device (decorative motif), 42, 115, 117, 118, 120, 121, 144
Wingate, Colonel Joshua, New Hampshire Militia Regiment, 186
Winter Hill, Massachusetts, 179, 180, 191
Wolcott, Joshua, powder-horn owner [30], 112, 113, 114; horn illustrated, 113
Wolfe, General James, death, 101
Woodbury, Connecticut, 182
Woodland Indians, 161; vignette of warfare (decorative motif), 199
Woodsman (decorative motif), with tomahawk and musket, 122
Wooster, Colonel David, 108, 115, 128, 131, 139, 142, 146, 149, 201
Work (fortification term), definition, 63
Wright, Captain, 137
Wyllys, Colonel Samuel, 3rd Connecticut Regiment, 2nd Company (Hartford), 162, 164, 197, 207
Wyman, Colonel Isaac, New Hampshire Regiment, 199

Y

Yale, Amasa, powder-horn owner [28], 111, 112; horn and detail illustrated, 111